African Urba

Gordon Prain · Nancy Karanja · Diana Lee-Smith
Editors

African Urban Harvest

Agriculture in the Cities of Cameroon,
Kenya and Uganda

International Development Research Centre
Ottawa • Cairo • Dakar • Montevideo • Nairobi • New Delhi • Singapore

Editors
Gordon Prain
Urban Harvest, CIP
Av. La Molina 1895
La Molina, Lima
Peru
gprain@cgiar.org

Nancy Karanja
Urban Harvest, CIP
Nairobi
ILRI Campus
Kenya
nancy.karanja@cgiar.org

Diana Lee-Smith
Mazingira Institute
Box 14550
00800 Nairobi
Kenya
diana.leesmith@gmail.com

A copublication with the

International Development Research
  Centre
P.O. Box 8500
Ottawa, ON, Canada K1G 3H9
info@idrc.ca / www.idrc.ca
e-ISBN 978-1-55250-492-5

and

International Potato Center (CIP)
Avenida La Molina, 1895
P.O. Box 1558
La Molina, Lima, Peru

ISBN 978-1-4419-6249-2 (hardcover)   e-ISBN 978-1-4419-6250-8
ISBN 978-1-4419-6571-4 (softcover)
DOI 10.1007/978-1-4419-6250-8
Springer New York Dordrecht Heidelberg London

Library of Congress Control Number: 2010932400

© International Potato Center (CIP), 2010
All rights reserved. This work may not be translated or copied in whole or in part without the written permission of the publisher (Springer Science+Business Media, LLC, 233 Spring Street, New York, NY 10013, USA), except for brief excerpts in connection with reviews or scholarly analysis. Use in connection with any form of information storage and retrieval, electronic adaptation, computer software, or by similar or dissimilar methodology now known or hereafter developed is forbidden.
The use in this publication of trade names, trademarks, service marks, and similar terms, even if they are not identified as such, is not to be taken as an expression of opinion as to whether or not they are subject to proprietary rights.

Cover illustration: Urban Harvest

Springer is part of Springer Science+Business Media (www.springer.com)

Jac Smit – In Memoriam

This book is dedicated to the memory of Jac Smit (1929–2009), who devoted his life and energy to the subject discussed in these pages (www.jacsmit.com). Jac helped bring the crucial role that urban agriculture plays in African cities to world attention in the early 1990s through his writings and tireless advocacy work.

Foreword

Within less than a generation a majority of Africans will live in urban areas. As the pace of urban growth outstrips job creation and the capacity of most governments to provide essential infrastructure, the face of African poverty is being transformed from what was once considered a primarily rural phenomenon to one including tens of millions of unemployed and underemployed living at the margins of the formal urban economy. Indeed, research has shown that the depths of income poverty, health risks and food insecurity in the informal settlements of some African cities are often worse than in stressed rural communities.

Urban agriculture is uniquely well positioned to respond to these challenges. Ironically, despite increasingly integrated global food systems, local production in urban and peri-urban areas remains essential to feeding African cities, due in part to systemic failures in markets linking rural and urban areas. This fact was dramatically demonstrated in the urban food riots provoked by the global food price spikes of 2007 and 2008. Historically discouraged and often prohibited by municipal governments, urban agriculture has attracted growing interest – and legitimacy – during the past decade and a half, largely due to a growing body of hard evidence showing that it can provide a sustainable motor for both human development and economic growth. The current volume represents a seminal contribution to that paradigm shift.

Drawing on the results of rigorous research conducted in Cameroon, Kenya and Uganda, the authors carefully explore the dynamics of urban and peri-urban agriculture through three complementary lenses – livelihoods, health and policies/institutions. From these scholarly studies, stronger typologies and analytical tools are emerging. The country case studies reveal the rich diversity of food and non-food agricultural systems that have emerged in different urban settings, and underline powerfully the importance of understanding agro-ecological, political, institutional and historical context before designing interventions. The case studies also reveal the complex and dynamic two-way demographic and economic interactions between rural and urban areas.

And contrary to much conventional thinking, the case studies show that urban agriculture is not simply a coping strategy for the poor, but is often a highly lucrative economic strategy pursued by the rich as well. Illuminated by insightful political economy analyses, power relations are exposed as critical factors in defining urban land tenure and land-use systems, water rights, the structure and operation of food markets, and the design and enforcement of health and food safety measures.

The complex and multi-level nature of power relations challenge simplistic silver bullet approaches to rationalizing urban and peri-urban systems, and to making them more efficient, equitable and safe. Careful, deliberate and inclusive political processes are key processes that include a wide range of actors and that are driven by hard evidence. In-depth case studies of multi-disciplinary and multi-agency platforms in Kampala and Nairobi presented in the final chapters point to ways forward, demonstrating how action research can contribute to policy reform by providing the right evidence, to the right audience and at the right time, and by catalyzing and framing negotiations between key stakeholders.

Urban Harvest, a system-wide initiative of the Consultative Group on Agricultural Research, must be congratulated for producing this marvelous volume. Similar congratulations are due to the International Development Research Centre, which supported, since 2000, the research process that produced the underlying evidence. Indeed, IDRC has been a pioneer in the field, having funded some of the earliest work in the 1980s that helped define the scope, scale and importance of urban agricultural systems globally. Since that time, IDRC helped create linkages with and between many regional and international organizations – including Urban Harvest, the RUAF Foundation, The Urban Agriculture Network (TUAN), UN-Habitat, the United Nations Development Programme (UNDP) and a host of local research and development organizations. The synergies created through these multi-dimensional partnerships created benefits that were out of reach for each working in isolation. The lessons of the IDRC experience for development programming more broadly are clear – persistence, continuity, and adaptation based on hard evidence can achieve transformational changes in thinking and approach. Rather than chasing fads, IDRC stayed with a problem for more than two decades, created and nurtured partnerships, and built local capacity that has now taken the lead. It is hoped that going forward the CGIAR has the same wisdom to retain a meaningful focus on urban agriculture and to build on the partnerships fostered by Urban Harvest.

It is fitting in many ways that this book is dedicated to Jac Smit, himself a pioneer and passionate advocate of urban agriculture. Often referred to as the "father of urban agriculture", starting in the 1960s Jac worked in more than 30 countries helping planners to understand and support urban agriculture in ways that directly improved lives and livelihoods of urban populations. Jac founded TUAN, and as the lead author of the classic reference book, *Urban Agriculture: Food, Jobs and Sustainable Cities,* published in 1996 by UNDP, Jac helped construct the intellectual framework within which urban agriculture is now perceived. A great teacher, Jac not only informed but inspired those of us privileged enough to have worked with him. Jac Smit died in November 2009. He would have loved this book, a book that in many ways reaffirms his vibrant legacy.

Adjunct Professor   Peter J. Matlon
Department of Applied Economics and Management,
Cornell University, Ithaca, NY, USA

Acknowledgments

An enormous number of people and institutions worldwide were involved in the production of this book, contributing to both the research and the ensuing production of the final manuscript. It is the work of Urban Harvest's Global Coordinator, Gordon Prain, Africa Regional Coordinator 2005 to present, Nancy Karanja, and Africa Regional Coordinator 2002–2005, Diana Lee-Smith. The period dealt with is mostly 2002–2006 but the text also refers to many research-related events that took place more recently. Thanks are due to all the many institutions and individuals whose collective effort went into the research-to-policy stakeholder meetings and processes that took place before, during and after the research itself and are described in Chapter 15. Most are not mentioned here but their contributions to the book's message are significant.

The Urban Harvest program is convened by the International Potato Center (CIP), one of the fifteen international agricultural research centres supported by the CGIAR – the Consultative Group on International Agricultural Research. The CGIAR is a global partnership of governments, multilateral organizations and private foundations that works to promote food security, poverty eradication and the sound management of natural resources in the developing world. We would like to acknowledge the vision and initiative of Dr. Wanda Collins, former Deputy Director General for Research at CIP, who led the development of the original proposal to the CGIAR on urban and peri-urban agriculture. We also gratefully acknowledge CIP's logistical and financial support, in particular to the lead editor, which made the publication of the book possible. The support of the Dalla Lana School of Public Health, University of Toronto, is also recognized, with special thanks to Donald Cole. Without a grant from the Ford Foundation (East and Southern Africa Office) most of the detailed editorial work could not have been done. The International Development Research Centre (IDRC) is not only funding the publication costs of this book and acting as a co-publisher, but members of its staff are also contributors to Chapter 14. Further, it has financially supported many of the research activities described in the book, including the Urban Harvest Program itself. As always with IDRC, our acknowledgement of their role goes way beyond the financial to include almost all aspects of the implementation of this research. Further financial support was received from the World Bank, which provided most of the Program funding for Urban Harvest.

As described in Chapter 14, many CGIAR centres contributed to the research, but three deserve special mention: IITA which led the research team in Yaoundé (special thanks to Christian Nolte); CIAT which led the team in Kampala (special thanks to Roger Kirkby); and ILRI which led the team in Nairobi (special thanks to Dannie Romney). Because the Kampala research took off in the direction of influencing policy and legal change as described in Chapter 15, the editors would also like to thank the Kampala Urban Food Security, Agriculture and Livestock Coordinating Committee (KUFSALCC), especially George Nasinyama and Abdelrahman Lubowa, for their contributions to the research in Kampala, while the role played by Kampala's CIP office must also be acknowledged.

The early chapters of the book concern urban agriculture in and around Yaoundé, Cameroon. The investigation of urban agriculture livelihoods in Yaoundé was designed and carried out by a team from IITA, IRAD and the University of Dschang. The nutrition study was executed by a team made up of scientists from CIRAD, IITA, ISSEA and the University of Yaoundé using data previously collected during seven surveys carried out in 1999–2000. Team members from ICRAF, IRAD and CIRAD made the survey of horticultural and tree-seedling nurseries, while the study of water pollution affecting and caused by urban agriculture in Yaoundé was designed and executed by team members from the University of Yaoundé, Ecole Nationale Polytechnique Supérieure (ENPS), IITA and ICLARM. A method for the quantification and spatial analysis of urban food supply flows was developed and applied to a test case of fresh cassava roots by team members from IITA, CIRAD, CARBAP and INC. The authors would like to thank the following persons specifically for their contributions to this chapter: Olivier David, François Damesse, Anne Degrande, Divine Foundjem Tita, Elie Foukou, Fernand Isseri, Emmanuel Ngikam and Valentina Robiglio. Funding for the studies into urban farming systems in Yaoundé that appear in Chapter 3 came from the World Bank managed by Urban Harvest. IITA and ICRAF contributed material resources and researcher time, and Diana Lee-Smith provided supervision and support for the work through regular workshops.

The research into crop–livestock integration in urban farming systems of Yaoundé, which forms the basis of Chapter 4, was led by researchers from IRAD, with additional participation by team members from IITA, ICLARM and University of Dschang, and was funded by Urban Harvest and IITA. The authors would specifically like to thank Jeanette Tchakounte, Andre Zoyuim, Randall Brummet and Herbert Lekane for their contributions to the work described in this chapter.

The institutional study in Chapter 5 relies on 20 years of research on Yaoundé by the lead author, and draws on data from a wide range of disciplines and sectors. It involved the collaboration of representatives of numerous institutions currently involved in the management of UA or engaged with it in some way: the *Communauté Urbaine de Yaounde* (CUY); the *Communes d'Arrondissemen*; NGOs and neighbourhood associations; provincial, departmental and divisional offices of agriculture; relevant government ministry services; researchers engaged in related

Acknowledgments

studies and local residents and media organizations. The authors acknowledge funding from Urban Harvest and the important inputs from Diana Lee-Smith, who also translated into English all three of the chapters on Cameroon. In addition, IDRC paid for the participation of both Dr. Bopda and Mr. Awono in a regional municipal meeting on urban food security and UA held in Nairobi in 2002, co-organized between UN-Habitat, IDRC and Urban Harvest.

Chapter 6 is based on the outcome of the Urban Harvest funded project "Strengthening UA in Kampala". Along with the International Centre for Tropical Agriculture (CIAT) as lead organization, collaborating institutions included the International Potato Center (CIP), the International Institute of Tropical Agriculture (IITA), National Agricultural Research Organization (NARO), the Department of Agricultural Extension and Education at Makerere University, Kampala City Council (KCC) and two NGOs, Environmental Alert and Plan International. Additional funds for the studies in this chapter were mobilised from CIAT and CIP and financial contributions from the International Network for Improvement of Banana and Plantain (INIBAP, a Division of Bioversity International) and the Department for International Development of the UK (DfID). The latter two grants made possible the household sample survey of farmers and the pilot listing survey of UA enterprises in two parishes in 2004, respectively. The authors would also like to thank the residents of Bukesa, Banda, Buziga and Komamboga parishes.

The study in Chapter 7, a component of the Urban Harvest funded "Strengthening UA in Kampala", was designed and led by researchers from the department of Agricultural Extension and Education at Makerere University, working with Kampala City Council, NARO, the Ministry of Agriculture, Environmental Alert – an NGO, Plan International – an NGO, and the International Center for Tropical Agriculture (CIAT). Additional funds were raised from CIP, BMZ and the "Farm Africa" Fund (DfID). The study of market opportunities for urban farmers in Kampala described in Chapter 8 was led by CIAT in collaboration with IITA (both of which provided additional resources), NARO, the Ministry of Agriculture and a farmers' network.

Much of the research in Chapter 9 was funded by the Canadian International Development Agency (CIDA) through a grant to Urban Harvest and the Dalla Lana School of Public Health, University of Toronto. In addition to the research team assembled for the Strengthening Urban Agriculture in Kampala Project, the International Livestock Research Institute (ILRI) was involved through separate grants from Urban Harvest and IDRC regional funds. As well as the CIDA funding and the additional Urban Harvest and IDRC grants, subsidiary grants were received from DfID and from IDRC's AGROPOLIS awards program, for three master's and doctoral degree studies linked to the research program.

The contribution of the following persons to the research in Uganda is also gratefully acknowledged: Grace Nabulo, Margaret Ssemwanga Azuba, Gertrude Atukunda, Maria Kaweesa, John Musisi Muwanga, Guy Blomme, Renee Sebastian, Popy Dimoulas, Frank Mwine, Regina Kapinga, Silver Tumwegamire, Prossy Musoke, Eva Birimukatonda, Pamela Businge, Elly Kaganzi, Denis Bisasem, A. Namagembe, C. Niringiye, B. Odongo, C. Owori and E. Olowo Onyango. Dr John

Aluma, former Deputy Director General of Research at NARO, made an inestimable contribution to moving forward the cause of urban agriculture policy in Uganda before his untimely death.

The research in Chapter 10 on nutrient cycles in Nairobi, Kenya was led by ILRI, with participation from ICRAF, CIP, Kenya Agricultural Research Institute (KARI), the University of Nairobi and an NGO, Kenya Greentowns Partnership Association. The study was funded by Urban Harvest using World Bank funds, with additional support from DfID, ILRI and CIP.

Investigations into crop-livestock-waste interactions in Nakuru, Kenya (Chapter 11) were jointly supported by Urban Harvest, DfID and IDRC funds, which supported research involving multiple stakeholders including the Nakuru Municipal Council (NMC), University of Nairobi, Egerton University and an NGO, Kenya Greentowns. Special thanks are due to the Municipality of Nakuru, particularly Simon Kiarie, former Director of Environment in the Municipality.

The study that culminated in Chapter 12 on the benefits and risks of urban dairy production in Nakuru, Kenya was led by scientists from the University of Nairobi in collaboration with the NMC and Kenya Greentowns. The work was funded jointly by DfID and Urban Harvest, which also provided research coordination, co-supervision of an MSc thesis and active team engagement.

Chapter 13 describes work carried out by scientists from ICRAF and the National Museums of Kenya on a topic related to Urban Harvest concerns and supported by an award from Urban Harvest and IDRC based on performance in the Anglophone Africa Training Course on UA in 2004 (see Chapter 14). The lead author, who would like to acknowledge Kisumu City Council for granting permission to conduct surveys in their urban and peri-urban markets, first became engaged with UA work as researcher and co-author of Chapter 10, reflecting the catalytic and networking role of Urban Harvest in encouraging UA work in the CGIAR.

For their personal support and guidance for the work in Kenya and the Urban Harvest programme in general, the Editors are grateful to Steve Staal, Amos Omore, Carlos Sere, Dennis Garrity, Bruce Scott, Bill Thorpe, John and Brigid Mac Dermott, Thomas Fitz Randolph, Delia Grace and, during his time as Mayor of Nairobi, Cllr. Dick Waithaka.

We would like to recognize the National Institute of Cartography, Cameroon, Kampala City Council, the Uganda Ministry of Agriculture, Animal Industries and Fisheries (MAAIF), Makerere University, University of Nairobi, Kenya Agricultural Research Institute, Institut des Sciences de L`Environment, Dakar Senegal, ENDA-RUP, Mazingira Institute and the Centre for Indigenous Knowledge Systems and by-Products (CIKSAP) for expertise regarding module development and delivery of the Anglophone Africa Training Course, 2004. Our gratitude is also extended to individuals such as the late Ato Yilma Getachew of Addis Ababa, Ethiopia. They all worked with experts from RUAF, IWMI, MDP and UMP-LAC as well as Urban Harvest to produce the learning materials.

Mary Njenga, who has worked as a researcher with Urban Harvest since 2002, deserves special mention and thanks. Mary has been a major contributor to many of the research activities reported in this book, and has advised and supported many

Acknowledgments

more where her name is not recorded. The editors also express their gratitude for the assistance provided to them in the production of the book by Cathy Barker, Ana Luisa Muñoz, Alison Light, Manisha Pahwa, Rosamond Rice and Alice Thomann, and to Charles Crissman, Paul Stapleton and Jan Low in the CIP Lima and Nairobi offices for help in making the production of this book possible. Grateful thanks also go to the anonymous chapter reviewers for their invaluable insights and detailed work on the text. A special thanks to Richard Stren for his helpful insights. All errors and omissions remain the responsibility of the editors.

Finally we want to thank Josephine, Marco, Andre and Anna Prain, Joseph Karanja Kiiru, Susan Karanja and Yvonne Karanja, Shermit Lamba and Davinder Lamba, for their patience during the production of the book, which took too much (by far) of our attention away from family life.

Lima, Peru
Nairobi, Kenya
June 2009

Gordon Prain
Nancy Karanja, Diana Lee-Smith

Contents

1. **The Institutional and Regional Context** 1
   Gordon Prain

2. **Urban Agriculture in Africa: What Has Been Learned?** 13
   Gordon Prain and Diana Lee-Smith

Part I   Cameroon

3. **Urban Farming Systems in Yaoundé – Building a Mosaic** 39
   Athanase Pr Bopda, Randall Brummett, Sandrine Dury,
   Pascale Elong, Samuel Foto-Menbohan, James Gockowski,
   Christophe Kana, Joseph Kengue, Robert Ngonthe, Christian
   Nolte, Nelly Soua, Emile Tanawa, Zac Tchouendjeu, and
   Ludovic Temple

4. **Crop–Livestock Integration in the Urban Farming
   Systems of Yaoundé** 61
   Thomas Dongmo, François Meffeja, Jean Marin Fotso, and
   Christian Nolte

5. **Institutional Development of Urban Agriculture – An
   Ongoing History of Yaoundé** 71
   Athanase Pr Bopda and Louis Awono

Part II   Uganda

6. **Changing Trends in Urban Agriculture in Kampala** 97
   Sonii David, Diana Lee-Smith, Julius Kyaligonza, Wasike
   Mangeni, Sarah Kimeze, Lucy Aliguma, Abdelrahman
   Lubowa, and George W. Nasinyama

7. **Can Schools be Agents of Urban Agriculture Extension
   and Seed Production?** 123
   Fred Baseke, Richard Miiro, Margaret Azuba, Maria
   Kaweesa, Moses Kalyebara, and Peter King'ori

8 **Identifying Market Opportunities for Urban
   and Peri-Urban Farmers in Kampala** 139
   Robinah Nyapendi, Rupert Best, Shaun Ferris, and John Jagwe

9 **Health Impact Assessment of Urban Agriculture in Kampala** ... 167
   George W. Nasinyama, Donald C. Cole, and Diana Lee-Smith

Part III Kenya

10 **Recycling Nutrients from Organic Wastes in Kenya's
   Capital City** 193
   Mary Njenga, Dannie Romney, Nancy Karanja, Kuria
   Gathuru, Stephen Kimani, Sammy Carsan, and Will Frost

11 **Crop–Livestock–Waste Interactions in Nakuru's
   Urban Agriculture** 213
   Nancy Karanja, Mary Njenga, Kuria Gathuru, Anthony
   Karanja, and Patrick Muendo Munyao

12 **Benefits and Selected Health Risks of Urban Dairy
   Production in Nakuru, Kenya** 229
   Erastus K. Kang'ethe, Alice Njehu, Nancy Karanja, Mary
   Njenga, Kuria Gathuru, and Anthony Karanja

13 **Urban Agroforestry Products in Kisumu, Kenya: A Rapid
   Market Assessment** 249
   Sammy Carsan, Dennis Osino, Paul Opanga,
   and Anthony J. Simons

Part IV Urban Agriculture and Institutional Change

14 **IDRC and Its Partners in Sub-Saharan Africa 2000–2008** 267
   Luc J.A. Mougeot, Francois Gasengayire, Diana Lee-Smith,
   Gordon Prain, and Henk de Zeeuw

15 **The Contribution of Research–Development Partnerships
   to Building Urban Agriculture Policy** 287
   Diana Lee-Smith and Gordon Prain

Acronyms 309

Index 313

Contributors

Lucy Aliguma Investing In Agricultural Innovations (Invagri) Ltd., P.O. Box 4959, Kampala, Uganda, lucyaliguma55@hotmail.com

Louis Awono Communaute Urbaine de Yaounde (CUY), B.P. 14702, Yaounde, Cameroon

Margaret Azuba Agriculture Department, Kampala City Council, Kampala, Uganda, msazuba@yahoo.com

Fred Baseke Department of Agricultural Extension, Makerere University, P.O. Box 7062, Kampala, Uganda

Rupert Best Independent Consultant, Calle 140 No. 6-57, Apartamento 102, Torre 5, Bogota, Colombia, rupertbest@gmail.com

Athanase Pr Bopda Institut National de Cartographie (INC), 779, Avenue Mgr Vogt, B.P. 157, Yaoundé, Cameroon, bopda20001@yahoo.com

Randall Brummett Worldfish, Yaoundé, Cameroon, r.brummett@cgiar.org

Sammy Carsan World Agroforestry Center, (ICRAF), Nairobi, Kenya; University of the Free State, Bloemfontein, South Africa, s.carsan@cgiar.org

Donald C. Cole Dalla Lana School of Public Health, University of Toronto, Toronto, ON, Canada, donald.cole@utoronto.ca

Sonii David Sustainable Tree Crops Program, International Institute of Tropical Agriculture (IITA), P.O. Box 135, Accra, Ghana, s.david@cgiar.org

Henk de Zeeuw ETC, P.O. Box 64, 3830 AB, Leusden, The Netherlands, h.dezeeuw@etcnl.nl

Thomas Dongmo IRAD, s/c B.P. 3214, Yaounde-Messa, Cameroon, thomas.dongmo@gmail.com

Sandrine Dury CIRAD UMR Moisa, TA C-99/15 34398, Montpellier, France, sandrine.dury@cirad.fr

Pascale Elong University of Dschang, B.P. 96, Dschang, Cameroon

Shaun Ferris Catholic Relief Services, US headquarters, 228, W. Lexington St., Baltimore, MD 21201-3413, info@crs.org

Samuel Foto-Menbohan University de Yaoundé, B.P. 812, Yaoundé, Cameroon

Jean Marin Fotso Institut de Recherche Agricole pour le Developpement (IRAD), s/c B.P. 3214, Yaounde-Messa, Cameroon, jeanmarinfotso@yahoo.fr

Will Frost formally with the World Agroforestry Center, Nairobi, Kenya

Francois Gasengayire IDRC, Eastern and Southern Africa, P.O. Box 62084, 00200, Nairobi, Kenya, fgasengayire@idrc.or.ke

Kuria Gathuru Kenya Green Towns Partnership Association, P.O. Box 69420-00400, Nairobi, Kenya, kuriagathuru@yahoo.com

James Gockowski IITA, PMB 5320, Ibadan, Nigeria, j.gockowski@cgiar.org

John Jagwe IITA, P.O. Box 7878, Kampala, Uganda, j.jagwe@iita-uganda.org

Moses Kalyebara Plan Uganda, Plot 126 Luthuli Avenue, Bugolobi, Kampala, Uganda

Christophe Kana ISSEA, B.P. 294, Yaoundé, Cameroon

Erastus K. Kang'ethe Department of Public Health, Pharmacology and Toxicology, Faculty of Veterinary Medicine, University of Nairobi, Nairobi, Kenya, ekiambi@yahoo.com

Anthony Karanja KEMRI-Welcome Trust, P.O. Box 43640-00100, Nairobi, Kenya, anthony.ngugi@shtm.ac.uk

Nancy Karanja Urban Harvest, c/o CIP, P.O. Box 25171, Nairobi 00603, Kenya, nancy.karanja@cgiar.org

Maria Kaweesa Environmental Alert, Kampala, Uganda, mkaweesa@envalert.org

Joseph Kengue Plant Genetic Resources Unit, IRAD, P.O. Box 2067, Yaoundé, Cameroon, irad-fruits@camnet.cm

Stephen Kimani Kilimo Trust Foundation, Plot 2, Serukuma Road, P.O. Box 71782, Kampala, Uganda, skimani@africaonline.co.ke

Sarah Kimeze Agency for Interregional Development (AFID), P.O. Box 28264, Kampala Plot 789, Bombo Road, Makerere Kavule, Kampala, Uganda, sarankimeze@yahoo.com

Peter King'ori Emergency Nutrition Consultant, UNICEF Rwanda, Kigoli, Rwanda, pkingori@gmail.com

Contributors

Julius Kyaligonza formally Urban Harvest, c/o Department of Veterinary Public Health and Preventive Medicine, Makerere University, Kampala, Uganda

Diana Lee-Smith Mazingira Institute, Box 14550, Nairobi, Kenya, diana.leesmith@gmail.com

Abdelrahman Lubowa College of Health Sciences, Makerere University, Kampala, Uganda, a.lubowa@yahoo.com

Wasike Mangeni Department of Sociology, Makerere University, P.O. Box 7062, Kampala, Uganda

François Meffeja Institut de Recherche Agricole pour le Developpement (IRAD), B.P. 2067, Yaoundé, Cameroon, meffeja@yahoo.fr

Richard Miiro Department of Agricultural Extension, Makerere University, P.O. Box 7062, Kampala, Uganda, rfmiiro@agric.mak.ac.ug

Luc J.A. Mougeot International Development Research Centre (IDRC), P.O. Box 8500, Ottawa, ON, Canada, lmougeot@idrc.ca

Patrick Muendo Munyao Urban Harvest, c/o CIP, P.O. Box 25171, Nairobi 00603, Kenya, patrick@farm-africa.org

George W. Nasinyama Department of Veterinary Public Health and Preventive Medicine, Makerere University, Kampala, Uganda, nasinyama@vetmed.mak.ac.ug

Robert Ngonthe The Institute for Statistics and Applied Economics (ISSEA), B.P. 294, Yaoundé, Cameroon, rngonthe@yahoo.fr

Alice Njehu International Livestock Research Institute (ILRI), P.O. Box 30709-00100, Nairobi, Kenya, anjehu@cgiar.org

Mary Njenga Urban Harvest, c/o CIP, P.O. Box 25171, Nairobi 00603, Kenya, m.njenga@cgiar.org

Christian Nolte Formally IITA, B.P. 2572, Yaoundé, Cameroon, cnoltemail@yahoo.co.uk

Robinah Nyapendi Programme Officer Governance and Trade Oxfam GB in Uganda, P.O. Box 6228, Kampala, Uganda, rnyaps@hotmail.com

Paul Opanga Elangata Wuas Ecosystem Management Programme, National Museums of Kenya, P.O. Box 40658, Museum Hill, Nairobi, Kenya, paul.opanga@bwint.org

Dennis Osino World Agroforestry Center (ICRAF), Nairobi, Kenya, d.osino@cgiar.org

Gordon Prain Urban Harvest, International Potato Center (CIP), Lima, Peru, g.prain@cgiar.org

Dannie Romney CABI Africa, ICRAF Complex, United Nations Avenue, Gigiri, P.O. Box 633-00621, Nairobi, Kenya

Anthony J. Simons World Agroforestry Center, Nairobi, Kenya, t.simmons@cgiar.org

Nelly Soua Humid Forest Station, IITA, B.P. 2008 (Messa), Yaoundé, Cameroon

Emile Tanawa Department of Geography, University of Yaoundé, B.P. 337, Yaoundé, Cameroon

Zac Tchouendjeu World Agroforestry Centre, B.P. 16 317, Yaoundé, Cameroon, z.tchouendjeu@cgiar.org

Ludovic Temple CIRAD UM R Moisa, 73 rue J.F. Breton, 34398, Montpellier, Cedex 5, France, ludovic.temple@cirad.fr

Chapter 1
The Institutional and Regional Context

Gordon Prain

How abundant is Africa's urban harvest, how much does it help feed and support the 250 million people now living in the continent's towns and cities? And how could it do this better? Crop cultivation and livestock raising have long histories in urban Africa, as in other urban areas of the world (Mumford 1961; Southall 1998), but broad awareness among researchers and policy makers of either the history or the contemporary facts of life in African urban development is much more recent. This book, which is a continuation of a research agenda on African urban agriculture begun more than 20 years ago, seeks to answer the two questions above with evidence and practical proposals for technical interventions and policy support.

The need to emphasize both technology and policy support for urban agriculture is a consequence of two other histories, of agricultural research on the one hand, and of urban institutions and policy in Africa on the other. With limited exceptions, urban agriculture has until recently been marginalized, ignored or proscribed in these histories. Before moving to a discussion of the urban agriculture research agenda in Africa and the contribution of this volume, it is important to at least touch on those two histories to help understand why the agenda has taken the form that it has.

Agricultural Research and Rural Bias

The Consultative Group on International Agricultural Research (CGIAR) is a global network of governments, multilateral organizations and private foundations that works to promote food security, poverty eradication and the sustainable management of natural resources in the developing world. Its research program is carried out through 15 commodity-based or natural resource-focused international agricultural research centres located mostly in developing countries. The earliest centres, conducting research on rice and wheat, and then the CGIAR itself, were established in the late 1960s and early 1970s, explicitly to bring agricultural science to bear on

G. Prain (✉)
Urban Harvest, International Potato Center (CIP), Lima, Peru
e-mail: g.prain@cgiar.org

the severe food crises and famines that were afflicting many rural areas of the global South at that time. Initially through productivity improvement in rice, wheat and later maize, three crops which then as now provided staple food to well over half the world's population (Plucknett & Smith 1982), the network sought to help small rural farmers feed themselves better through a science-based "Green Revolution". Through changing the input response and harvest index of the major food crops, the revolution succeeded, but perhaps in ways that were not fully intended. The successes were most significant in the high-potential production areas such as the Indo-Gangetic plain, the irrigated lowland valleys and plains of Southeast and East Asia and the maize-producing valleys of central Mexico.

These systems were the proverbial bread baskets of some of the most densely populated areas of the developing world where food shortages and famines had been severe (Sen 1981).[1] Major beneficiaries of productivity increases were thus the urban consumers, who had easier access to cheaper staple foods, and their governments, for whom food shortages and high prices represented a political challenge, especially from the more powerful urban constituency. This phenomenon became known as "urban bias" in national food policies and was used to explain "why poor people stay poor" (Lipton 1977; Bates 1981). The poor in this analysis were those in rural areas farming beyond the high-potential production areas. They benefited much less from the Green Revolution and were penalized by national food policies. This analysis became an "intellectual cornerstone" of international efforts to stimulate development in Africa and continued to shape the ideas of economic planners and policy-makers into the 1990s (Maxwell 1998). In a way, much of the subsequent history of international and national agricultural research can be seen as an attempt to rectify perceived urban bias, especially in low potential, stressed agricultural systems in Africa.

The polarization of rural and urban has even come to characterise the broader investment by public sector agencies in technology generation, perhaps to the detriment of rural areas as well as urban. Whereas agricultural technology development has been almost exclusively oriented toward addressing rural needs, research on manufacturing processes, product transformation, infrastructure and sanitation have been typically focused mainly on urban needs. The World Bank, up to quite recently, organized its major development initiatives and investments around these two poles: a Vice-Presidency for Environmentally and Socially Sustainable Development (ESSD), which included agriculture and rural development and a Vice-Presidency for Infrastructure, which included Urban Development. The World Bank is one of three principal sponsors of the CGIAR and its Chairperson has always been the Vice-President of ESSD.

Whilst agricultural research and other international development efforts seem to have been hardening the boundaries between rural and urban "sectors", those boundaries have been gradually eroded on the ground by the changing circumstances facing poor populations in the developing world. In Africa, demographic

---

[1] Sen's analysis is concerned with the failure of food entitlements among specific sectors of the developing world as an explanation for famine, rather than simply with food supply or availability

change has been historically unprecedented, with an urban growth rate averaging 5 percent over the past two decades (Kessides 2006). Currently, Africa is on average one-third urbanized, but the biggest urban growth is still to come. It is projected that 367 million people will be added to the urban population between 2000 and 2030, more than twice the level of projected rural growth, leading to a majority of the population classified as urban in about 20 years.

It is often assumed that one way rural-to-urban migration is the major cause of urban growth, but in fact two other factors are as or more important: natural increase of the urban population and the reclassification of rural areas as cities and towns expand and engulf their hinterlands. Migration itself is a complex process in Africa, more commonly seasonal and cyclical than one-way (Ellis & Harris 2004). As is discussed in detail in this book, for very many households in Africa, rural–urban boundaries are quite artificial and household members move regularly between the two, creating multi-locational households in the process (Chapter 2). In fact, there is a multitude of rural–urban linkages involving not only labour, but agricultural inputs, marketing chains, micronutrients, social ties and political obligations which have been increasingly analyzed and documented (cf. Tacoli 1998; Satterthwaite & Tacoli 2002). These are also described in the different chapters of this book.

The movements of people between town and countryside reflect poverty-driven livelihood strategies and there is indeed clear evidence of urban poverty increasing as towns and cities grow, the so-called urbanization of poverty (Haddad et al. 1999). This relationship is complex though, and estimates of urban poverty are probably underestimated because of the higher costs of non-food items and services (Amis 2002). As is to be expected, rural poverty rates continue to exceed urban rates on all measures in Africa. However, reviewing data from 19 African countries, Kessides notes that what is surprising is not that rural rates are higher, but how close urban poverty is to rural poverty in many cases. This despite the presence of factors in urban areas which should lower poverty levels (Kessides 2006, p. 17). With urban populations spending as much as 80 percent of their household budgets on food (Maxwell 1998), poverty inevitably challenges their food security and leaves limited funds for other necessities such as housing, healthcare and schooling. These are important drivers underlying cultivation of crops and livestock raising, which contribute additional food for household consumption as well as releasing cash that would otherwise be used for food for other purchases needed in the household economy.

Although the International Development Research Center (IDRC) had been a solid supporter of research and development on urban agriculture since the 1980s (see below), the United Nations Development Program was one of the first multilateral agencies to recognize the trends described above, commissioning an authoritative study on the current status of urban agriculture in the 1990s (UNDP 1996). This publication pulled together available information on the role of urban agriculture and generated new estimates of the numbers of people involved in agricultural production and marketing in specific cities and the quantities of foodstuffs consumed in cities that came from urban and peri-urban production. Several of these estimates are cited in Chapter 2. The study concluded that as many as 800 million people worldwide were involved in urban agriculture-related activities.

This publication and international agreements of the United Nations – the Habitat Agenda of 1996 and the Earth Summit Agenda 21 of 1992 (UN-Habitat 2001; UNEP 2002) – also called for agricultural research for the development of technology tailored to the specific needs of urban agriculture. Urban agriculture is not simply rural agriculture done in cities. Different constraints and opportunities that are absent or of limited significance in rural contexts affect urban agriculture. Some examples include: the need to make use of micro-spaces; poor soil quality or lack of soil; opportunities to use vertical space; abundant availability of recyclable nutrients for soils and animals; availability of waste water; need for bio-management of insect and disease pests; need to manage high risk urban contaminants, etc. Some pioneer research on some of these issues has been conducted in Cuba for the widespread urban agriculture present there and results have been widely applied (Altieri et al. 1997; Cruz & Sánchez Medina 2003).

When the CGIAR underwent its Third System Review 2 years after the UNDP publication, with the external evaluation examining past performance, current priorities and issues in need of attention, one of the UNDP authors, Jac Smit, was involved in the wide consultation process with CGIAR partners and critics. As a result, the panel's final report noted the almost complete absence of attention in the CGIAR research program to agricultural production systems in urban and peri-urban areas, and the review recommended the launch of a program to coordinate the contribution of the different centres to agricultural production and marketing in these environments. The CGIAR responded by calling for proposals for a cross-cutting research initiative on this theme the following year, eventually accepting a proposal by the International Potato Center for a Strategic Initiative on Urban and Peri-urban Agriculture (SIUPA), later renamed Urban Harvest.

The outputs and outcomes that have been produced by Urban Harvest and its partners working in Sub-Saharan Africa are reported in this book. However, as of 2009, the commitment of the CGIAR to conducting research for development on agricultural systems in and around cities was in doubt. A more recent restructuring gave no indication of continuing to address this type of agriculture but, on the contrary, renewed the commitment to commodity research on major staple crops, reminiscent of the earlier Green Revolution perspective.

The Treatment of Agriculture in African Urban Institutions and Policy

It is probable that the absence of attention to urban and peri-urban agriculture in international and in national agricultural research[2] has also been influenced by a

---

[2] Although not discussed here, attention to urban agriculture has also been absent from most national or regional agricultural research organizations and networks in Africa. One notable exception is CORAF/WECARD (West and Central African Council for Agricultural Research and Development), a network of national agricultural research institutes which identifies one of its

broader set of notions about the meaning of rural and urban. These notions – which evolved in the social and cultural history of Northern Europe and its former colonies over several centuries – have left their mark on how urban agriculture is handled by institutions and policy in Africa.

"The City in History" (Mumford 1961) has been described as the integrated development of cities and agriculture on a global scale. Cities had not only to manage links with the rural hinterland to ensure adequate supplies of food for non-agricultural city folk, but support and manage the agriculture going on within and around the city that assured its survival (Steel 2008). Bopda and Awono's study of Yaoundé in this volume describes how African cities are strongly influenced by the agricultural background of their populations, by the way agricultural products and people flow back and forth between centre and hinterland on a daily basis and by the way most cities grew up from small semi-rural settlements in an unplanned way, with agriculture continuing to be practiced on available public lands.

The history of many cities in Africa has also been strongly influenced by European settlement and colonial government (Lee-Smith & Lamba 2000). This dynamic was studied empirically in relation to food in the 1980s (Guyer 1987). Yaoundé began as a German military garrison and trading station, self-sufficient in food thanks to local production. But in common with Kenyan and other colonial authorities, both policy and repression were used in an effort to chase agriculture out of cities. This was at least partly to replicate the vision of urban society emerging from northern Europe's industrial revolution, that is, cities moving away from agrarian society toward industrialization and the wealth created through capital investment. Rapid urban growth in Europe occurred around manufacturing and service industries. It consisted of dense, low-cost housing for the industrial workforce – the future inner city slums – together with elite suburban settlements occupied by the "captains of industry" and the professional classes (Fishman 1987). Yet in terms of food, this division was as much ideological as real. Because transport systems failed to keep pace with urban growth, food supply to cities remained a problem. In England and other European countries, municipal authorities were obliged to "allot" small plots to workers' families for food production (Burchardt 1997). Reduced in size or changing location, these allotment gardens have never left European cities.

Driven by the search for sources of raw materials as well as for new consumer markets, European economies exported the divide between "rural" and "urban" to their colonies, with efforts made to keep "rural" local populations out of towns except for the provision of services to the colonists (Tibaijuka 2004). Yet the legislation enacted during the colonial period and after independence is by no means clear-cut. Dick Foeken has pointed out several contradictions in the legislative situation in East Africa, sometimes supporting food production for obvious practical reasons, but always allowing for its control or elimination, most often on

---

outputs as strengthening peri-urban systems. In another indication of a changing perspective, the Forum for Agricultural Research in Africa (FARA) co-hosted a side event on urban horticulture during its 2008 annual meeting.

health grounds, leading to many internal contradictions among different legislations (Foeken 2005, p. 6). And as David and colleagues report in Chapter 6, urban agriculture in Kampala was more tolerated there because it was an institution – like 'mailo' land tenure – that partly survived because of its association with the royal house of Buganda, enlisted in the 'indirect' system of colonial rule. Yet in Uganda as in Kenya and Cameroon, the confused statutes or regulations resulted in a marginal form of urban agriculture, sometimes tolerated by city authorities, but practiced with insecurity and uncertainty by producers, who faced the regular possibility of harassment and intimidation. As the authors of the Yaoundé study put it: "urban agriculture has been playing hide-and-seek with urban management for a century".

Official attitudes to urban agriculture in the region seemed to change during the world economic crisis in the 1980s and early 1990s. The implementation of the IMF's structural adjustment programs following the crisis led to reduced subsidies, decreased investments in infrastructure, lower farm incomes and increased urban unemployment among public sector employees. Agriculture became a recognized survival strategy for many families, including government employees, some now out of work (Maxwell 1994; Bopda & Awono, this volume). Both Maxwell and later Page (2001) have noted that during this period, with households facing individual crises, the practice of urban agriculture in effect substituted many of the state's social security functions and to some extent blunted political opposition to many of the economic "adjustments". Chapters 2 and 15 discuss the extent to which stakeholder dialogue and platform-building overcomes this lack of participation in the political process through empowerment of producers and involvement in policy development and change.

An Urban Agriculture Research Agenda for Africa

The research for development agenda on urban agriculture to which this book contributes began in the 1980s, coinciding with the earliest responses to structural adjustment among local urban populations (Maxwell 1998). The collection of summary papers brought together under the title "Cities Feeding People" (Egziabher et al. 1994) – mostly based on research supported by the International Development Research Centre (IDRC) – represented a large part of what was known about the practice in Sub-Saharan Africa at that time. The aim of that early work was to understand the extent and type of agricultural practices in cities, the relative involvement of women and men and the degree to which production was for subsistence or commercially motivated.

That volume worked with a straightforward definition of urban agriculture as "not merely the growing of food crops and fruit trees but ...also...the raising of animals", while acknowledging the difficulty of defining the terms "urban" and "peri-urban" (Tinker 1994). Plenty of ink has been spilt since on the issue of defining urban agriculture. A more elaborate definition was provided by (Luc Mougeot 2000, p. 10):

1 The Institutional and Regional Context

> Urban agriculture is an industry located within (intra-urban) or on the fringe (peri-urban) of a town, a city or a metropolis, which grows or raises, processes and distributes a diversity of food and non-food products, (re-) using largely human and material resources, products and services found in and around that urban area, and in turn supplying human and material resources, products and services largely to that urban area.

Industry is here understood in a sectoral sense involving the totality of production, processing and use. It includes the idea of production units consuming their production as well as products being sold in the market. In other words, it does not prejudge commercial versus subsistence as aims of urban agriculture. A second important point is the idea of an integration of urban agriculture activities with the human and material resources of the urban area through recycling of inputs and outputs. The spatial dimension of urban agriculture is the third key point. It is located along a continuum from the peri-urban fringe or interface, where it blends into rural type agriculture in some locations but is significantly affected by urban processes and actions in others (brick-making, leisure centres, dormitory towns, etc).

Many of the early studies were particularly concerned with the contribution of local food production to urban food security. At the time, "feeding the cities" was a major policy concern, albeit from a food supply perspective and particularly looking at rural–urban food flows rather than local production, which in the 1980s was "an absolute anathema to local planners" (Maxwell 1998, p. 15). These early studies disturbed the simplistic picture of a rural–urban divide.

The policy and institutional dimension of urban agriculture was one of two subsequent themes prioritized within this research agenda during the latter part of the 1990s and the advances and continuing challenges were summarized in a conference in Havana in 1999 with the title, "Growing cities, growing food: getting urban agriculture on the policy agenda" (Bakker et al. 2000). Throughout the past decade there has continued to be considerable focus on opening institutional space for agriculture in urban administrations, making broad policy declarations, reviewing and revising legislation at local and national level, and understanding the kinds of governance mechanisms which can achieve and support these changes.

The second theme was health. One of the grounds for local government proscription or repression of urban agriculture had always been the health risks it posed to urban populations through being an enabling environment for disease (moisture, dirt, animals) or a direct pathway through plants taking up pathogens or poisons (Birley & Lock 1999; Lock & de Zeeuw 2003). Yet conversely, one of the major potential contributions of urban agriculture was food and nutrition security and for both of these potential outcomes, evidence was needed. The companion volume to this book (Cole et al. 2008) describes the recent history of this research (see Chapter 9). It is worth pointing out that during the 1990s food security and especially nutrition has fallen off urban planning and development priorities, with almost all the substantial work having been done by nutritionists and other researchers outside of urban management or planning (Maxwell 1998).

Perhaps partly because of this body of work conducted during the 1990s, including the conclusion that urban food security could become the "greatest humanitarian challenge of the next century" (Atkinson 1995, p. 152), urban food and nutrition security did appear to return to the priorities of policy makers in Africa during the first years of the 21st century. At the level of international policy-making, food security and urban agriculture were highlighted in the Habitat Agenda emerging from the UN HABITAT II meeting in Istanbul 1996, but failed to make it into UN-Habitat's program at that time. A session on food security was however part of the World Urban Forum in Nairobi in April 2002, and a month later a workshop on "Urban Policy Implications of Enhancing Food Security in African Cities" was also convened in Nairobi and attended by mayors and city officials from all over the continent. It reviewed the role of urban and peri-urban agriculture (UPA) and rural to-urban food flows in confronting urban food insecurity and how to strengthen these through policy.

At these and other meetings on the same issue which took place in Africa at that time (see Chapter 14 below), declarations were made and many officials returned to their cities with a changed view of urban agriculture, determined to strengthen local food production as a strategy for urban poverty alleviation (FAO 2002). Yet the circumstances of agriculture in African cities in 2002 were not well known. The pioneering studies carried out in the 1980s had provided vital new information from major African cities, but important knowledge gaps remained, along with the possibility that significant changes might have occurred in the following decades. As the CGIAR Strategic Initiative on Urban and Peri-urban Agriculture (SIUPA, later Urban Harvest) was established in 2000, it was seen as important that such forums fed into and were linked to the setting of the research agenda. The numerous initiatives by IDRC and the partners it supported (see Chapter 14) also fed into the setting of the research agenda. Especially important was an emerging integrated vision of urban agriculture as technical practices offering economic benefits and entailing potential health impacts – both positive and negative – in an institutional and policy context that could make or break it.

As it set out to harness the technical skills and capacities of the international agricultural research centres for improving urban and peri-urban agriculture, Urban Harvest set its African research agenda through an analysis of the research gaps and needs during a multi-stakeholder planning meeting in Nairobi in 2000 (Chapters 2 and 14). Representatives of six African cities, as well as international and national researchers and urban development specialists, attended that meeting and three countries – Cameroon, Uganda and Kenya – answered the call for proposals that followed. This book describes that process, containing the results of the research that was undertaken and the story of the interventions that were made in urban governance.

The book is divided into five sections, the first setting the context and summarizing the research findings, then a section on the research from each country, Cameroon, Uganda and Kenya, followed by the final section addressing the way forward in the light of potential institutional change. Following this introduction on the institutional and regional context, Chapter 2 provides an analysis of the results from all the research activities in a common framework. This is essentially the framework provided by the three major research themes established through the stakeholder meeting in 2000:

1   The Institutional and Regional Context

- Urban agriculture, livelihoods, and markets;
- Urban ecosystem health; and
- Policy and institutional dialogue and change.

Chapters 3, 4, and 5 contain the results of the research undertaken in Cameroon and led by scientists from the International Institute for Tropical Agriculture (IITA). The data are from Yaoundé, but a national perspective informs the institutional study in Chapter 5. Chapters 6, 7, 8, and 9 report the results of the collaborative research for development that was undertaken in Uganda. The results in Chapters 6, 7, and 8 are all from work carried out in Kampala led by scientists from the International Centre for Tropical Agriculture (CIAT), while Chapter 9 summarizes a major study – also carried out in Kampala – on health and urban agriculture. While initiated by Urban Harvest and the University of Toronto, this health study generated a research-to-development institution whose history is also analysed in the book's conclusions. The section on Kenya (Chapters 10, 11, 12, and 13) presents data from three cities, Nairobi, Nakuru and Kisumu, supported by the Urban Harvest Programme. The Nairobi study in Chapter 10 was led by the International Livestock Research Institute (ILRI), while Chapters 11 and 12 represent follow-up research in Nakuru. Chapter 13 describes a study supported by Urban Harvest as a result of its Anglophone Africa Training Course on Urban Agriculture held in 2004, and led by the World Agroforestry Centre (ICRAF), which was also involved in the Nairobi study described in Chapter 10.

In the concluding section, Chapter 14 begins by returning to the international institutional context of urban agriculture. It looks back in some detail at the urban agriculture research agenda as implemented in the Sub-Saharan African region over the past 6 years or so by IDRC, Urban Harvest, the Resource Centre for Urban Agriculture and Food Security (RUAF) and other partners. Finally, Chapter 15 provides an in-depth analysis of the kinds of policy and institutional change related to urban agriculture that have been achieved in the three countries through the partnership platforms and stakeholder dialogue established in the different cities.

References

Altieri, M, Rosset, P & Nicholls, C 1997, 'Biological control and agriculture modernization: towards resolution of some contradictions', *Agriculture and Human Values*, vol. 14, pp. 303–310.

Amis, P 2002, 'Thinking about chronic urban poverty', *Chronic Poverty Research Centre (CPRC) Working Paper no. 12*, World Bank, Washington, DC.

Atkinson, S 1995, 'Approaches and actors in urban food security in developing countries', *Habitat International*, vol. 19, no. 2, pp. 151–163.

Bakker, N, Dubbeling, M, Gundel, S, Sabel-Koschela, U & de Zeeuw, H (eds) 2000, *Growing cities, growing food: urban agriculture on the policy agenda. A reader on urban agriculture*, Deutsche Stiftung fur internationale Entwicklung (DSE), Germany.

Bates, R 1981, *Markets and states in tropical Africa*, University of California Press, Berkeley, CA, USA.

Birley, M & Lock, K 1999, *The health impacts of peri-urban natural resource development*, Liverpool School of Tropical Medicine, Liverpool.

Burchardt, J 1997 'Rural social relations, 1830–50: opposition to allotments for labourers', *Journal of Agricultural and Rural History*, vol. 45(part II), pp. 165–175.

Cole, DC, Lee-Smith, D & Nasinymama, GW (eds) 2008, *Healthy city harvests: Generating evidence to guide policy on urban agriculture*, CIP/Urban Harvest and Makerere University Press, Lima, Peru.

Cruz, M & Sánchez Medina, R 2003, *Agriculture in the city. A key to sustainability in Havana, Cuba*, Ian Randell, IDRC, Kingston, Jamaica.

Egziabher, AG, Lee-Smith, D, Maxwell, DG, Memon, PA, Mougeot, LJA & Sawio, CJ 1994, *Cities feeding people: an examination of urban agriculture in East Africa*, IDRC, Ottawa, ON.

Ellis, F & Harris, N 2004, 'New thinking about rural and urban development', *Keynote paper for department of international development sustainable development retreat*, London.

Fishman, R 1987, *Bourgeois utopias: the rise and fall of suburbia*, Basic, New York, NY.

Foeken, D 2005, Urban agriculture in East Africa as a tool for poverty reduction: a legal and policy dilemma? *ASC Working Paper 65/2005*, African Studies Centre, Leiden, Netherlands.

Food and Agriculture Organization (FAO) 2002, *Declaration: feeding cities in the Horn of Africa*, FAO Newsroom, www.fao.org/english/newsroom/news/2002/4820-en,html [Accessed 9 May 2009].

Guyer, J (ed) 1987, *Feeding African cities: studies in regional history*, Manchester University Press, Manchester.

Haddad, L, Ruel, M & Garrett, J 1999, 'Are urban poverty and undernutrition growing? Some newly assembled evidence', *World Development*, vol. 27, no. 11, pp. 1891–1904.

Kessides, C 2006, *The urban transition in Sub-Saharan Africa. Implications for economic growth and poverty reduction*, Cities Alliance, SIDA, World Bank, Washington, DC.

Lee-Smith, D & Lamba, D 2000 'Social transformation in a post colonial city: the case of Nairobi', with Lamba, D in Polese, M & Stren, R (eds) *The social sustainability of cities: diversity and the management of social change*, University of Toronto Press, Toronto, ON.

Lipton, M 1977, *Why poor people stay poor*, Harvard University Press, Cambridge, MA.

Lock, K & De Zeeuw, H 2003, 'Health and environment risks associated with urban agriculture' in *Annotated Bibliography on Urban Agriculture*, SIDA & ETC-RUAF, CD-ROM, RUAF, Leusden.

Maxwell, D 1998, *The political economy of urban food security in Sub-Saharan Africa*, FCND Discussion Paper No. 41, International Food Policy Research Institute, Washington, DC.

Maxwell, DG 1994 'The household logic of urban farming in Kampala', in Egziabher, AG, Lee-Smith, D, Maxwell, DG, Memon, PA, Mougeot, LJA & Sawio, CJ (eds) *Cities feeding people: an examination of urban agriculture in East Africa*, IDRC, Ottawa, ON, pp. 45–62.

Mougeot, LJA 2000, 'Achieving urban food and nutrition security in the developing world: the hidden significance of urban agriculture', *A 2020 Vision for Food, Agriculture, and the Environment, Focus 3, Brief 6 of 10*, International Food Policy Research Institute (IFPRI), Washington, DC.

Mumford, L 1961, *The city in history: its origins, its transformation and its prospects*, Harcourt, Brace and World, New York, NY.

Page, B 2001, 'Urban agriculture in Cameroon: an anti-politics machine in the making?' *Geoforum*, vol. 33, pp. 41–54.

Plucknett, DL & Smith, NJH 1982, 'Agricultural research and Third World food production', *Science*, vol. 217, pp. 215–220.

Satterthwaite, D & Tacoli, C 2002, 'Seeking an understanding of poverty that recognizes rural-urban differences and rural-urban linkages', in Rakodi, C & Lloyd-Jones, T (eds) *Urban Livelihoods: a people-centred approach to reducing poverty*, Earthscan, London, pp. 52–70.

Sen, A 1981, *Poverty and famines: an essay on entitlement and deprivation*, Oxford University Press, UK.

Southall, A 1998, *The City in time and space*, Cambridge University Press, Cambridge.

Steel, C 2008, *Hungry city: how food shapes our lives*, Chatto and Windus, London.

Tacoli, C 1998, 'Rural-urban interactions: a guide to the literature', *Environment and Urbanization*, vol. 10, no. 1, pp. 147–166.

Tibaijuka, AK, 2004, *Africa on the move: an urban crisis in the making*, a submission to the Commission for Africa, Commission for Africa, Nairobi.

Tinker, I 1994, 'Urban agriculture is already feeding cities', in Egziabher, AG, Lee-Smith, D, Maxwell, DG, Memon, PA, Mougeot, LJA & Sawio, CJ (eds), *Cities feeding people: an examination of urban agriculture in East Africa*, IDRC, Ottawa, ON.

UNDP 1996, 'Urban agriculture: food, jobs and sustainable cities', *United Nations Development Program, Publication Series for Habitat II, volume1*, UNDP, New York, NY.

UNEP (United Nations Environment Programme) 2002 *Agenda 21: global programme of action on sustainable development*, http://www.un.org/esa/dsd/agenda21/.

UN-Habitat 2001, *The Istanbul declaration and the Habitat Agenda*, HS/441/97/E, Nairobi. http://www.unhabitat.org/downloads/docs/2072_61331_ist-dec.pdf.

Chapter 2
Urban Agriculture in Africa: What Has Been Learned?

Gordon Prain and Diana Lee-Smith

Introduction

Though the crisis in world food prices exploded during 2008, the problem of urban food insecurity in Africa has been a fact of life for many low-income urban dwellers for decades, and especially since the period of structural adjustment in the 1980s (Maxwell 1995). It is not that there is no food; it's that poor urban consumers cannot afford it. This is the stark but simple truth lying behind much of the agriculture that is widespread within and around African cities. What urban households have known and practiced for generations, urban decision-makers have begun to recognize much more recently: urban agriculture is a livelihood strategy.

The series of meetings convened on this theme around 2002 as described in the previous chapter began to discuss policies that might give more support to the practice. But as the earlier regional meetings convened by Urban Harvest had made clear, information was lacking about urban agriculture in the early 21st century. Which socio-economic groups were now most actively involved? Was food security or wealth generation the major goal for urban families? What was the relative contribution of livestock and crop production to the urban economy and society? How was agriculture combined with other occupations in household livelihoods and what were the roles of men and women? To what extent did it mobilize urban, peri-urban, and rural natural resources for productive ends, and what conflicts existed, especially for land and water? Was urban agriculture safe? Did its health benefits outweigh any possible risks?

Furthermore, anachronistic legislation prohibiting agricultural activities of different kinds continued to be on the books of many African cities, keeping open opportunities for harassment and corruption on the side of the authorities and insecurity on the side of the producers. Those seeking to change the legal and administrative frameworks toward more enabling regulations and by-laws needed greater

---

G. Prain (✉)
Urban Harvest, International Potato Center (CIP), Lima, Peru
e-mail: g.prain@cgiar.org

evidence of the positive contribution of urban agriculture to poverty alleviation and assurances that it was not a major pathway for health hazards (Cole et al. 2008).

It is in this context that the stakeholder consultations in which Urban Harvest participated between 2000 and 2002 led to the identification of three key areas for its research:

- *Urban agriculture, livelihoods, and markets.* What is the contribution of agriculture to urban livelihoods? How does it contribute to household food security, savings on food purchases, or generating income? How does it vary in terms of production, processing, marketing, and household consumption systems along the rural–urban transect? What technology interventions can enhance agricultural contributions to livelihoods?
- *Urban ecosystem health.* What are the positive or negative contributions of urban agriculture to the urban ecosystem and human nutrition and health? Does it recycle urban and peri-urban liquid and solid wastes, thus contributing to a healthier, more productive urban ecosystem? Does it provide pathways for diseases or their vectors to enter the urban ecosystem? What are the nutritional benefits? What are the feedback mechanisms between people's actions and population, community and environmental health?
- *Policy and institutional dialogue and change.* What national or local policies influence the practice of agriculture in cities? Which institutions are involved? What methods can be developed for building communication and consensus among the different stakeholders? What institutional alliances or platforms can support more sustainable urban agriculture? How can policy and regulation be improved and agriculture institutionalized in local governments?

The research presented in this book advances understanding of agriculture in selected African cities in relation to this thematic structure. In complex city ecosystems, which include informal economies and social networks, poor households depend on multiple income sources and a wide range of non-material assets to ensure their livelihood. Inadequate assets can leave households vulnerable to economic, environmental, health, and political stresses and shocks. This is the vulnerability context. Drawing on existing conceptual frameworks in sustainable rural livelihoods and urban livelihoods research (Farrington et al. 1999; Rakodi & Lloyd-Jones 2002) and in the area of ecosystem health, a research framework was elaborated linking the three themes to the major areas of urban development (Fig. 2.1).

This chapter reviews what has been learned about urban agriculture in Cameroon, Kenya, and Uganda, in relation to the three themes of livelihoods, ecosystem health, and policy and institutional dialogue and change. In relation to the third theme, we focus here on some outcomes of policy dialogue, but return to the subject in Chapter 15, especially to describe the change process. And, while we review here the learning from the three countries on ecosystem health, it should also be noted that the topic of health and urban agriculture is dealt with in greater detail in our companion book (Cole et al. 2008).

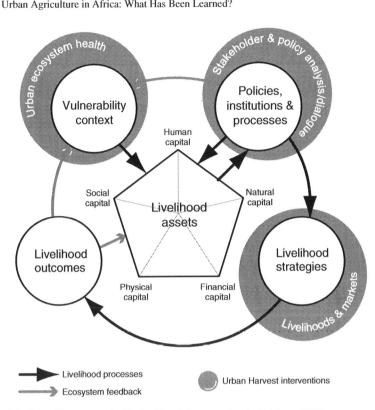

Fig. 2.1 Urban Harvest sustainable livelihoods framework, adapted from DFID

Urban Agriculture, Livelihoods, and Markets

Conceptual Approach

As seen in Fig. 2.1, five types of capitals or assets are distinguished, from which households develop their livelihood strategies:

Natural capital mainly involves the amount and quality of accessible land, water, and biodiversity. Households access these in a variety of ways ranging from formal land titles or membership in irrigation associations to casual cultivation of public spaces and illegal or informal use of wastewater and solid organic wastes.

Physical capital includes buildings, equipment, domestic animals, transport, seeds, and other inputs. Livestock-raising depends on the physical capital of the animals themselves and the housing which affects their health and productivity. Crop production needs equipment and sometimes structures like screen houses. Means of transport can dramatically change access to inputs and the marketing possibilities of crops and animals.

Human capital includes manual labour, different types of practical skill, different types and sources of knowledge, and good health or wellness. The human capital of a household includes the different knowledge and skills of women and men, the indigenous knowledge of the older generation, and the modern education of the young.

Financial capital is made up of available income and savings, and also formal and informal credit access. Because this type of capital depends on relations of trust, it is closely related to the fifth type of asset, social capital.

Social capital includes support acquired through formal or informal membership in networks and groups, often involving different kinds of reciprocal services, including the exchange of psycho-social welfare. Trust is the currency needed to enter these arrangements and it is strengthened or weakened through participation in them.

Households deploy these assets in livelihood strategies – petty trading, self or hired employment, agricultural production as well as migration, joining organizations, seeking formal education or training, and so on. Household members engaging in these strategies must also engage with the structures and processes that make up the public life of the city. These include institutions such as markets, local government, educational systems, policies and regulations about doing business, about keeping animals, about using land or water resources, or about the handling of wastes. The types of resources a household commands will determine how far these structures and processes can be influenced favourably – through a social connection or financial resources to buy compliance – and how far they impede the household from implementing livelihood strategies and achieving livelihood outcomes, for example, improved food security or more disposable income to use in education of children.

These livelihood processes and outcomes can in turn exert positive and/or negative ecosystem feedback on the livelihood assets – through increasing or decreasing certain capitals – and on the vulnerability context.

Livelihoods in Space

The spatial dimension is an important focus of the studies reported in this book. There are two aspects to this. First, livelihoods vary along the continuum from rural to inner urban location because of differences in the presence of and access to assets; differences in the vulnerability context; and differences in the institutions, policies, and processes that households must deal with. Secondly, urban households may have spatially dynamic livelihood strategies, with members drawing on assets widely distributed within and beyond the urban areas and seeking opportunities to deploy household assets in different places. In some cases, this leads to multi-locational households (Baker & Akin Aina 1995).

The study of Kampala by David et al. (Chapter 6) analyzes how livelihoods and the role of agriculture vary along the continuum from low-density peri-urban areas around the city to heavily populated settlements near the city centre. Kampala City Council classifies areas where agriculture is found in four types: peri-urban (peripheral), peri-urban transition, urban new (dense slum), and urban old, using criteria of population density, land availability, and the prevalence of crop and livestock production. These types were found to differ with respect to natural capital assets, specifically the amount of land available for farming and access to water surfaces. Occupants of new slum areas of Kampala can access nearby wetlands as their main farming location whereas three-quarters of cultivation in the inner city is done on very small plots around the homestead. In peri-urban areas, plots are bigger and there is greater choice of location. Local races of livestock are likewise more common in peri-urban areas where they can be free-range or grazed, whereas zero-grazing and bird cages are essential in the space-constrained urban old and new areas where improved breeds do better.

The availability of physical capital along the rural–urban continuum also shapes agriculture. The further one moves into Kampala city, the more common it is for producer households to live in rented accommodation, with correspondingly higher levels of instability and greater likelihood of limited cultivation of fewer, shorter duration crops. The exception is the cultivation of banana, widely found in inner urban areas, not only in Kampala but also in Yaoundé.

The various studies in Yaoundé analyzed urban agriculture spatially in relation to its changing political boundaries, although there was no consistent or official classification of the continuum as in Kampala. The fertile inland valleys (a type of wetland locally referred to as *bas-fonds*) are important agricultural systems in urban Yaoundé, where they act as sinks for urban wastes used as nutrients and are closely linked to urban markets. The dominant production systems identified are also correlated with a spatial distribution related to livelihood assets. Commercial crop farming dominates in the inner urban setting with its access to the *bas-fonds*. Nearly all the household-based livestock enterprises had a commercial orientation, the larger ones in the peri-urban areas and the smaller ones in the urban areas. Like Kampala, Yaoundé is a tropical highland city. As the density of housing, businesses, and roads declines in the peri-urban areas, households farm larger mixed plantings of maize, leafy vegetables, and root crops as well as keeping livestock.

The second dimension of a spatial analysis of livelihoods and urban agriculture concerns the way producing households seek to access different assets and opportunities to deploy their resources in different locations as part of their livelihood strategies. In Nairobi, market gardeners in urban and peri-urban areas access compost produced from domestic organic wastes by urban recyclers and manure from Maasai cattle-herders in Nairobi's rural hinterland (Chapter 10). In Nakuru, Kenya (see Chapter 11), livestock-keepers use the manure from urban-raised livestock to fertilize crops on both urban and rural plots, underlining the importance of multi-locational households and casting doubt on the usefulness of maintaining strict distinctions between urban and rural livelihoods (Satterthwaite & Tacoli 2002, p. 55; Simon et al. 2006).

Very often agriculture-related and other types of livelihood strategies explore opportunities in a geographical region or sub-region, accounting for the flows of cash, food, social support, and cultural commitments across this space. In Yaoundé, one of the most common forms of association among urban women producers is based on common ethnic links to home villages, a pattern commonly found right across East and Central Africa (Chapter 3). The same chapter describes how similar patterns of trust (social capital) drive the market flows of cassava into Yaoundé.

Who Is Using Agriculture as an Urban Livelihood Strategy?

So far we have been talking about urban livelihoods and agriculture in relation to households in general, but through these studies we understand that both low- and high-income urban households are involved in agriculture, for different reasons and with different strategies. We also recognize that household members have different degrees of involvement in agriculture and that women contribute in a major way to livelihoods through farming.

Relatively few studies of urban agriculture have provided accurate quantitative assessments of the urban population involved, or the proportion of those farming who are women. Figure 2.2 presents most of the available data on this issue from previous studies in Sub-Saharan Africa, though one needs to be cautious with the very different types of data sources and their reliability. Aside from a couple of outliers on the lower and upper limits (Accra and Addis), there is nevertheless some consistency across these studies, which have a median percentage of those farming of 35 percent, of which around two-thirds are women.

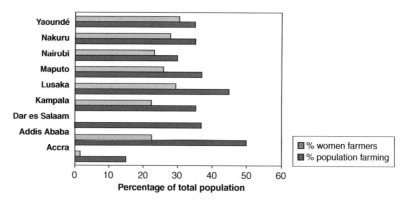

Fig. 2.2 Proportion of urban population farming in selected cities of Sub-Saharan Africa
Source: Yaoundé: UNDP (1996); Nakuru: Foeken (2006); Nairobi: Mwangi and Foeken (1996); Lee-Smith (2001); Maputo: UNDP (1996); Lusaka: UNDP (1996); Kampala: Nabulo et al. (2004, 2006); Dar es Salaam: Sawio (1998); Addis: Yilma (2003); Tegegne (2004); Accra: Obuobie et al. (2004); Maxwell et al. (2000)
Note: the 15% figure for Accra is based on inner urban areas and the estimate for peri-urban is thought to be much higher

The studies reported in this volume on Yaoundé and Kampala throw more detailed light on these earlier estimates. While the Yaoundé study did not sample both farming and non-farming populations to estimate the prevalence of agriculture as a livelihood strategy, it did examine official statistics. Although these show that farming increased almost 10-fold during 1957–1987, this was far smaller than the growth of the overall population (almost a 100-fold increase), meaning the farming population fell from about 18 to 2 percent of the economically active population. As the authors point out, this grossly underestimates the situation in a city where 60 percent of the 150 km^2 urban area is still dedicated to agriculture and where farming is a component of mixed livelihood strategies, mostly carried out by women. Indeed, the empirical studies by these authors indicate that women account for 87 percent of the urban farming population, including 79 percent of those primarily growing commercially. So, the facts that agriculture is often part of mixed livelihoods, that it is mostly women's work (often invisible to official statistics), and that in Yaoundé as in many other African cities its legal status is ambiguous, combine to make these official statistics highly unreliable. The UNDP figure, given in Fig. 2.2, was more likely correct.

The Kampala study provides us with a more fine-grained understanding of the significance of agriculture within the overall urban population. As discussed above, Kampala City Council (KCC) and the team of researchers working with them have viewed urban agriculture in spatial terms and recognized that the phenomenon will be different at different points along the peri-urban to urban continuum. Drawing on a KCC census of selected urban and peri-urban parishes and case studies of selected locations along the continuum, the authors were able to confirm this insight. They found that the involvement of the population in farming varied from just over 25 percent in a newly urbanized area of Kampala to 96 percent in one peri-urban zone. Whereas a second urban zone sampled showed a similar proportion of households involved in agriculture (28 percent), suggesting similar levels of agriculture in more built-up areas of the city, the second peri-urban area sampled was found to have 38 percent of households in farming. The big variability in the importance of farming at the peri-urban interface underlines the enormous range of activities in this transition zone (Simon et al. 2006).

The KCC census found other demographic differences, such as larger households in the peri urban compared with the urban areas, the peri-urban households being closer in size to rural households, which have on average 4.9 members in Uganda compared with 4.2 in urban areas (Uganda Bureau of Statistics [UBOS] & ORC Macro 2001). Yet the survey of urban farming households showed a much larger size of seven persons in all four areas studied, urban and peri-urban, consistent with Maxwell's earlier findings of a statistically significant difference between farming and non-farming urban households in Kampala (Maxwell 1995). In Yaoundé, the size of producing households was also found to be large, averaging 7.9 persons compared to 6.6 for all households in the city. The median farming household size in Nakuru, Kenya, was also found to be 5–7 persons compared to 2–4 persons for non-farming households (Foeken 2006, p. 181). It is not clear whether households that produce are poorer and need to farm for food security, or whether farming enables them to support more people.

We learn more about this when we consider the socio-economic status of producing households. Then it becomes clear that, while most urban producing households are indeed poor, this is mainly because of greater family size. Earlier studies in the region have shown that in some cities such as Nakuru, the poor are proportionally less represented among urban agricultural producers than the better-off (Foeken 2006) and that this is particularly marked among households raising larger livestock such as cattle. Given the investment and maintenance costs of large livestock, this is understandable. Thus in the Nakuru studies on crop–livestock interactions and the health risks associated with urban dairy production, the socio-economic profile of the sample was biased toward better-off households because of their greater involvement in this sector (Chapters 11 and 12). In Kampala, farming households in the peri-urban areas are better off than those in the outer and inner urban areas, measured by income, access to land, and house ownership.

Contribution of Agriculture to Household Income and Savings

There is now clear evidence that agriculture in rural areas is no longer the single activity of families, nor even, in many cases, the main activity (Ellis 2000; Bebbington 1999). In peri-urban and urban areas of the developing world, the diversity of livelihoods is even more in evidence. Keith Hart, a British anthropologist who coined the term "informal sector" to describe the employment situation in African cities, provided the following description from 1960s Accra:

> Mr. A. D. worked as a street-cleaner…as an afternoon gardener…and as a night watchman…In addition to this annual income of approximately £320, he grew vegetables on his own plot of land which brought in another £100 or so. (Hart 1973, p. 66)

This person had been 20 years in Accra, showing such behaviour is not restricted to short-term migrants. In fact, it is a way of life for millions of urban Africans. Hart points out the rarity of a single income stream for low-income urban families. With the urban population in Sub-Saharan Africa more than four times what it was when Hart conducted his study, the urban employment situation has become even more acute and the practice of multiple livelihoods strategies more intense (Kessides 2006).

In all the cities studied in this volume, agriculture is clearly only a part of diverse livelihoods (Table 2.1) while still providing a significant contribution to income. Formal employment, business, and trade tend to dominate, except for the commercial producers in the inland valleys of Yaoundé, who can sell dry-season vegetables grown using waste-water irrigation for more than double the wet-season price. Few employment alternatives are as lucrative and incomes are estimated to be about 50 percent above the minimum wage. Likewise, the Kampala study notes that its findings, of 70 percent of heads of farming households earning more than

2 Urban Agriculture in Africa: What Has Been Learned?

Table 2.1 Principal income source for households involved in urban agriculture

	Yaoundé		Kampala	Nakuru
	"Commercial"	"Subsistence"		
Farming	70	33	22.9	46.6
Non-farming	30	67	77.3	53.3
Business/trading			35.8	15.5
Casual employ			14.5	15
Formal employ		67	27	22.8

national annual income per capita (US$330), are consistent with other studies of urban farming.

The spatial pattern in Yaoundé, with intensive commercial vegetable and maize production in the valleys running through the inner city and less commercial activity in the peri-urban uplands, is another example of how location within the urban environment affects opportunity and strategy and the types of households involved. Agriculture is the major livelihood strategy for producers in the well-endowed urban inland valleys, whereas for those dependent on rain-fed upland agriculture in peri-urban areas it is more of a food security strategy that supplements other income sources. On the other hand, Kampala is much wetter than Yaoundé and has a more abundant rain-fed agriculture in the peri-urban areas including many low-lying wetlands next to Lake Victoria. Consequently, agriculture plays a bigger part both in the deployment of household labour and in generating income than in the built-up urban areas (Figs. 2.3 and 2.4).

At first glance the Yaoundé pattern of income from agriculture seems to reverse that of Kampala described above, but further analysis is needed. As has already been mentioned, in both these systems in Yaoundé, women are the primary household members involved in farming, but in the uplands, households have not only more

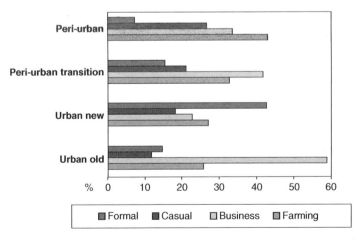

Fig. 2.3 Primary activities of farming households along peri-urban to urban continuum, Kampala (% of households surveyed)

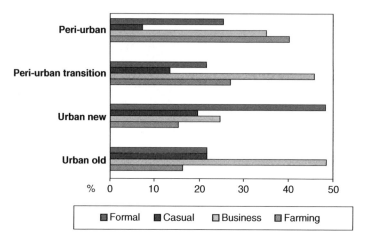

Fig. 2.4 Main sources of income of farming households along peri-urban to urban continuum, Kampala (% of households surveyed)

income streams but also higher education levels than those cultivating the inland valleys. This latter variable has been strongly associated positively with lower levels of poverty in Cameroon and elsewhere (World Bank 2004) and underlines the need to look at crop and animal production in the broader context of income, food security, and social capital.

For almost all farming households in the cities examined in this volume, agricultural produce is for eating as well as selling. Commercial producers of traditional leafy vegetables in the inland valleys of Yaoundé still consume about 25 percent of these vegetables themselves, about 50 percent of their "commercially produced" maize, and raise other crops specifically for home consumption. In Nakuru, 50 percent of the kale which is eaten – kale is one of the most important vegetables that accounts for 21 percent of the value of foods consumed – comes from their own plots. This illustrates the importance of local food production not only as a direct food source for enhancing food security but also as a means of saving income for other purposes. The authors of the Kampala study make this same point: "food production as a form of savings was still clearly the main purpose for which households were farming in 2003."

The Kampala data suggest that this trend is more marked in peri-urban than urban areas, where commercial sales account for 19 percent in one urban case and 52 percent in another. In the peri-urban areas surveyed, commercial sales accounted for only 24 and 13 percent. As already noted, especially in Kampala and Nakuru, commercialization was more associated with livestock production, with its opportunities for sale of products such as milk and eggs in addition to meat. Livestock production is also associated with higher income households as discussed above. Some data from Kampala suggest further investigation is needed to explore a possible link between urban and peri-urban location and the keeping of livestock.

Women in Sub-Saharan Africa are associated with agricultural production for subsistence rather than commerce (Hovorka & Lee-Smith 2006; Chapters 3 and 12 above). Nevertheless, the data from both Yaoundé and Nakuru in Chapters 3 and 12 suggest that this relationship is a cultural norm that could be changing in the practice of urban agriculture. Women frequently carry out the majority of urban farm labour, including tasks culturally assigned to men, such as managing livestock, but may not control the income generated, as noted in Nakuru. But in Yaoundé, it is women who dominate both subsistence and commercial production. The very rich picture painted in these studies of the patterns of farming and their purposes and outputs raise possibilities for defining further research that may help improve incomes and offset hunger.

Urban Agriculture and Markets

Products need to be marketed if farmers are to derive income from agriculture, and while much effort has been invested in finding ways to "link farmers to markets" in rural settings as a key strategy for poverty reduction, there has been less attention to marketing urban produce. Several chapters in this book investigate the extent to which urban and peri-urban producers take advantage of their closer links to markets while two chapters (8 and 13) deal exclusively with marketing of crop, agro-forestry, and/or livestock products in Kampala and Kisumu, respectively. The continuing high levels of home consumption reported in these studies – even if for generating savings – suggests that urban farmers in all three countries studied could take much more advantage of their proximity to market than they currently do.

The authors of Chapter 8 seek to identify best-bet marketing opportunities, taking into consideration current demand and supply as well as production conditions in both urban and peri-urban areas of Kampala. They identified 40 products in high demand across some or all categories of outlet, from supermarkets and catering services to small groceries or kiosks. This demand was not being met, even in the small kiosks. Of the seven most important products for generating income, only one – poultry – was found right across the peri-urban to urban transect, while seven of the eleven most commonly found products were primarily food security crops. Two strategies were identified for taking better advantage of market opportunities:

- Focusing on increasing volumes of existing products with low risk, such as poultry and mushrooms;
- Focusing on products with higher returns (fruit, vegetables, and pig-raising) but developing strategies to reduce the higher risks, such as collective action through producer associations.

The authors of Chapter 13 examined value chains for agro-forestry products in Kisumu, Kenya, looking at the origins of products, marketing channels, and price margins. Products coming from urban, peri-urban, and rural production systems were differentiated, as well as those going to wholesale and retail urban markets and those being sold on to other urban centres. Three types of value chain were found, with different kinds of linkages between producers and consumers and highly variable margins between value chain actors in the three channels. The study found that peri-urban farmers were less motivated to grasp new opportunities than those in urban areas. This important difference, also found in Kampala, has been noted elsewhere (Warnaars & Pradel 2007). Thus farming populations in and around cities include both risk-averse behaviour typically associated with low-income marginal rural agriculture and the more risk-taking enterprise attitude associated with informal urban businesses. While the Kampala data link these attitudes respectively to peri-urban and urban farmers, suggesting behaviour linked to a spatial continuum, this may well vary and needs to be the subject of specific research.

Like Kisumu, Yaoundé presents opportunities for production and marketing of agro-forestry products in urban and peri-urban areas, as described in Chapter 3. In both places indigenous plant species used in culturally specific ways are involved, whether for food or medicine, and this raises other issues in relation to how livelihoods and marketing interact with complex social and environmental concerns.

The 19 types of medicinal plants being traded in Kisumu were mostly collected wild from surrounding peri-urban areas because of cultural beliefs against their cultivation.[1] Wilderness and nature conservation are important aspects of natural capital. As in many parts of Kenya, the study found a dearth of supply of fuel or construction materials, partly due to logging bans implemented in response to severe deforestation. However, the authors note the existence of urban and peri-urban open spaces in Kisumu which could be taken advantage of for tree-planting, representing a largely untapped opportunity for livelihood and market development.

It has been demonstrated for Nairobi (Basweti et al. 2001) that seedling nurseries offer an important source of income for significant numbers of people in urban and peri-urban areas and make a contribution to urban greening. The Kisumu study notes that the authorities are "comfortable" with these enterprises that essentially "squat" on public space along the main access roads, because of their contribution to the environment and to beautification.

The Kisumu and Yaoundé agroforestry studies reveal not only competition but also collaboration along market chains, including exchange of technical information and skills such as grafting. The Yaoundé study identified opportunities for seedling nursery owners in rural and urban areas to benefit economically from collaboration. Previously unidentified demand for indigenous tree species could end up linking urban enterprises, with limited access to these species but with ready access to markets, with rural enterprises, which have access to a diversity of indigenous species but poor connection to markets. The market for African leafy vegetables has been demonstrated in Yaoundé (Gockowski et al., 2003 and Chapter 3) as also in a

---

[1] Although not elaborated in the chapter, these beliefs appear to relate to the power of these medicines deriving from their wild growth and its absence in cultivated versions (Schippmann et al. 2002).

similar study in Nairobi (Mwangi et al. 2007). The similar development of market-chain information and collaboration could help in these production systems, for example in the case of nutrients from waste as agricultural inputs. This is dealt with in the section on ecosystem health below.

Unlike urban and peri-urban agriculture in general, nurseries were found to be dominated by men. In Kisumu, young adult men aged 25–40 run these businesses and men account for 90 percent of ownership in Yaoundé, a city where women otherwise are responsible for 87 percent of agricultural enterprises. These men were often well-educated, owned more than one nursery, and employed labour. It appears that the interdependence of nurseries in terms of technical support and even marketing collaboration involves networking among men from which women are largely excluded.

Urban Ecosystem Health

Conceptual Approach

Because urban agriculture is connected to so many different natural, physical, and human-designed systems, a very broad approach is needed to understand its relationship to health, including the health of livestock and the environment as well as of humans (Cole et al. 2008, p. 34). A number of the chapters in this volume address the subject within this broad approach, using a range of concepts. The term "ecohealth" itself has emerged in recent years as researchers explored some of these multiple system connections, assisted by IDRC's programme on "Ecosystem and Human Health," or "Ecohealth." Focusing on interacting social, political, economic, and ecological parameters, ecohealth broadened approaches to human health away from simply bio-medical concerns (Cole et al. 2008, p. 9). It encompasses the concept of a healthy, well-functioning ecosystem, in which human health is dependent upon ecosystem function, since humans are part of the ecosystem.

The livelihoods framework (Fig. 2.1) links the urban ecosystem health theme to the vulnerability context of households. Although external economic and political stresses and shocks such as price rises, drastic policy changes, or political upheavals often characterize households' vulnerability context, stresses and shocks are also related to the health of the ecosystem. Poor sanitation, the accumulation of wastes, disease prevalence including zoonoses, sudden epidemic outbreaks, and low levels of micronutrients in the local food system, all increase the vulnerability of households, weaken their ability to accumulate and deploy household assets and so make moving out of poverty more difficult and moving into poverty more likely.

Different chapters in this book examine the way that urban agriculture impinges on urban ecosystem health, especially its potential to increase or decrease ecosystem health risks and thus vulnerability. The role of livestock in ecosystem health is especially interesting and is addressed in Chapter 12. While livestock can introduce vital micronutrients into local food systems, they also can be the source of disease affecting humans. Chapter 9 is a summarized version of our companion book addressing how to ensure "Healthy City Harvests" from urban agriculture (Cole et al. 2008). While drawing on concepts and perspectives in the field of public health, it is rooted in an urban ecosystems health approach. For example, an important public health tool utilized in the studies reported in Chapter 9 is Health Impact Assessment (HIA), which has often focused heavily on risk assessment. But studies of urban agriculture, including those reported in this chapter, have retained a balanced examination of both health benefits – such as from food security and improved nutrition – and health risks – such as from chemical and biological contaminants (Lee-Smith & Prain 2006).

Many of the chapters of this book seek to understand the positive and negative impacts on ecosystem health of different urban agricultural activities, such as the case of livestock mentioned above. Research in Nairobi and Nakuru (Chapters 10 and 11) not only report the potential environmental and economic benefits of recycling solid wastes for composting, but also show how their use in agriculture can be a pathway for negative human health effects, especially when the wastes contain heavy metals. These chapters take a broad ecosystem perspective of the production and recycling of these wastes, viewing them as potential resources with multi-directional flows along the continuum among rural, peri-urban, and urban areas. They are at once a wealth of nutrients potentially benefitting urban agro-enterprises economically and contributing to solving the city's environmental problems (Smit & Nasr 2001) while at the same time representing a flow of contaminants potentially posing health risks to producers and consumers. This is also examined in Yaoundé in Chapter 4. Another concept related to urban ecosystem health is "greening" of the city. As mentioned in the previous section, increasing the areas of vegetation in a city through agriculture or urban forestry helps reduce heat island effect by increasing levels of evapotranspiration (Ohmachi & Roman 2002, p. 172). While not directly addressed by the studies in this volume, this is one of the positive spill-over effects of many individual livelihood decisions reported.

Benefits and Risks of Livestock-Raising

The capacity of urban agriculture to provide good food to households and alleviate hunger, thus contributing to meeting one of the main Millennium Development Goals, is arguably its greatest potential benefit. However, rigorous studies showing the impact of urban agriculture on food and nutrition security in cities, such as the one contained in this book, are rare. The statistical study undertaken in Kampala between 2003 and 2005 (Sebastian et al. 2008; Yeudall et al. 2008) and summarized in Chapter 9 of this volume confirms the validity of farmers' statements that keeping urban livestock benefits them in terms of nutritional as well as income contributions to their livelihoods (Cole et al., 2008, pp. 104). While numerous studies show that the major factor influencing household food security (HFS) is wealth, this study confirms that land for urban farming, urban livestock-keeping (especially pig-raising), and women's education also contributed significantly. And with regard to child nutrition, the study also clearly showed that consumption of animal source foods (ASF) was associated with better nutritional status, strongly suggesting the positive role of urban livestock-keeping.

Both benefits and risks to human health of urban livestock-keeping are examined in Chapter 9 for Kampala as well as in Chapter 12 for Nakuru. There is a wide range of potential health hazards that can be transmitted from livestock to humans, through a variety of pathways, and the level of health risk needs to be assessed and managed in relation to the potential benefits. The detailed studies of Kampala and Nakuru in this book contribute to a growing body of knowledge on this topic that is based on empirical studies in cities of the global South. The Nakuru study is particularly important in giving insight into the gender dimension of urban livestock-keeping and stressing the need for better farmer education on risk mitigation. A useful output from Kampala was understanding what urban farmers already know and do about mitigating health risks and what conditions encourage them to act on this or constrain them. Access to resources, including water, are important, as is secure tenure and the right to farm. Together with the learning about the nutritional benefits of keeping urban livestock mentioned already, these are really crucial findings in terms of how urban agriculture may impact hunger and poverty alleviation.

Benefits and Risks of Horticulture

Work in Yaoundé reported in Chapter 3 indicates the importance for the diets of low-income urban households of year-round availability of traditional leafy vegetables. They are an important source of nutrients for urban consumers, providing for example 8 percent of protein and 40 percent of calcium intake. For the very poor, with low consumption of animal-source foods, they are even more important. About 27 percent of consumption of these vegetables by poor, Yaoundé households comes from their own home gardens. Overall, Yaoundé households get 10 percent from their own home gardens but another 20 percent of their overall consumption is in the form of gift exchanges with relatives and friends. The inland valley horticulture in Yaoundé described earlier is a crucial source of traditional leafy vegetables for low-income consumers during the dry season.

As well as being pathways for micronutrients, horticultural crops are also potential pathways for biological and chemical contaminants, negatively affecting the health of the urban ecosystem, including human health. This is especially so in intensive urban production systems, where the uptake of soil nutrients can be mixed with pathogens and chemicals. Studies described in Yaoundé in Chapter 3 and Kampala in Chapter 9 differentiate between biological and chemical hazards. The former mostly arise from contamination by human wastes due to inadequate sanitation, while the latter arise mainly from discharges into water, soils, or the air from industries or combustion (including from vehicles). The former are pathogenic and can cause infectious disease while the latter are toxic, can bio-accumulate over time, and cause chronic disease.

Both chapters explore the origins and pathways of these contaminants with respect to the two cities. The Kampala studies are considerably more detailed, exploring many types of contaminants and different pathways. Chapter 9 looks at complex organic chemical compounds, which have hardly been examined in relation to urban agriculture previously, and derives policy guidelines for urban crop production in situations of air, soil, and water pollution based on empirical study. The study of water pollution affecting and caused by urban agriculture in Yaoundé found that the few large industries are the main sources of chemical contamination. Preventing farming is not the solution, both studies recommending better sanitation as a major way to mitigate biological health risks and emission controls for chemical risks. But public and farmer education are the crucial immediate measures required.

Solid Wastes: Understanding Rural–Peri-urban–Urban Resource Flows

Like agriculture-based livelihood strategies, the recycling of nutrients for urban agriculture needs to be understood in spatial terms. The natural resource linkages involving liquid and solid wastes can impact both positively and negatively on rural and urban spaces where wastes are taken up as part of producer households' livelihood strategies.

The majority of solid waste in developing world cities is organic – about 70 percent in Nairobi or almost half-a-million tons every year (JICA 1997) – and this is mostly not managed by the authorities. In Nairobi less than half is collected and reaches city dumps, mostly from commercial or higher income residential areas, the rest piling up in poorer areas or dumped in streams and rivers. The authors of Chapter 10 estimate that the nitrogen, phosphorus, and potassium locked up in this resource represent a fertilizer value of about $2 million.

There are significant marketing flows of animal manure between the rural hinterland of Nairobi and gardeners, landscapers, and horticultural producers in and around the city. There is also urban, peri-urban, and rural use of urban manures in Yaoundé and Nakuru. By contrast, less than 1 percent of organic solid waste generated in Nairobi is reused. In Nakuru, whereas over 90 percent of domestic waste generated by farming households is recycled, the vast majority of waste from non-farming households is not, and ends up dumped in open spaces or, less commonly, removed to the municipal landfill.

Nevertheless, despite the concentration of nutrients in manure, the recycling systems are imperfect. In Nakuru, on average 46 percent of urban and peri-urban produced manure is recycled, mostly for crop production within the same area, with a small amount being carried to rural farms or sold and about 14 percent being used in a variety of other ways including for biogas. Still, more than 50 percent of manure is dumped. These average figures disguise very large differences in management of manure along the peri-urban to urban continuum however. In the inner urban areas, more than 80 percent of manure is dumped, creating significant environmental problems. In the more peri-urban areas of the city, only about 13 percent is dumped, with 60 percent recycled in urban and rural plots. The picture in Nairobi is similar, in that intensive urban marketing of rural manures compares with 60 percent of manure from urban-grazed cattle being dumped. Urban manure producers are quite disconnected from the marketing system based on rural supply from Maasai cattle.

Yaoundé provides a more optimistic scenario of the recycling of nutrients in agriculture. Apart from the intensity of horticulture in the urban inland valleys described in Chapter 3, the authors of Chapter 4 estimate that, of the approximately 10 000 tons of poultry and pig manure annually produced in the city, about 60 percent is being recycled in the urban and peri-urban areas with a further 10 percent being sold elsewhere, mostly in Bamenda, the capital city of another province. Most of the remaining 30 percent being dumped is probably pig manure, considered problematic as a fertilizer by many farmers in Yaoundé, as elsewhere. Recent reevaluation of the use of composted pig manure in Lima suggests that the level of reuse could be increased in Yaoundé through technical support. In all cases there seem to be opportunities to increase the level of recycling of manures within urban and peri-urban areas to further benefit city environments and enhance agricultural productivity.

Actual and Potential Benefits of Waste Recycling

These experiences in Nairobi, Nakuru, and Yaoundé suggest that urban agriculture is currently absorbing a much more significant quantity of animal waste than vegetative waste, whether crop residues or food wastes. Taking Nakuru as an example, much larger volumes of both crop residues and domestic food wastes are being used as animal feed than for composting (Fig. 2.5). This may reflect the higher value and higher costs associated with livestock-raising compared to crop production and households' need to reduce feed costs by using domestic or farm wastes. But it also

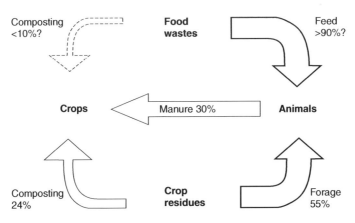

Fig. 2.5 Agricultural recycling of organic wastes in Nakuru

seems to indicate that the direct use of waste as livestock feed is preferred to the necessity of processing it for compost.

A further point not directly dealt with in these studies is that raising livestock also absorbs significant quantities of the domestic waste of non-farming households. A recent study of pig-raisers sampled from six parishes across the peri-urban to urban continuum in Kampala (Lubowa et al. forthcoming) found that the 144 pig-raisers recycled nearly 70 tonnes of organic wastes per week, or half a tonne per enterprise. These wastes – mostly from domestic and restaurant food leftovers, with farmers' own household wastes forming a significant but limited part, and market wastes – account for 1.3 percent of Kampala's estimated weekly generation of 5535 tonnes of organic waste. If extrapolated to the 98 parishes in proportion to their pig-raising enterprises, as much as 20 percent of the city's organic wastes might usefully be recycled as high-quality protein-rich feed.

The low level of recycling through composting compared to total volumes of organic wastes in Nairobi and Nakuru is only one part of the story. In Nairobi, because of the uneven distribution of waste production and collection, many informal settlements suffer disproportionately from the public sector incapacity to manage sold waste. Direct action on this problem by local community groups, often youth-led, resulted in significant environmental improvements in shanty areas where these groups were active, because composting was only one part of a whole process of collection, sorting, selling, or processing and reuse, of both inorganic and organic wastes. Consumers of compost in Nairobi include seedling nursery enterprises, ornamental gardens, landscapers, and farmers in the urban area and small farmers and large horticulture enterprises in the peri-urban and rural areas in central Kenya. The study suggests that demand outstrips supply for compost producers closer to the city centre with strong networks of institutional customers, while those located on the periphery, especially in informal settlements, experienced limited demand and small sales. While lack of information about points of sale was a factor, quality and potential health risks were as well. The pricing of compost in relation to its nutrient content emerged from the study as a key constraint on its use. Improving quality standards was found to be needed for both manure and compost.

Overall, the lessons from these studies on recycling of organic wastes for feed or soil amendment is that there is a huge potential for both. Manure does better than compost in marketing terms, while using food waste for feed is currently more feasible for producers than is composting. While all of these benefit environmental management, composting in particular needs to address quality issues and to increase productivity and marketing networks in order to tap into an apparently growing demand.

Policy and Institutional Dialogue and Change

Given that institutional and policy change were objectives of the research work described in this book and a necessary part of making agriculture a key component of sustainable cities, the activities undertaken necessarily intervened in local urban governance. Governance is defined here as the relationship between civil society and the state, between rulers and the ruled, the government and the governed (Gore 2008, p. 57). This concept, derived from political science, examines the power relations between the various actors, and can usefully be applied to urban agriculture (Lee-Smith & Cole 2008; Gore 2008). The notion of "dialogue" included in the name of this research theme adds a broader dimension. It derives from participatory and anthropological research approaches that have also strongly influenced the framing of the theme (Chambers et al. 1989; Röling & Wagemakers 1998; Prain 2006). It highlights the micro-level interactions, sensitizations, negotiations, and shifts in ideas from which mezzo- and macro-level policy change flows. It also underlines the role of multi-disciplinary and multi-agency platforms (Röling & Jiggins 1998, pp. 301–304) that provide a space for these interactions. Dialogue also includes the notion of social capital, meaning the degree of association, trust, mutual confidence, and social interaction characterizing relationships (Bebbington 1999; DfID 1999). Dialogue not only helps build social capital but is more fruitful and creative when partners to negotiation enjoy greater levels of social capital. Because policy and institutional change through dialogue is a key outcome for ensuring the sustainability of livelihoods and ecosystem health benefits, it is dealt with in the final chapter of this volume. Here we make some brief observations about the process of platform formation and the results achieved.

Most local governments want to see greater levels of food security and poverty reduction among their constituents and if this can be partly achieved through productive use of urban natural resources rather than through costly social programs, why not? The "why not" derives both from the complexity of urban agriculture itself – "at once a form of land-use, an economic subsector of agriculture and an expression of the multiple ways in which the urban and rural worlds intersect" (Bopda and Awono Chapter 5) – and the complexity of urban governance.

In the cases described in this book, local and international researchers partnered with representatives of government, universities, and civil society organizations to try to address both the complexity of urban agriculture itself and of local governance. The regional stakeholder meeting held in Nairobi in 2000 to define the Urban Harvest program deliberately sought to invite representatives of different organizations from cities in the region as a first step in building platforms for research-development collaboration in those cities. The formation of platforms was a prelude to policy and institutional change. Writers such as Röling and Jiggins, who have been at the forefront of developing the notion of platforms, have pointed out that these entities can be very variable. They can be "one-time meetings, elected committees, formally appointed boards or councils or even parastatal or government bodies" (1998, p. 303). Indeed the "morphology" of the stakeholder platforms which have formed in the three countries studied here has been quite different, especially in the range of interests represented and the extent to which the platform has been perceived as a formal mechanism for resolving conflicting interests. In some respects these platforms have functioned as think-tanks to resolve differences among higher level decision-makers, generate ideas about what next steps are needed and facilitate those steps.

The variability of policy and institutional outcomes across the three countries is described in Chapter 15, which tries to map those outcomes and understand their variability. What we have learnt is that, however committed the participation in stakeholder platforms may be, to some extent all participants are constrained by their own institutional histories and even by the historical circumstances of agriculture within the particular city. And as was indicated in Chapter 1, the history of urban institutions and policy has frequently included the marginalization or proscription of urban agriculture.

Nevertheless, despite the negative treatment of agriculture in many cities of Africa, policy change can be seen as having at least begun and to be in process in the three countries studied in this book. Perhaps because of the specific link of agriculture to the Kingdom of Buganda, local institutions in Kampala collaborated in urban agriculture management and in support to farmers. Kampala City Council seems to be unique in having developed not only a Department of Agriculture but a typology of urban farming systems and land types within its boundaries by the end of the 20th century. Such institutions were markedly absent in the other places studied, although Nakuru in Kenya had environmental initiatives addressing urban agriculture in the 1990s, through international assistance projects (Foeken, D 2005).

Conclusion

This book tries to assess the extent to which urban agriculture is contributing to household livelihoods and to the health of the urban ecosystem on the basis of case studies in three countries and five cities. Because agriculture in cities is intricately bound up with use of and competition for resources and with regulations on public health and other sectors, we consider the relation of urban agriculture to urban governance as critical and this has been one thematic area of research. The mechanism for achieving institutional and policy recognition of agriculture has been the stakeholder platform. The most significant advances by such a platform — including its formalization – have been in Kampala, where new ordinances or laws governing urban agriculture were put in place, the City's Department of Agriculture was empowered and policy was drafted. In Nakuru likewise, new laws governing urban agriculture were developed (although they had not been put in place at the time of publication) and research support had strengthened the urban agriculture activities of the Municipal Environment Department. There was considerable progress in Kenya at the national level, with a policy process beginning partly as a result of a national workshop empowering the Kenya Agriculture Research Institute (KARI). Progress in Nairobi City Council was poor however, although a policy forum started by civil society and with a vibrant farmers' network also led to the policy process (see Chapter 15 below). The final project workshop and exposition in Yaoundé augured well for policy and institutional change there, but was not followed up by any form of assistance.

It is the sheer pervasiveness of urban agriculture, both geographically and sectorally, which makes it such an ideal vehicle for institution-building and this is the conclusion of the Yaoundé study. As the urban dimensions of food crises and climate change become increasingly apparent, this opportunity will have to be grasped and agriculture integrated in the socio-economic and environmental planning of cities.

References

Baker, J & Akin Aina, T (eds) 1995, *The migration experience in Africa*, Nordic Africa Institute, Uppsala.
Basweti, C, Lengkeek, A, Prytz, L & Jaenicke, H 2001, 'Tree nursery trade in urban and peri-urban areas. A survey in Nairobi and Kiambu Districts, Kenya', *RELMA Working Paper* No. 13, Regional Land Management Unit (RELMA), Nairobi, Kenya.
Bebbington, A 1999, 'Livelihoods, capitals and capabilities: a framework for analyzing peasant viability, rural livelihoods and poverty', *World Development* vol. 27, no. 12, pp. 2021–2044.
Chambers, R, Pacey, A & Thrupp, LA (eds) 1989, *Farmer first*, Intermediate Technology Publications, UK.
Cole, DC, Lee-Smith, D & Nasinymama, GW (eds) 2008, *Healthy city harvests: Generating evidence to guide policy on urban agriculture*, CIP/Urban Harvest and Makerere University Press, Lima, Peru.
DfID 1999, *Sustainable livelihoods guidance sheets*, Department for International Development, London, http://www.livelihoods.org/info/guidance_sheets_rtfs [Accessed 23 June 2009].
Ellis, F 2000, *Rural livelihoods and diversity in developing countries*, Oxford University Press, Oxford, UK.
Farrington, J, Carney, D, Ashley, C & Turton, C 1999, 'Sustainable Livelihoods in Practice: Early applications of concepts in rural areas,' *ODI – Natural Resource Perspectives* vol. 42, p. 13.
Foeken, D 2005, 'Urban agriculture in East Africa as a tool for poverty reduction: a legal and policy dilemma?', *ASC Working Paper* 65/2005, African Studies Centre, Leiden, Netherlands.
Foeken, D (ed) 2006, *To subsidise my income: urban farming in an East-African town*, Africa Studies Centre, no. 7, The Netherlands.
Gockowski, J, Mbazo'o, J, Mbah, G & Fouda Moulende, T 2003, 'African traditional leafy vegetables and the urban and peri-urban poor', *Food Policy*, vol. 28, no. 3, pp. 221–235.
Gore, C 2008, 'Healthy urban food production and local government', in Cole, DC, Lee-Smith, D & Nasinymama, GW (eds) *Healthy city harvests: generating evidence to guide policy on urban agriculture*, CIP/Urban Harvest and Makerere University Press, Lima, Peru, pp. 49–65.
Hart, K 1973, 'Informal Income Opportunities and Urban Employment in Ghana.' *Modern African Studies* vol. 11, no. 1, pp. 61–89.
Hovorka, A & Lee-Smith, D 2006, 'Gendering the UA agenda', in van Veenhuizen, R (ed) *Cities farming for the future: urban agriculture for green and productive cities*, RUAF Foundation, IDRC & IIRR, Ottawa, ON, pp. 125–144.
JICA (Japan International Cooperation Agency) 1997, *Master plan study of Nairobi*.
Kessides, C 2006, *The urban transition in Sub-Saharan Africa. Implications for economic growth and poverty reduction*, Cities Alliance, SIDA, World Bank, Washington, DC.
Lee-Smith, D 2001, 'Crop production in urban/peri-urban agriculture in Kenya', in Kahindi, JP, Karanja, NK, Alabaster, G & Nandwa, S (eds) *Proceedings of a workshop on enhancement of productivity and sustainability of UPA through efficient management of urban waste*, 8–9 October, 2001, Nairobi
Lee-Smith, D & Cole, D 2008, 'Can the city produce safe food?', in Cole, DC, Lee-Smith, D & Nasinymama, GW (eds), *Healthy city harvests: Generating evidence to guide policy on urban agriculture*, CIP/Urban Harvest and Makerere University Press, Lima, Peru pp. 3–13.
Lee-Smith, D & Prain, G 2006, 'Urban agriculture and health', in Hawkes, C & Ruel, MT (eds) *Understanding the links between agriculture and health*, 2020 Focus no. 13, Brief

13 of 16, IFPRI, Washington, DC, www.ifpri.orghttp://www.ifpril.org/2020/focus/focus13.asp [Accessed 23 June 2009].

Lubowa, A, Prain, G & Kyomugisha, E 2010, 'The use of commercial food and other organic wastes for urban and peri-urban pig-production in Kampala, Uganda', *Urban Harvest Working Paper 5*, Lima, Peru, (Forthcoming).

Maxwell, D 1995, *Labour, land, food and farming: a household analysis of urban agriculture in Kampala, Uganda*, Unpublished PhD Dissertation, University of Wisconsin-Madison, United States of America.

Maxwell, D, Levin, C, Armar-Klemesu, M, Ruel, M, Morris, S & Ahiadeke, M 2000, *Urban livelihoods and food and nutrition security in greater accra, Ghana*, IFPRI – International Food Policy Research Institute, Washington, DC.

Mwangi, AM & Foeken, D 1996, 'Urban agriculture, food security and nutrition in low-income areas of Nairobi', *African Urban Quarterly* vol. 11, no. 2/3.

Mwangi, S, Kimathi, M, Kamore, M, Karanja, N, Njenga, M & Farm Concern International 2007, 'Creating market opportunities for poor women farmers in Kenya', *Urban Agriculture Magazine 17*, Leusden.

Nabulo, G, Nasinyama, G, Lee-Smith, D & Cole D 2004, 'Gender analysis of urban agriculture in Kampala, Uganda', *Urban Agriculture Magazine* vol. 12, pp. 32–33.

Nabulo, G, Oryem-Origa, H & Diamond, M 2006, 'Assessment of lead, cadmium, and zinc contamination of roadside soils, surface films, and vegetables in Kampala City, Uganda', *Environmental Research* vol. 101, pp. 42–52.

Obuobie, E, Dreschel, P & Danso, G 2004, 'Gender in open-space irrigated urban vegetable farming in Ghana', *Urban Agriculture Magazine, no. 12*, RUAF, Leusden.

Ohmachi, T & Roman, ER (eds) 2002, *Metro Manila: in search of a sustainable future: impact analysis of metropolitan policies for development and environmental conservation*, Japan Society for the Promotion of Science (JSPS) Manila Project, in collaboration with University of the Philippines Press, Quezon City.

Prain, G, 2006, 'Participatory technology development for urban agriculture: collaboration and adaptation along the urban-rural transect', in van Veenhuizen, R (ed) *Cities farming for the future – urban agriculture for green and productive cities*, RUAF Foundation, IDRC and IIRR, Leusden, pp. 273–312.

Rakodi, C & Lloyd-Jones, T 2002, *Urban livelihoods: a people-centered approach to reducing poverty*, Earthscan, London.

Röling, NG & Jiggins, J 1998, 'The ecological knowledge system', in Röling, NG & Wagemakers, MAE (eds) *Facilitating sustainable agriculture*, Cambridge University Press, Cambridge, pp. 283–311.

Röling, NG & Wagemakers, MAE (eds) 1998, *Facilitating sustainable agriculture*, Cambridge University Press, Cambridge.

Satterthwaite, D & Tacoli, C 2002, 'Seeking an understanding of poverty that recognizes rural-urban differences and rural-urban linkages', in Rakodi, C & Lloyd-Jones, T (eds) *Urban livelihoods: a people-centred approach to reducing poverty*, Earthscan Publications, London, pp. 52 70.

Sawio, C 1998, 'Managing urban agriculture in Dar es Salaam', *Cities feeding people, Report 20*, IDRC, Ottawa, ON.

Schippmann, U, Leaman, DJ & Cunningham, AB 2002, 'Impact of cultivation and gathering of medicinal plants on biodiversity: global trends and issues' in *Biodiversity and the ecosystem approach in agriculture, forestry and fisheries. Satellite event on the occasion of the ninth regular session of the Commission on Genetic Resources for Food and Agriculture*, 12–13 October 2002, Inter-Departmental Working Group on Biological Diversity for Food and Agriculture, FAO, Rome.

Sebastian, R, Lubowa, A, Yeudall, F, Cole, DC & Ibrahim, S 2008, 'The association between household food security and urban farming in Kampala', in Cole, DC, Lee-Smith, D & Nasinymama, GW (eds), *Healthy city harvests: Generating evidence to guide policy on urban agriculture*, CIP/Urban Harvest and Makerere University Press, Lima, Peru, pp. 69–87.

Simon, D, McGregor, D & Thomson, D 2006, 'Contemporary perspectives on the peri-urban zones of cities in developing areas', in McGregor, D, Simon, D & Thomson, D (eds) *The peri-urban interface. Approaches to sustainable natural and human resource use*, Earthscan, London, pp. 3–12.

Smit, J & Nasr J 2001, 'Agriculture – urban agriculture for sustainable cities: using wastes and idle land and water bodies as resources', in Satterthwaite, D (ed.) *Sustainable Cities*, Earthscan, London, pp. 221–233.

Tegegne, A 2004, 'Urban livestock production and gender in Addis Ababa', Ethiopia', *Urban Agriculture Magazine, no. 12*, pp. 30–31.

Uganda Bureau of Statistics (UBOS) & ORC Macro 2001, *Uganda demographic and health survey 2000–2001*, UBOS and ORC Macro, Calverton, MD.

UNDP 1996, *Urban agriculture: food, jobs and sustainable cities*, United Nations Development Program Publication Series for Habitat II, vol. 1, UNDP, NewYork.

Warnaars, M & Pradel, W 2007, 'A comparative study of the perceptions of urban and rural farmer field school participants in Peru', *Urban Harvest Working Paper Series*, paper 4, International Potato Center (CIP), Lima, Peru.

World Bank 2004, *Enquête Camerounaise auprès des ménages, 1996. Synoptique des résultats d'enquête normalizes*, Survey Databank – World Bank Africa Region, Washington, DC.

Yeudall, F, Sebastian, R, Lubowa, A, Kikafunda, J, Cole, DC & Ibrahim, S 2008, 'Nutritional security of children of urban farmers', in Cole, DC, Lee-Smith, D & Nasinymama, GW (eds) *Healthy city harvests: Generating evidence to guide policy on urban agriculture*, CIP/Urban Harvest and Makerere University Press, Lima, Peru, pp. 89–103.

Yilma, G 2003, 'Micro-technologies for congested urban centers in Ethiopia', *Urban Agriculture Magazine*, http://www.ruaf.org/node/327 [Accessed 18 June 2009].

Part I
Cameroon

Cameroon Overview

The three chapters in this section present research undertaken in Yaoundé, the capital city of Cameroon, from 2002 to 2004 and set in a national perspective. Although there have been one or two previous studies of different aspects of urban agriculture in the city, particularly concerning the growing of green leafy vegetables for the urban market, this research constitutes the first coordinated national effort to document the activity and to link this to public policy processes. The research contained in this section is also notable for its attempt to introduce a spatial perspective to the subject, both in the way of understanding how different urban and peri-urban agriculture activities function, and in the way of linking this to physical planning and local government administration.

Led by researchers from the International Institute for Tropical Agriculture (IITA) but incorporating a wide range of governmental and non-governmental institutions, the Urban Harvest-supported research team carried out a series of detailed investigations to build up the mosaic picture presented in Chapter 3. This describes how perishable items produced by urban farming contribute to livelihoods and nutrition, how urban farming affects and is affected by water pollution, and what the patterns of production and flow are of tree seedlings and cassava, an important local food crop. They are framed within a long-term perspective which is both spatial and historical, thanks to the studies over many years by one team member who is also an official of the National Institute of Cartography.

Chapter 4 is a study of crop–livestock farming in the capital, a common and popular form of urban farming in this and other cities because of its productivity. The study includes a detailed investigation of waste production and utilization for this type of farming system at city scale.

Chapter 5 returns to the spatial and historical framework to present the study of institutions for urban agriculture, examining why and how strong institutions have not emerged to deal with food production in the city, at the same time presenting this as a challenge for effective urban government. The chapter concludes with a summary of how the findings of all these studies were dealt with in a public forum attempting to move forward the urban agriculture agenda in Cameroon in early 2004.

Chapter 3
Urban Farming Systems in Yaoundé – Building a Mosaic

Athanase Pr Bopda, Randall Brummett, Sandrine Dury, Pascale Elong,
Samuel Foto-Menbohan, James Gockowski, Christophe Kana,
Joseph Kengue, Robert Ngonthe, Christian Nolte, Nelly Soua,
Emile Tanawa, Zac Tchouendjeu, and Ludovic Temple

Background to the Set of Studies

Urban agriculture is prevalent in Cameroon, the first country examined in this book of case studies, yet its role in urban life was little studied until the 1990s. At that time researchers began to look at some aspects of this complex phenomenon, such as the role of traditional leafy vegetables in the diet and incomes of the urban poor (Gockowski & Ndoumbé 1999). Following their attendance at a regional stakeholder meeting organized by Urban Harvest in late 2000, scientists from different institutions came together in 2001 to move forward work they were pursuing independently on different topics related to urban agriculture in Yaoundé. This interdisciplinary collaboration produced the original empirical studies contained in this chapter and the two that follow, which aim at a deeper understanding of some of the complexities of urban farming in the country and indicate directions for further work, both in research and the development of public policy.

This chapter is by way of an overview, briefly setting the scene and then presenting a mosaic of five empirical studies on different aspects of urban agriculture to develop a composite picture. This continues in the next chapter, which is an empirical study of crop–livestock farming systems in the capital. Each of these presents not only data but also their implications for policy intervention. Chapter 5, the last on Cameroon, is a study of institutional aspects of urban agriculture set in the context of Yaoundé's history. Besides being a critique of how the topic is, or is not, addressed through the institutions of government, this study is based on an original, extensive socio-spatial analysis of Cameroon's capital city that forms the background to, and points the way forward for, the other studies, not just for Yaoundé but for the whole of Cameroon.

In February 2004, the results of all these studies were displayed in an exhibition at Yaoundé's Town Hall, and presented at a public meeting to an audience of around 50 participants from sectors and institutions concerned. The Government

---

A. Pr Bopda (✉)
Institut National de Cartographie (INC), 779, Avenue Mgr Vogt, BP 157, Yaoundé, Cameroon
e-mail: bopda20001@yahoo.com

Delegate to the Urban Community of Yaoundé expressed a commitment to further public action on urban agriculture, which, in spite of its problems, is a source of employment and improved quality of life. The outcome of this meeting, and its concrete policy recommendations for urban agriculture in Cameroon in general, are discussed in the conclusion to Chapter 5.

This chapter presents the first five studies. Data on livelihood strategies of low-income urban crop farmers and nutritional contributions of perishable products to urban households are complemented by assessments of the potential for producing fruit tree seedlings in town and the dangers posed by water pollution affecting and generated by urban farming. Finally in this first chapter, a method for the quantification and spatial analysis of urban food supply flows is presented through its application to cassava, an important staple crop in Central Cameroon.

The Origins and Nature of Urban Agriculture in Yaoundé

Like many large cities in Sub-Saharan Africa, Yaoundé is relatively new and rapidly growing. Urban at its centre and rural at its periphery, stretching 24 km North–South and 16 km West–East, Yaoundé has had its boundaries extended decade after decade in an effort to control and manage orderly development. Urban agriculture is found in all the villages and quarters of the town and in each ring of concentric development, reflecting not only people's attachment to their rural origins but also their poverty and need for food and income.

Agriculture played a role in the histories of both Yaoundé and Douala, the other main urban centres of Cameroon, both having taken shape around existing agricultural land-use. Now a major port on the Gulf of Guinea and the largest urban centre in the country, Douala was a town before the Europeans arrived whereas Yaoundé was founded in 1888 in the interior and only gradually became an important town. In fact, confrontations over access to agricultural land in Douala contributed to Yaoundé becoming the capital around the beginning of World War I. As Douala grew during the German occupation of 1884–1916 new crops were introduced, but the local population struggled both against eviction from their land and prohibitions against urban farming, trade and commerce (Geschière & Konings 1993). Yaoundé was originally not only a military station but also an agricultural research station, as shown in the map by Franqueville (1984) in Fig. 3.1 below. The German botanist Zenker who ran this station ensured that agriculture remained a part of the increasingly urban landscape. Between the two wars, the town was a focal point in the push to develop smallholder cocoa production; in the 1940s Yaoundé was a leading centre of fish farming and in the 1960s a wide belt of market gardens evolved to serve the growing population (Laburthe-Tolra 1985).

Yaoundé's spatial organization follows the classic form of a colonial urban centre evolving into a metropolis, with an old town centre, forbidden to black people, surrounded by a sanitary cordon and then successive rings of unplanned indigenous settlement. During the colonial period, residences of black and white people had

3 Urban Farming Systems in Yaoundé – Building a Mosaic

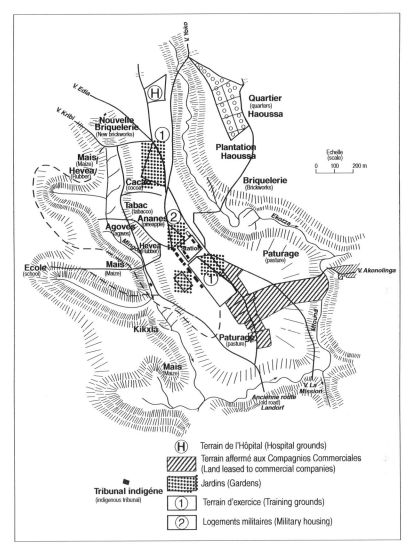

Fig. 3.1 Agriculture in the emerging city of Yaoundé in 1890

to be 800 m apart – the distance a mosquito carrying malaria from black people was supposed to be able to fly and infect a white person. Such overtly racist public health regulations, widespread in colonized Africa, were implemented in 1908 by the Germans, upheld in 1923 under French administration and reinforced as a planning principle in the early 1950s (Bopda & Awono 2003).

Indigenous settlement first grew in a discontinuous ring around the colonial town in the 1920s, with the highest densities at the places near where local people were allowed entry. In the following three decades two further kinds of settlement developed, the first resulting from the policy of forced eviction and resettlement as the colonial town itself grew, the second, which became more pronounced, resulting from people moving away from denser areas near the centre to the more rural periphery, either on their own initiative or that of the administration. Beyond this lay indigenous settlements considered by all concerned as land reserves for the growing urban centre.

Thus Yaoundé grew, and grows, by assimilating its rural periphery. The most spectacular way this occurs is through the resettlement of people subject to forced eviction from the most dense or otherwise out-of-date parts of the town. Between 1970 and 1990, 650 ha of new settlement were created in this way, to accommodate people displaced from 587 ha closer to the centre (Bopda & Awono 2003).

Despite such measures, the official urban plans of Yaoundé, of 1952, 1963, 1980 and 2005, have been largely ineffective since squatter communities and private developments without proper permits build up rapidly. The hilly nature of the city constrains development on steeper slopes, while the valley bottoms prone to flooding are only built upon close to the town centre where land is most valuable. The resulting pattern is characterized by variations in density ranging from 320 persons per hectare in the old indigenous quarters to less than 80 in the town centre (MINPAT 1993).

Yaoundé's annual rate of population growth has been between 6 and 9 percent for nearly a century, while its spatial expansion has been about 6 percent per annum for about half a century. And while migrants from all over the country form the majority, 40 percent of the city's residents are now born in town, meaning the population is becoming ever more urban in its habits and expectations (Bopda & Grasland 1996).

As the political capital, Yaoundé's economic life is dominated by bureaucracy and the civil service and its people dream of escaping manual labour as peasants and becoming white-collar workers. Yet everywhere in the town one sees people, mostly women, cultivating gardens. Although the land of Yaoundé is being converted into real estate, as institutions of government expropriate land for civic use and the functions of urban life, agriculture constantly reappears on undeveloped plots and in the interstices of the town, mainly the lowland marshes along waterways. These green spaces are cultivated by those who need food and work, even though this is illegal within the city limits.

The following sections of this chapter annotate this picture in greater depth by presenting the results of the empirical studies undertaken by the interdisciplinary team in 2002–2003. We begin with a methodological note on the five studies.

Methods Used for the Five Studies

The administration of urban space in Cameroon is confused and complex, as is the case in many Sub-Saharan African countries. Yaoundé is administratively defined as an Urban Community of six sub-divisions, which is surrounded by Mfoundi

Division, encompassing an additional seven, each having a separate administration. Mfoundi is often taken as the peri-urban area, while the term urban is generally used to refer to the six sub-divisions of the Urban Community of Yaoundé (CUY). Three of the studies described here took their samples from the CUY, while the two that studied movement of tree seedlings and cassava sampled more widely.

The investigation of livelihoods was based on a purposive sample of 121 crop cultivators interviewed in 2003. Respondents were selected from eleven sites typical of both the central areas of Yaoundé, mostly the valley bottoms prone to flooding, and areas toward the periphery, which are less built-up. The sample was stratified by commercial or household food production based on a preliminary qualitative appraisal. Crop producers were classified as commercial if they sold at least half of one of their products in the market. Sixty food producers and 61 commercial producers were purposively selected for interview. While not allowing for statistical extrapolation, the sampling procedure provided indicative data and allowed meaningful comparisons to be made about the relationship of urban crop production to the well-being of producers' households.

The nutrition study used previously collected data from seven surveys carried out in 1999–2000 using dietary recall to examine the intake of nutrients, including the contribution of urban agriculture to the nutrition of urban dwellers. A random sample of 183 households representative of the Yaoundé population was stratified into high-, middle- and low-income groups. The results of this data analysis are also discussed in Gockowski et al. (2003) to which we refer.

The survey of horticultural and tree seedling nurseries was done in the six urban and seven peri-urban sub-divisions of the Mfoundi division and three rural villages beyond this boundary in order to investigate their characteristics and links to the flow of seedling production into urban areas. Places with nurseries were chosen based on information collected from extension agents and producers and a list made of all those with nurseries. A representative set was selected for more detailed investigation of the spatial pattern of rural, peri-urban and urban nurseries. Owners of all 39 selected nurseries were interviewed and the quantities of seedlings produced estimated.

The study of water pollution affecting and caused by urban agriculture in Yaoundé used both new and existing data to identify and measure point and nonpoint sources of pollutants affecting surface water quality, which in turn impacts agricultural production in bottom lands used to grow vegetables and proposed for fish farming. Both chemical and biological pollution levels of surface waters were measured and considered. Suggestive data on the human health impacts of accumulated contaminants in valley bottoms were also obtained through a survey of farmers.

A method for the quantification and spatial analysis of urban food supply flows (Temple 2001) was developed and applied to a test case of fresh cassava roots. Data were collected from cassava producers, transporters and retailers involved in bringing fresh cassava roots to Yaoundé's markets and applied to the model. Some 130 cassava sellers and 55 drivers were interviewed in July 2002 at the four most important of the eight multiple-function markets in Yaoundé. The number of vehicles seen transporting food items into Yaoundé was first counted between 5.30 and 10.30 am, the period when most vehicles ply the village roads, on seven of the supply axes identified in a preliminary survey as important in terms of food supply. The busiest axis was identified for use in quantification exercises. On three consecutive days at each market, the number of sacks of fresh cassava roots unloaded from each vehicle arriving between 6 and 10 am (when most arrive from the supply areas) was counted. The owners of the products were also identified and asked the number of sacks transported and their villages of origin in order to cross-check the volume. More structured interviews were held with drivers; appointments having being made during unloading. After cleaning and constructing tables from these qualitative data, they were applied to maps of supplying villages, markets and routes using GIS positioning. Trip times in the dry and wet seasons were also factored in as variables.

Livelihoods of Urban Crop Growers in Yaoundé

The cultivation of leafy green vegetables and maize on vacant lots, unused municipal lands and valley bottoms is the most visible aspect of urban agriculture in Yaoundé, others being the ubiquitous small beds of commonly used herbs, stands of plantains and bananas (Lemeilleur et al. 2003), and avocado, African plum (*Dacryodes edulis*), mango, and guava trees, all growing next to houses or lining the streets. In addition, small animal husbandry of guinea pigs, rabbits, chickens, goats and pigs is widespread. Animal husbandry is the subject of the next chapter. While they represent only a part of the city's urban and peri-urban agriculture of Yaoundé (Temple & Moustier 2004), these crop growers were selected for study based on their economic importance to farming households, the numbers of persons involved, their importance in supplying urban markets and their potential for improved productivity.

Three main types of cropping systems were distinguished:

- Mixed crop systems dominated by improved varieties of maize in the upland areas;
- Mono-crop systems of improved maize grown in valley bottoms;
- Intensive horticultural systems in valley bottoms.

Households were classified according to their objectives regarding cultivation (Table 3.1):

- To contribute to the household food supply;
- To generate cash income (selling at least half of one of their products).

We observed a high level of consistency between where a producer operated, the way the enterprise looked and the household's objective vis-à-vis urban agriculture.

Table 3.1 Selected characteristics of urban cultivators of two types

	Food consumption		Commercial		Total	
	Number	Percent	Number	Percent	Number	Percent
Woman farmer	57	95.0	48	78.6	105	86.8
Married farmer	41	68.3	48	78.6	89	73.6
Primary education	28	47.6	17	27.9	45	37.2
Western origin	34	56.7	32	52.5	66	54.5
Valley bottom location	21	35.0	53	86.9	74	61.2
Upland location	48	80.0	24	39.3	72	59.5
Renting land	20	33.3	40	65.6	60	49.6
Inherited land	14	23.3	7	11.5	21	17.4
Bought land	10	16.7	8	13.1	18	14.9
Squatting	12	20.0	5	8.2	17	14.1
Borrowed land	9	15.0	6	9.8	15	12.4
Association member	40	66.7	37	60.7	77	63.6
Total cultivators	60	100.0	61	100.0	121	100.0

Producers doing intensive horticulture and growing monoculture maize in the valley bottoms were more likely to be pursuing a commercial objective than were those cultivating a mixture of crops on upland fields, who produced primarily to augment household food supplies.

Gockowski and others have shown that green leafy vegetables are important sources of nutrition and employment among urban and peri-urban households, contributing a significant share of essential nutrients for the urban poor in particular. Over 32 000 households were engaged in producing and marketing of leafy vegetables in Yaoundé in 1998–1999, most of the farmers being women using an extensive mixed crop system (Gockowski et al. 2003).

Most urban crop cultivators were women (87 percent), even 79 percent of the commercial producers being women, and 95 percent of the food producers. The few men in the sample produced mostly for commercial objectives. The large majority of cultivators were married, although the proportion was lower (69 percent) among the household food producers, among whom there were more widows, divorced, separated and single people, most being women. Our results showed that on average, cultivators had been living in Yaoundé for 21 years and practicing agriculture for nine.

Household food producers had higher levels of education than commercial producers. Forty-eight percent of household producers had secondary or even higher education, compared to only 28 percent of commercial producers. Half the farmers employed child labour, some in hazardous activities such as application of pesticides. Commercial producers used higher rates of chemical inputs and had a much higher reported rate of morbidity and greater health-care expenses.

Over half the cultivators of both types were immigrants from the densely populated and impoverished western highlands of Cameroon (55 percent of the total).

About half the cultivators rented the land they were using, this being the case for many more of the commercial cultivators (two-third) who were mainly in the valley bottoms and fewer of the food producers (one-third) who farmed more in the uplands. A minority of each type of producer owned the land they were cultivating. The unstable and transitory nature of urban cultivation is evidenced by the fact that over half the farmers interviewed (55 percent) said that they had been forced to abandon fields because of land-tenure problems.

The plots of land cultivated were small, averaging about 400 m^2, although farmers generally cultivated more than one (1.8 plots for food producers and 4.1 plots for commercial producers). Both home food producers and commercial farmers concentrated heavily on maize and traditional green leafy vegetables, the most common of which were *amaranthus* spp. and *Solanum nigrum*. Food producers consumed more than 80 percent of all their crops, selling or giving away the rest in equal proportions. The commercial producers consumed about a quarter of their traditional leafy vegetable production themselves and nearly half their maize production, but they sold almost all the lettuce and condiments that they grew. It is worth noting that the commercial producers also grew various other crops mainly for their own use, and that they also gave away some of their produce. Thus urban produce reaches consumers through a variety of pathways – the households of cultivators, their friends and neighbours and consumers who shop in markets – potentially benefiting them all through increased food intake, especially of micronutrients.

Commercial producers used more purchased inputs than did the food producers. Eighty percent or more of commercial cultivators used chicken manure, inorganic fertilizer and insecticides, compared with 28, 37 and 23 percent of the household food producers who used each of these items respectively. Fungicides were used by 29 percent of commercial and 10 percent of household food producers. It would seem that commercial producers can afford more inputs due to their produce sales. It is possible they may be less concerned about the effects of overuse of chemicals because they are not the main consumers but further research would be needed to explore this relationship.

A high proportion of the cultivators (67 percent of household food producers and 61 percent of commercial producers) were members of associations (Table 3.1), another characteristic associated with women. Studies elsewhere in Africa show rural women have a higher incidence than urban women of being involved in formal and informal collectives, although such membership persists among urban women (Lee-Smith 1997). While social solidarity was the most common reason for belonging, these groupings present opportunities for organizing urban cultivators to participate in public programs. More than half the associations identified in this study were linked to the cultivator's village of origin – that is, they were ethnic in orientation – and just under half were savings associations.

The household size of urban cultivators was large, 7.9 persons, suggesting either that urban food production can support larger families or that such families need to produce food for their survival (or both). Whether, and why, cultivators' households may be poorer than others needs investigating. A random sample of 150 urban households in Yaoundé in 1998 found that the overall family size was 6.6 persons with a mean of 7.9 persons among the poorest 20 percent of households (Gockowski et al. 2003).

Further, we estimated the monthly income derived from commercial traditional leafy vegetable production was 36 000 FCFA (Franc Communauté Financière Africaine), approximately 50 percent above the minimum wage, suggesting that cultivators' households are, on the contrary, not among the poorest. The data certainly suggest that cultivation is a strategy of poor households to alleviate poverty and feed more people in the process. Most commercial producers (70 percent) said agriculture was their principal source of income, whereas most household food producers (67 percent) cited a formal sector job as their principal source of income, with farming as their second most important source of revenue.

Our findings reveal an important gender dimension to urban cultivation, which is clearly dominated by women. The data also support the frequently-made assertion that women pursue it more often as a strategy for feeding their households, while men pursue it more often as a means of earning income, following the conventional assignment of gender roles in Sub-Saharan Africa (Hovorka & Lee-Smith 2006). However, this is by no means an exclusive division, with women also forming a large majority of commercial urban cultivators and food traders; more than 4100 women were engaged in green vegetable marketing in Yaoundé in 1998 (Gockowski et al. 2003). Women also constituted 94 percent of traders in fresh cassava root, as described below in this chapter.

Contribution of Perishable Products to Household Nutrition

Overall in Yaoundé we found no major nutritional shortages as the diet was rich and diversified, and the target of 2500 kcal/person/day was easily met for all three income groups. However, about 10 percent of the population was found to suffer from a chronic deficit of carbohydrates and protein, based on more detailed breakdown of the low-income group. Therefore, it is important to examine the role played by urban and peri-urban food production in meeting nutritional needs for vulnerable groups, and to determine if it could meet this shortfall to a greater extent.

The very poor obtained their protein and micronutrient needs from leafy vegetables and groundnuts. Leafy vegetables are critical to calcium intake as access to dairy products is limited in Yaoundé. There is an enormous diversity of traditional leafy vegetables grown in the city (Poubom et al. 1999). Three leafy vegetables (cassava leaves, *Vernonia* and *Amaranthus*) provide about 8 percent of the protein and 40 percent of the calcium intake of all urban consumers. Their nutritional contribution was found to be more significant for the urban poor than for wealthier families, as these traditional foods are rich on a per unit cost basis, supplying a disproportionate share of protein, minerals and vitamins. The protein of amaranths is particularly noted for its high level of lysine, an essential amino acid that is often lacking in diets based on roots and tubers (Schippers 2000).

It has been found that consumption of the less-nutritious exotic *Brassica* species, such as cabbage, is lower in Yaoundé than in other African urban centres, accounting for only 7 percent of expenditures on leafy vegetables.[1] Forty-two percent of households were found to be consuming *Brassica* species, compared to 100 percent consuming traditional leafy vegetables. This contrasts with markets such as Nairobi, Accra and Harare where *Brassica* species have replaced traditional vegetables. Expenditures on traditional vegetables, such as *Gnetum* spp. and cassava leaf, decline among wealthier households.

There are different types of vegetable production systems in Yaoundé, complementing each other in the food supply. Peri-urban growers in the uplands capture most of the rainy-season market, while producers in the valley bottoms who rely on wastewater irrigation dominate the dry season market when prices are more than double. The dry season decline in supply and corresponding increase in price affects poor consumers negatively and is a food security concern for the very poor.

Traditional leafy vegetables are relatively easy to grow and are among the most commonly found crops in home gardens, due to their low capital requirements (Poubom et al. 1999). The large number of sellers and their widespread dispersion among the numerous open-air markets of Yaoundé also serve as a substitute for a lack of refrigeration, allowing consumers to get fresh vegetables generally on the same day they are harvested. This is important, as the nutritional quality of these crops declines rapidly without refrigeration, which in general is lacking among the poor.

Twenty-seven percent of the leafy vegetables consumed by low-income households in Yaoundé come from their own home gardens. For all income groups, 10 percent of the total consumed was home-grown, while gifts from family and friends contributed more than 20 percent. Urban consumers themselves also produce 10 percent of the fresh cassava roots and fresh cassava leaves they consume. These consumers also produce between 5 and 8 percent of the plantain, cocoyam (taro), banana, processed cassava products (*bâton*), and papaya that they consume.

Seedling Production for Urban Horticulture

Fruit trees and other horticultural crops are produced in all thirteen sub-divisions of urban and peri-urban Yaoundé as well as in surrounding rural areas, although the geographical distribution is very unequal. Thirty-nine nurseries were studied in detail, the owners interviewed forming the study population (see "Methods used for the Five Studies", above). In Yaoundé city, most of the nurseries are located along major roads. Some are in busy squares, others near the administrative centre or up-market residential neighbourhoods. In order of importance, nursery owners said their location criteria were market access (31 percent), water availability (26 percent), accessibility (23 percent) and land availability (15 percent). Bushy wetland areas in valley bottoms are used most often, and most nursery owners

---

[1] The following paragraphs draw heavily on Gockowski et al. (2003)

(57 percent) placed their site alongside existing nurseries for reasons of convenience, publicity and security.

Around 70 percent of the nurseries had been operating for between 3 and 8 years, with one being 48 years old, originating from the time when horticulture was promoted in and around Yaoundé. Over 95 percent of all nurseries are private enterprises. Horticultural nurseries were most common (41 percent), followed by mixed nurseries (36 percent) and fruit tree nurseries (23 percent). Nurseries reproduced up to 50 plant species, but five species per nursery was most typical. An estimated 32 258 plants were reproduced, 24 864 of which were sold. Individual numbers ranged from about 1500 to 5000 produced and sold by a single nursery. The nine existing fruit tree nurseries were estimated to have the potential to reproduce about 38 000 plants and sell 28 000, compared to the fourteen mixed nurseries, which could produce and sell about 47 000 and 37 000 respectively. The most important fruit trees were avocado and mango, followed by citrus and guava.

There were three types of nursery enterprise: family businesses, women's groups and individual male ownership. Over one-third of the owners had at least one other nursery in a different part of Yaoundé. Only 10 percent of the owners were women. While education among fruit and vegetable producers was generally low, 60 percent of nursery owners had professional training in gardening or nursery keeping, and a quarter had engineering qualifications, of which a good 14 percent were agronomists. Eleven percent of the owners belonged to a professional organization. It was apparent that only a few of the nursery owners live exclusively on revenues of the nursery. Most nursery owners (77 percent) employed other people, usually 1–4 persons, although up to 12 employees were found. Clients were mostly civil servants, entrepreneurs, farmers, retirees and diplomats.

The nursery owners' principal source of technical and price information was from other nursery operators (32 percent) or special publications (32 percent). Sixteen percent obtained information from extension programs whilst 13 percent obtained it from the Institut de Recherche Agricole pour le Developpement (IRAD) and 6 percent from professional schools. Only 30 percent of the nursery owners actively tried to promote their enterprise through any type of public relations, and this mainly came about through media reports or occasional visits by interested groups.

At a meeting in 2003 between urban and rural nurseries, rural owners mentioned their lack of market access and market information as problems, whereas the urban owners complained about the limited number of species at their disposal and the lack of multiplication techniques. Seed material is available at IRAD, but not in sufficient quantity. This meeting was a landmark in identifying the potential benefits of collaboration between rural and urban nurseries to overcome several of the obstacles to production and marketing. There is scope for collaboration with benefits for both sides. The urban nursery owners saw that creating links with rural nursery owners would open possibilities for increased production. This was seen as of particular potential for marketing local fruit trees such as safoutier (*Dacryyodes edulis*) and kola (*Garcinia kola*). Both of these are used in cooking popular traditional foods and would create new urban markets.

Water Pollution Affecting and Caused by Urban Agriculture

Urban agriculture can be an environmental polluter and at the same time be affected by harmful substances derived from other sources. In addition, farmers working in and around polluted surface water supplies risk their health. Surface, run-off and sub-surface water from a large number of small streams in and around Yaoundé converge into the M'fou drainage of the Nyong River watershed and are subject to two major pollution mechanisms:

- Point pollution (industries, fuel stations, septic tanks, non-functional sewers, markets);
- Non-point or diffuse pollution (e.g., dumped household refuse; household latrines; agricultural inputs such as pesticides, organic and inorganic fertilizers; informal small animal production units; soil erosion).

According to our investigations in 2003 Yaoundé had 2687 known industrial enterprises discharging their effluent (an estimated 296 000 m^3 per month) and solid wastes (an estimated 4600 tonnes per month) untreated, directly into surface waters (Table 3.2). More than half the effluents come from large manufacturing plants. These are mainly six enterprises including two breweries, slaughterhouses and dairies. The next largest polluters are hotels and restaurants, hospitals and educational institutions, and garages or car repair workshops. Other sources contribute less, with animal production adding about 0.1 percent of the total. The load of each effluent with specific chemical compounds or biological substances is known for some but not all of the large plants identified. For example, effluents from breweries were found to contain 2080 mg/l Chemical Oxygen Demand (COD),

Table 3.2 Estimated amount of effluent discharged by known industrial polluters

Type of industrial unit	Total number	Number surveyed	Total effluent (m^3)	%
Manufacturing	224	44	171 192	57.8
Hotels and restaurants	152	15	27 686	9.4
Hospitals and social services	150	30	24 368	8.2
Education	152	15	20 594	7.0
Garages and car and repair workshops	712	71	13 809	4.7
Public administration	232	24	10 512	3.6
Markets, sewers, etc.	141	14	10 037	3.4
Furniture and light manufacturing	459	46	4764	1.6
Construction	62	7	4156	1.4
International organizations	111	5	4029	1.4
Banks and insurance companies	161	16	2046	0.7
Water, electricity and gas supply	24	3	1690	0.6
Transport and communication	90	10	888	0.3
Animal production	17	3	258	0.1
Total	2687	303	296 029	100.0

Source: Authors' survey

259 mg/l metal sulphide (MES), 1650 mg/l Biochemical Oxygen Demand, 5-Day (BOD_5), 41.5 mg/l Phosphate (PO_4), and 71.5 mg/l Ammonium (NH_4), along with 1000 colony-forming units (CFU) 100 ml^{-1} fecal coliforms and 50 CFU 100 ml^{-1} fecal streptococci.

Pollution levels are generally higher in lower parts of the watershed rather than higher up. For example, in the *Abiergué* watershed it was found that BOD_5 levels were 150 mg/l in the lower part versus 35 mg/l in the upper part. This is largely a result of increasing density of households, most of which have inadequate toilet facilities, as one moves downhill. This lack of adequate sanitation is of concern with respect to fecal contamination. Previously unpublished data from two neighbourhoods in 2001 reveal that only one-third of households had either a septic tank or regulation latrine, while two-thirds had insecure latrines in direct contact with the upper groundwater. This appears to be a major factor in the biological contamination of surface water. Water-quality assessments of all stream flows in Yaoundé were rated strongly to very strongly polluted, notably with fecal bacteria (Table 3.3 and Fig. 3.2).

Table 3.3 Mean water pollution levels of 12 Yaoundé streams

Parameter	Units	Mean	S.E.	CV	Min.	Max.
pH		7.1	0.80	0.11	6.5	9.9
Temperature	°C	26.0	0.81	0.03	24.3	27.4
Conductivity	µs cm^{-1}	321.2	158.58	0.49	127.0	716.5
MES	mgl^{-1}	115.2	234.32	2.03	7.0	956.0
O_2	mgl^{-1}	2.7	0.76	0.29	1.1	3.5
BOD_5	mgl^{-1}	48.7	66.73	1.37	7.4	271.0
COD	mgl^{-1}	95.2	182.11	1.91	5.3	739.0
NO_3	mgl^{-1}	1.9	2.19	1.14	0.6	8.9
NH_4	mgl^{-1}	10.4	14.78	1.42	1.0	59.4
PO_4	mgl^{-1}	9.8	12.46	1.27	1.7	50.0
Fecal coliform	CFU 100ml^{-1}	1.7E+06	2.5E+06	1.44	1500	8.2E+06
Fecal streptococci	CFU 100ml^{-1}	9.7E+05	3.3E+06	3.42	100	1.3E+07

Source: Authors' survey

In addition, concentrations of fecal bacteria and heavy metals, (Chromium (Cr), Cadmium (Cd), Mercury (Hg), Lead (Pb) and Zinc (Zn)) were measured in fish, mud and water from Yaoundé's two urban lakes. Predatory fishes had higher concentrations of both types of coliform bacteria than omnivorous species, and a human source for the contamination was indicated (Table 3.4). All the heavy metals were found in lake sediments, but only lead and zinc were found in water and fish, with concentrations in fish tissue generally below recommended limits (Demanou & Brummett 2003).

Apart from drinking water, which should conform to higher standards, the World Health Organization (WHO) recommends that water coming into direct contact with humans should contain less than 1.0E+04 CFU 100 ml^{-1} of total fecal coliform

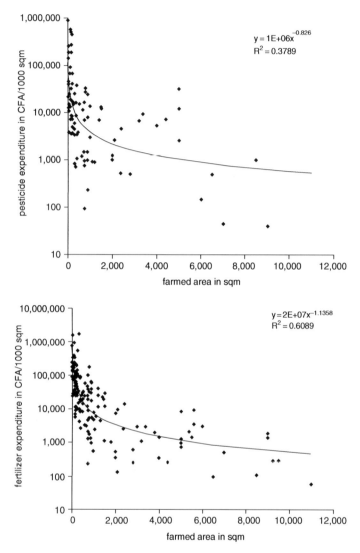

Fig. 3.2 Pesticide use by urban and peri-urban farmers in relation to land source
Source: Endamana et al. (2003)

bacteria. These levels were exceeded in 10 out of the 12 rivers and both of the two major lakes in central Yaoundé.

Contamination can occur through direct contact when fetching irrigation water or bathing as well as by consuming fresh vegetables irrigated with surface water. Pathogens in contaminated water include cholera, typhoid and dysentery. In a survey of 60 urban farmers we found that those who cultivate the bottomlands complain

Table 3.4 Geometric mean densities of fecal coliform (FC) and fecal streptococcus (FS) bacterial in the intestinal contents of four fish species from two lakes in urban Youndé, Cameroon (CFU/100 g). No *H. niloticus* were captured in Lac Central

	Lac central		Lac melen	
	FC	FS	FC	FS
Oreochromis niloticus (herbivore)	1.2×10^7	5.1×10^6	2.3×10^6	2.3×10^5
Hemichromis elongatus (carnivore)	3.6×10^7	2.7×10^6	5.3×10^6	4.4×10^5
Clarias gariepinus (omnivore)	2.7×10^7	3.2×10^6	3.4×10^7	1.3×10^7
Heterotis niloticus (herbivore)			3.1×10^7	3.4×10^6

Source: Authors' survey

more often (>50 percent farmers) about health problems and spend more money on health than those working at the slopes and uplands (20 percent).

Urban farmers not only are affected by water pollution but also contribute to it by using excessive or wrongly applied fertilizers and pesticides. The average intensive vegetable grower in Yaoundé uses 20 bags of chicken manure on 393 m² of land, corresponding to 18 tonnes per hectare (t/ha) versus 10 t/ha used by the average extensive vegetable grower (Gockowski et al. 2003).

In a survey of 296 urban and peri-urban farmers Endamana et al. (2003) found that small landholders use relatively more fertilizers and pesticides than larger landholders. They also determined that 51 percent of urban and peri-urban inland valley farmers applying pesticides use Methyl-Parathion (a highly toxic insecticide), 16 percent use Maneb (a non-toxic fungicide), 16 percent use Deltamethrin (a pyrethroid), 6 percent Benomyl (a non-toxic fungicide), and 2 percent use Carbofuran (a highly to moderately toxic insecticide). Unfortunately, no data are available about the impact of pesticide use on water quality.

The implications of this enquiry for public policy are taken up in the conclusions to this chapter.

Rural–Urban Linkages: Measuring Supply Flows of Fresh Cassava Roots

Urban food security depends on both *in situ* production and the inflow of food from surrounding areas, both rural and peri-urban (Guyer 1987). With increasing urban population, it is necessary to understand the supply system of foodstuffs to towns and how this can be maintained or improved. In the last two decades of the 20th century, rapid urbanization brought changes in food production and commercialization in southern Cameroon, where growing urban demand has created opportunities for increasingly intensive commercial agriculture of maize, bitter cassava, yam and potato. An increase in cocoa prices in 1998, however, created disincentives to the production of bulky staple crops (e.g., plantain, fresh cassava root and cocoyam) in favour of high-priced cash crops, resulting in rising prices for staples and serious food access problems for the poor.

The main constraint seems to be the cost of transport and handling caused by poor infrastructure, both roads and markets, while oil price increases explain the higher urban prices. Thus the structural factor of demographic transition (more urban for less rural population) is the main cause of shortages of these staples for the urban poor, with every small shift in cocoa or oil prices exacerbating the effects as the supply system reaches its limits (Dury et al. 2004).

This study aimed at an in-depth understanding of the case of fresh cassava roots in Yaoundé, by devising a method for the identification, quantification and spatial analysis of urban flows. Altogether, it is estimated that about 50 000 cubic metric tonnes of fresh cassava roots were consumed in Yaoundé in 2002, a value of 2.5 billions FCFA and about 30 percent of the household expenditure for root and tubers (Dury et al. 2000).

Unlike other food items like plantains and cocoyams, there are no specialized wholesalers involved in the marketing of fresh cassava roots. The supply of the product is guaranteed by about 30 000 producers who commercialize in small quantities and rely on public transport in small vehicles that carry both passengers and farm produce. Fresh cassava roots are sold in three different types of bags (30, 75 or 100 kg). Average individual quantities handled per trip range from 250 kg to about 500 kg, depending on the market.

The most common type of vehicle used (65 percent) was the Toyota Corolla, which can carry seven persons when typically overloaded and 390 kg of cassava roots. The Liteace minivan (22 percent) was the next most common, carrying 14 persons when overloaded and about 550 kg of cassava root. Trucks, which generally travel at night, also transport small amounts of foodstuffs to Yaoundé in addition to their main cargoes of drinks, building materials, timber and other items. Most of the small vehicle drivers (60 percent) did not mind whether they were transporting persons or foodstuffs, even the very heavy and bulky cassava, although a few (12 percent) said they make more profit in transporting foodstuffs while 11 percent thought their vehicles would last longer by transporting people rather than such cargo.

The drivers transporting food move to and from the villages in search of passengers who are either picked up in front of their houses or along the roadsides. Before harvesting, farmers make appointments with drivers at least a day before travelling, or else notify passing vehicles by means of a flag made of a cloth tied to a stick. After the passenger has been picked up the flag is removed so as not to confuse the next passing driver. Most drivers ply routes to and from their home villages and do not go further, explaining that they operate through social ties, dealing with the people they know instead of risking their security with strangers. Most drivers make an average of two return trips a day, but some can go up to four, depending on the distance of the village from Yaoundé and factors such as the condition of the roads and the number of police checkpoints. The more the police checkpoints the fewer trips can be made. All drivers said these checkpoints are their main problem, with police compelling them to pay bribes without which they would not be allowed to continue.

Our study also found that there are no specialized wholesale markets for fresh cassava roots in Yaoundé, with eight markets performing several activities simultaneously: wholesaling (in the sense of unloading products coming from rural areas), semi-wholesaling (selling to retailers from other markets) and retailing (selling in small quantities to final consumers). Seven others markets are purely for retailing by the producer/sellers. Our study concentrated on four of the eight markets with multiple functions, which constitute about 80 percent of all retail sales in Yaoundé. Information was gathered from drivers and producers as they arrived from the numerous supply villages.

It was found that all of the sellers interviewed in our surveys were also the farmers, that 94 percent of them were women and that they came from 61 different villages. For example, 26 cassava producers were interviewed at the Ekounou market from 8 to 10 July 2002, all of them women and each selling fresh cassava roots in local primary exchange units known as "Sack Mbanjock" (about 32 kg each). Each handled an average of nine sacks per trip, approximately 300 kg. The minimum number of sacks handled per trip by these women was three (approximately 100 kg) and the maximum 20 (approximately 650 kg). The pattern was similar in the other markets studied, although the other units of sale, the "Farine" (76 kg) and the "Sack Aliment" (about 100 kg), were found in addition to the Mbanjock in the markets where trucks operated and a few men were trading cassava roots (see Plate 3.1).

The small size of vehicles that are not adapted to carrying goods explain why producers prefer the smaller Mbanjock sacks, while, conversely, the high numbers of small vehicles has been attributed to the small size of unit quantities coming

Plate 3.1 Typical scene unloading cassava roots as a Youndé market

from the production areas. In support of this, it has been said that the larger vehicles like pick-ups and trucks specializing in transporting food usually return from the production areas with more than 50 and 70 percent respectively of their capacity empty (Dongmo 1985).

The findings are similar to those in other countries including Chad, namely that, unlike other products such as plantains with marketing chains involving many intermediaries, the marketing chain of fresh cassava is short, with the retailer being the only intermediary between producer and consumer (Duteurtre 2000). The main suppliers of the product to Yaoundé are the producers themselves with their characteristic small supplies per trip. The producers do not specialize in cassava but deal with a combination of products. The absence of specialized wholesalers leaves the sub-sector less efficient. This absence of specialized wholesalers can be explained by the bulky and perishable nature of cassava, which does not allow for it to be stored for bulk transport to the urban market. Taking a gender perspective, our findings suggest that women are involved in less profitable crops that simply dispose of household surplus, and generate little cash profit.

Among the 110 supply villages that we identified, 61 had significant quantities of cassava on hand and 50 did not. The most important supply villages identified were located within a radius of a 1–4 h drive to Yaoundé. However, the atomistic supply of the product complicates the task of determining which supply villages are most important, and why. The number of observations in each market was small, taking into consideration the numerous supply villages. In fact, cassava may leave any of the villages on any particular day for the urban market. Further studies based on time series data would contribute better data on the quantities flowing from the supply areas. In-depth studies of the social, economic and spatial factors that influence producers to sell or transform cassava roots would also be of benefit, and a gender analysis would be particularly useful.

However, it can be concluded from our study that, in the case of this product, distance from the road and distance from Yaoundé have a real impact on food supply, that the condition of the road and the season (wet or dry) has an effect on the amount of food that gets to town, and that decision makers could increase food supply and food security by improving the condition of roads that serve villages which produce food such as cassava root (Temple & Moustier 2004). The main advantage of this type of study is that it mixes qualitative and quantitative data and analyses them spatially, allowing for the development of tools that can be applied in planning.

Conclusions of the Five Studies

Urban agriculture in Yaoundé may be viewed from a variety of disciplinary viewpoints and from the perspectives of a variety of different stakeholders. Our studies confirm that it is an activity widely pursued by households as a strategy for generating income and providing food, and call for public sector intervention to guide farmers and consumers in the interest of food safety, while ensuring food supplies are maintained. It is mainly women who engage in both the production and sale of food crops, most of which are crucial to the nutrition of residents of the town, especially those with low incomes. The different studies presented here indicate a number of directions for improving the production and supply of different foodstuffs consumed by the population of Yaoundé.

Better support is needed for the use of the land for urban cultivation, in an effort to meet the shortfall in energy and protein identified in an affordable way. This would make possible increased and more stable production of leafy vegetables and groundnuts for protein and cassava and other energy-rich foods for carbohydrates. Urban farmers also need better support for marketing. Institution building has to relate to the current ethnically based associations linked to cultivators' villages of origin and to women as the majority of traders as well as cultivators. Chapter 5 will address further some of the institutional problems related to marketing.

The current level of water pollution affecting agricultural activities in Yaoundé and its impact on human health raises questions for public policy. Up to now there have been no studies available that would establish the relative effect of this risk *vis-à-vis* all the other factors affecting human health in an African city like Yaoundé, although an attempt to do this is in Kampala, Uganda, is described in this volume.

As an immediate step in risk mitigation, urban agriculturalists should be educated about the dangers of using and getting into contact with urban river water. Furthermore, consumers should be informed that the use of fresh vegetables without bacteriological decontamination is dangerous. This can be achieved relatively easily and cheaply by washing vegetables to be eaten raw and treating them with small amounts of household bleach. Alternatively, cooking vegetables before consumption kills germs that might affect human health. Currently, fish appear safe to eat in terms of heavy metal risk. In terms of risk from biological contaminants, cooking removes such risks from food, but raw fish must be handled with care in the kitchen as contaminants may be released on cutting, especially the larger fish, and the livers of such fish should be avoided. Surfaces where they are cut should be cleaned before other foods to be eaten raw touch them (Demanou & Brummett 2003).

With regard to contamination of water by agro-chemicals, extension information and better regulation of pesticide-use are both clearly needed to protect farmers' households and consumers alike from the harmful effects of overuse of chemicals.

Policy-making in the new millennium should clearly distinguish such risks from the benefits that growing food in urban areas brings to the alleviation of hunger in households in addition to the alleviation of poverty through sales and income. Women, more than men, are involved in urban and peri-urban agriculture, implying the need for specific policy targeting when strategies of meeting the Millennium Development Goals on hunger and poverty are being implemented. The data presented here should aid in the development of such targeted programs to increase the safe supply of urban food through support to urban and peri-urban farmers and traders. Despite its complexity, the subject of urban agriculture can no longer be neglected by public policy and planning, due to its scale and therefore its role in meeting these targets.

References

Bopda, A & Awono, J 2003, *L'Agriculture urbaine et péri-urbaine à Yaoundé. Aspects institutionnels. Rapport final*, 88 pp. SIUPA, Yaoundé.

Bopda, A & Grasland, C 1996, 'Noyaux régionaux et limites territoriales au Cameroun. Migrations et structures par âge de la population en 1987', in Bocquet-Appel, JP, Courgeau, D & Pumain, D (eds) *Spatial analysis of biodemographic data*, Eurotext/INED, Paris, pp. 131–156.

Demanou, J & Brummett, RE 2003, 'Heavy metal and faecal bacterial contamination of urban lakes in Yaoundé, Cameroon', *African Journal of Aquatic Science*, vol. 28, no. 1, pp. 49–56.

Dongmo, JL 1985, *L'approvisionnement alimentaire de Yaoundé*, IRAD, Yaounde, Mimeo.

Dury, S, Gautier, N, Jazet, E, Mba, M, Tchamda, C & Tsafack, G 2000, *La consommation alimentaire au Cameroun en 1996: données de l'enquête Camerounaise auprès des ménages (ECAM)*, Direction de la Statistique et de la Comptabilité Nationale (DSCN), Centre de coopération Internationale pour la Recherche Agronomique en Développement (CIRAD) & International Institute of Tropical Agriculture (IITA), Yaoundé.

Dury, S, Medou, JC, Tita, DF & Nolte, C 2004, 'Le système local d'approvisionnement alimentaire urbain atteint-il ses limites en Afrique sub-saharienne? L'exemple des amylacés dans les villes du Sud Cameroun', *Cahiers Agricultures*, vol. 13, no. 1, pp. 116–124.

Duteurtre, G 2000, *Une méthode d'analyse des filières*, Synthèse de l'atelier du 10–14 avril 2000, LRZV, N'Djaména: document de travail. N'Djamena: PRASAC, 2000. 46 pp.

Endamana, D, Kengne, IM, Gocokwksi, J, Nya, J, Wandji, D, Nyemeck, J, Soua, NN & Bakwowi, JN 2003, 'Wastewater re-use for urban agriculture in Yaoundé, Cameroon: opportunities and constraints', *Johannesburg Symposium 2003: Water, poverty and productive uses of water at the household level*, Multiple Use Water Services Group, Muldersdrift, South Africa, pp. 84–92.

Franqueville, A 1984, 'La population rurale africaine face a la penetration de l'economie moderne: le cas du Sud-Cameroun', *Le developpement rural en questions: paysages, espaces ruraux, systemes agraires : Maghreb, Afrique noire, Melanesie*, Paris, ORSTOM, pp.433–446.

Geschière, P & Konings, P (eds) 1993, *Itinéraires d'accumulation au Cameroun/Pathways of accumulation in Cameroon*, Paris, Karthala.

Gockowski, J & Ndoumbé, M 1999, 'An analysis of horticultural production and marketing systems in the forest margins ecoregional benchmark of southern Cameroon', *Resource and Crop Management Monograph Research No. 27*, International Institute of Tropical Agriculture (IITA), Ibadan, Nigeria, p. 61.

Gockowski, J, Mbazo'o, J, Mbah, G & Fouda Moulende, T 2003, 'African traditional leafy vegetables and the urban and peri-urban poor', *Food Policy*, vol. 28, no. 3, pp. 221–235.

Guyer, JI 1987, 'Feeding Yaoundé, capital of Cameroon', in JI Guyer (ed) *Feeding African cities: studies in regional history*, Manchester University Press, Manchester, pp. 112–153.

Hovorka, A & Lee-Smith, D 2006, 'Gendering the urban agriculture agenda' in *Cities farming for the future*, IDRC, Ottawa & ETC-RUAF, Leusden.

Laburthe-Tolra, PH 1985, 'Yaoundé d'après Zenker', *Annales de la faculté des lettres et sciences humaines*, no.2.

Lee-Smith, D 1997, *'My house is my husband': a Kenyan study of women's access to land and housing*, Thesis 8, Lund University, Sweden.

Lemeilleur S, Temple L & Kwa M 2003, Identification of banana production systems in urban and peri-urban agriculture in Yaoundé, *InfoMusa*, vol. 12, no. 1, pp.13–16.

MINPAT 1993, *Migrations et urbanisation: Le cas de Yaoundé et de Douala*, Yaoundé, DRDI/République du Cameroun.

Poubom, CF, Mboussi, AM & Fonkem, C 1999, 'The biodiversity of traditional leafy vegetables in Cameroon', in Chweya, JA & Eyzaguirre, PB (eds) *The biodiversity of traditional leafy vegetables,* IPGRI, Rome, Italy, pp. 23–49.

Schippers, R 2000, *African indigenous vegetables: an overview of cultivated species*, Natural Resources Institute/ACP-EU Technical Centre for Agricultural and Rural Cooperation, Chatham.

Temple L 2001, 'Quantification des productions et des échanges de fruits et légumes au Cameroun', *Cahiers Agricultures*, vol. 10, no. 2, pp. 87–94.

Temple, L & Moustier, P 2004, 'Les fonctions et contraintes de l'agriculture périurbaine dans quelques situations africaines (Yaoundé, Cotonou, Dakar)', *Cahiers Agricultures*, vol. 13, pp. 15–22.

Chapter 4
Crop–Livestock Integration in the Urban Farming Systems of Yaoundé

Thomas Dongmo, François Meffeja, Jean Marin Fotso, and Christian Nolte

Introduction

As a result of rapid rates of urban growth, especially in the larger centres of Douala and the capital, Yaoundé, about half the Cameroon's population lives in urban areas (World Bank 2003). Yaoundé's population increased at the rate of about 10 percent per annum from 0.64 million in 1987 to 1.5 million in 2000 (DSCN 2000), and is projected to be as much as 4 million by the year 2020. The city's population faces enormous problems of poverty, with unemployment registering around 25 percent (DRSP 2003). To sustain themselves, households resort to urban agriculture (UA), which also provides some income. Statistics taken from just one aspect of UA show that at least 32 000 households sell traditional leafy vegetables, most of which are produced by the women in those households, using extensive mixed crop-farming systems (Gockowski et al. 2003).

Integrated crop and livestock production on small mixed farms is one of the oldest agricultural practices typical of surrounding rural areas, and is an important component of the urban farming systems of Yaoundé. The rearing of small livestock in backyards in the city is a common supplementary economic activity that also constitutes an important source of animal protein for human nutrition. Livestock contribute to urban dwellers' incomes, health and nutrition, as well as being an integral part of their culture when used as gifts or for religious purposes. Yet little research has been done to document and evaluate the potential of urban crop–livestock mixed farming to become an integral part of Cameroon's agriculture sector. The study described in this chapter – part of the work of the interdisciplinary teamwork described in Chapter 3 above – was done in Yaoundé in 2002–2003 to address this deficiency.

Our specific objective was to characterize the technical, social and economic aspects of pig and poultry production in both urban and peri-urban areas of Yaoundé, and to explore the potential of these small mixed farms to become a viable part of

---

T. Dongmo (✉)
IRAD, s/c B.P. 3214, Yaounde-Messa, Cameroon
e-mail: thomas.dongmo@gmail.com

the country's agriculture sector. The resulting data describe the distribution channels for manure and offer estimates of the total quantity of manure produced, the quantity used in urban areas and the amount exported to other provinces. The constraints and opportunities for an integrated crop–livestock agriculture system in urban areas are also identified.

The study's findings generate a picture of a highly productive but relatively undeveloped part of the agricultural sector where opportunities exist for further intensification. The benefits would be increased alleviation of hunger and poverty among the urban low-income population, better development of small-scale enterprises in the urban economy, and better management of the urban environment. However, constraints and problems that stem from urban livestock keeping would need to be managed as part of this scenario. Manure management is a priority.

Methods Used for the Study

Cameroon's capital, Yaoundé, is located in the central highlands of the country. It is administered by the Governor of the Central Province and managed by an urban council located physically within the Division of Mfoundi, which has an area of 297 km^2. According to national statistics, 70 percent of the chickens and 50 percent of the pigs in Cameroon's Central Province are produced in Mfoundi Division, which is to say they are products of urban and peri-urban agriculture mostly on household small mixed farms. This heavy concentration of the provincial animal production around the capital is attributable to the insufficient and decrepit road network in rural areas, and the large potential market and better access to vaccines in the city.

Yaoundé's boundary has been expanded seven times since its establishment as a colonial settlement in 1889, the most recent being in 1992. In reality, it is an agglomeration of numerous settlements that are becoming more and more densely settled due to the high rate of urban population growth. The city is divided administratively into quarters that correspond to old villages, with the quarters grouped into subdivisions. For the purposes of this study, the 1992 boundary of the city was used as the urban area and the boundary of Mfoundi Division as inclusive of the peri-urban area.

A preliminary descriptive inquiry was carried out in Yaoundé and its surroundings in 2002 in order to identify the different pig and poultry production systems and to set research priorities. It was observed that in general, the more one approaches the centre of Yaoundé, the less frequently animal production is found. Livestock-keeping activities decline as one gets closer to densely populated areas. The preliminary survey was used to design the sample and the survey questionnaire, and to pre-test the latter before final implementation. Data were also collected from secondary sources through agricultural census reports, export statistics, interviews with key informants, and published and unpublished reports.

For the structured survey, a stratified random sample was selected, taken along seven major axes crossing Yaoundé. The starting point for each was the Central Post Office, going out toward the periphery. Primary data were collected in 2003 from 150 chicken and pig producers, 75 in each category, in 11 quarters purposively selected from the 16 sub-divisions of Yaoundé, as well as in four peripheral zones, so as to cover urban and peri-urban producers in accordance with the preliminary survey.

The piggeries and poultry farms eligible for the study were ones that had been established for at least 1 year. Interviews with these poultry and pig farmers were complemented by key informant interviews with feed producers, veterinarians, butchers and animal retailers.

Data were collected on farmers' socio-demographic characteristics and livestock production activities, including constraints and integration of livestock in the whole mixed farming system. Health problems, the socio-economic importance of the agriculture in the household's livelihood activities, and the technical and economic aspects of livestock production were also included.

The Findings on Livestock Enterprises in Yaoundé

More than 80 percent of animal production in the sampled areas was geared toward chicken (broilers and layers) and pigs, with about the same proportion of all three categories of enterprise being found in urban and peri-urban areas. The sizes of enterprises differed, with more small production units (less than 500 broilers or hens, 20 or fewer pigs per production cycle) found in urban areas, while bigger enterprises with more than 500 animals were found only in the peri-urban areas. Farm sizes varied from 200 to more than 1000 broilers, and between 1 and 41 pigs per herd. Small production units accounted for 80 percent of the broiler, 70 percent of the pig and around 10 percent of the hen-production units (Table 4.1).

Table 4.1 Number of animals per farm surveyed

# of Broilers	up to 200	200–500	500–1000	>1000	
% of Farms surveyed	30	50	15	5	
# of Pigs	1–10	11–20	21–30	31–40	>40
% of Farms surveyed	51	19	12	10	8

There were more women involved in keeping chickens and more men involved in keeping pigs. The majority of farmers interviewed were aged between 21 and 51 years and most were married (78 percent of those raising chickens and 95 percent of those raising pigs) suggesting that some woman-headed households were keeping chickens, because being a married farmer indicates a male-headed household. Around half the farmers had a secondary school education, and almost half had no previous experience in livestock production, suggesting they were urban dwellers who had decided to take up farming. However, over 75 percent of those interviewed had by then been involved in livestock production for at least 5 years. Chicken production was done by women in 65 percent of households, 80 percent of them being aged between 20 and 50 years of age, suggesting it was their main contribution to the workforce. By contrast, in 76 percent of the households pig production was carried out by men, 55 percent of whom were older than 50 years, and had retired from other work. The low participation of women in pig production was said to be due to the difficulties of the work. However, since family labour was almost universally used in both chicken and pig livestock enterprises, the high proportion of married couples keeping pigs where men run the business indicates that women were probably also participating in these enterprises (see Plate 4.1).

Plate 4.1 In Yaoundé, although pig production is said to be a man's task, women often do much of the work as "family labour"

The surveyed households engaged in livestock production mainly as a source of revenue, and both types of enterprise generated a good income. The sale price of a broiler chicken of 1.8 kg was between 2000 and 2500 FCFA, while a 40 kg bag of poultry manure from the laying hens brought in between 1000 and 1500 FCFA, and the broiler manure could be sold for 800–1000 FCFA. The chickens were sold mostly to buyers who came to the farm (60 percent) while the rest were sold in the local market or to restaurants and hotels. Some restaurant and hotel buyers also came to the farm to buy. Selling a kilogram of pork meat brought in 1300 FCFA, while a 40 kg bag of pig manure brought in 500 FCFA. Most of the pigs (80 percent) were sold exclusively at the farm gate. The buyers of the pigs were of three main categories: butchers, meat roasters and other households.

Many farmers sold their animals to generate liquidity (55 percent) and a significant proportion of all sales (30 percent) were for the specific purpose of paying children's school fees. A very high proportion of all livestock sales were said to be the result of demand during the period of festivities such as Christmas and Aid-el-Kebir (77 percent).

Crop–Livestock Integration – the Use of Manure in Yaoundé

The integration of crop and livestock production is part of the everyday reality of Yaoundé, with manure being a crucial part of the urban farming system for the production of garden crops. This study revealed that 90 percent of the mixed farmers also producing crops in the gardens of Yaoundé were aware that manure improves soil fertility in the same way as do purchased fertilizers and that poultry manure is richer than other manures. Some crop farmers we talked to when identifying our sample stated they were interested in keeping livestock as well, in order to generate their own manure supply, but said they were constrained by lack of family labour.

In Yaoundé the big poultry farmers are the main producers of poultry manure. Some of the manure produced in Yaoundé is sold in the Western Province, and this is mostly the higher quality manure of layer hens. Broiler manure is of lower quality than layer manure, and pig manure is of a lower quality than that of either type of poultry. Pig manure has a low dry matter content, which is a serious problem for its effective utilization, and is why much is not used by the farmer and instead becomes a source of pollution. Our study, however, identified some farmers who were using the pig manure as part of their production system, which suggests that further study is needed in order to better valorize that manure.

The use of manure by urban and peri-urban farmers in Yaoundé is systematic. Some farmers use manure alone, while some apply a mixture of manure and fertilizer. Because of the importance of this activity to the farming households, as well as to the economy in general, the study team undertook to calculate the total production and use of livestock manure in urban and peri-urban Yaoundé. We have used the following data and assumptions to estimate manure production.

The total parent stock for poultry in the urban and peri-urban area of Yaoundé is 70 000, according to recent official estimates. A breeder hen usually produces 160 eggs to be incubated, with a hatching rate of 80 percent, giving a yearly production of 8.9 million chickens. Half of these, 4.5 million birds, become broilers and about 2 million are layers, the remainder being cockerels. Chickens in Central Province constitute 20 percent of national production or 1.3 million broiler and layer chickens (Brilleau 1993). Poultry production in the Mfoundi Division and its environs make up 70 percent of poultry production in the Central Province, representing 0.91 million birds. As a proportion of this, layers represent at most 200 000 birds, with 710 000 broilers.

Cameroon had about one million pigs at the time of this study (DSCN 2000). According to the conclusions of a commission on animal production on the long-term planning of agricultural research (MINREST 1995), the forest zone where

Yaoundé is located has 18 percent of the animals, or 180 000 pigs. The demand for pork meat in relation to population distribution leads to a rough estimation of 50 000 pigs in Mfoundi Division.

Broilers and layer hens, fed mainly with maize and soybean, supplemented by fishmeal etc., produce 57 and 63 kg of fresh excrement per year, respectively (Salichon 1975; Oluyemi & Roberts 1979). A pig mainly fed with household refuse, supplemented by brewery by-products, produces 600–700 kg of fresh excrement a year (Eusebio 1980).

Using these data, along with those presented in Table 4.2, the following calculations can be made:

$$710\,000 \text{ broilers} \times 57 \times 30\% = 12\,141 \text{ tonnes of dry manure a year.}$$

$$200\,000 \text{ layer hens} \times 63 \times 30\% = 3780 \text{ tonnes of dry manure a year.}$$

$$50\,000 \text{ pigs} \times 650 \times 14.4\% = 4680 \text{ tonnes of dry manure per year.}$$

Table 4.2 Composition of animal manure according to source

	Dry matter %	Nitrogen (N)%	Phosphorus (P)%	Potassium (K)%
Pigs	14.4	5.6	2.5	1.4
Poultry	30	6.9	4.6	2.1

Source: Ministère de la coopération (1991)

Combining the totals, the total annual production of poultry and pig manure in Mfoundi is estimated to be 20 601 tonnes dry weight.

Several recent studies were used to estimate the quantity of manure utilized in urban and peri-urban crop gardening. The total valley area in Yaoundé is estimated at 6000 ha (Gockowski & Ndoumbé 1999) while the area used by cultivators varies from 228 ha (Temple 2003) to 245 ha (Damesse et al. 2003). The typical urban farmer uses 40 kg of manure for two vegetable beds on 12 m^2 of land (Dongmo & Hernandez 1999), leading to a conservative estimate that 8166 tonnes of manure are used by urban farmers if we take the lower figure of 245 ha of the valleys that are cultivated. There are two production seasons each year for all crops that are grown, although only just over half the urban farmers work on their farms all the time. This allows us to roughly estimate that approximately 12 250 of the 20 601 tonnes produced annually are used in Mfoundi Division, or around 60 percent.

However, the manure produced in Yaoundé is also exported to other provinces, part of it being sold off in the Northwest Province, notably Bamenda, where chicken manure fetches more than double the price: 1500–2500 FCFA per bag versus 700–1200 FCFA in the Yaoundé market. We estimate that about 10 percent of manure produced (about 2000 tonnes) is exported in this way.

Knowing that the total quantity of manure used is 12 250 tonnes in Mfoundi and 2000 tonnes elsewhere, it appears that the total level of manure utilization including areas outside Mfoundi Division is about 69 percent, and the amount of manure being thrown away in the environment is about 31 percent of the total or approximately 6350 tonnes. Based on the elements composing the manure and using the median value for nutrient content given in Table 4.2 above, discarding this amount of manure corresponds to a discharge of 400 tonnes of nitrogen, 229 tonnes of phosphorous and 114 tonnes of potassium as waste, which also constitutes serious environmental pollution.

Constraints and Opportunities for Crop–Livestock Production in Urban Areas

Pig production in town near residential areas is a source of many problems, particularly the unpleasant smell and noise, and also causes pollution, for example through contamination of the water table by nitrates. For these reasons the large pig farms are more often located at the urban periphery. Poultry farms can also have similar problems, although smell is less severe than in the case of piggeries.

Free grazing of animals – a common phenomenon in traditional livestock production – can be a source of infestation, with parasites like *Taenia solium* or *Trichinella Spirillis* for the live pigs and even for humans. Also, the social conflicts that arise from this type of traditional livestock production are not negligible, as neighbours may complain of noise or smell. The lack of technical training for farmers is a real problem. Officials of the extension services of the Ministry of Livestock are aware of the inadequate information and support that they are able to offer to livestock farmers, due to shortages of personnel, especially trainers, and of financial resources.

Other problems experienced by Yaoundé livestock farmers are difficulties in feeding the animals, the lack of finance, the unskilled labour force, marketing problems, and the lack of space for building piggeries and poultry houses. The harassment of livestock producers by municipal authorities is another constraint faced by the farmers.

There is a need to look into these difficulties in order to improve the conditions of livestock production in urban zones to achieve higher productivity and better performance in overcoming the constraints. Effective use of crop residues and other potential resources for pig and poultry feed in urban areas and market research based on farmers' conditions and available market opportunities are some of the immediate follow-up studies that are required. Agro-industries located in urban and peri-urban areas have an extensive potential market for their by-products in urban agriculture. During our study for example, it was estimated that 15 000 tonnes per year of brewery wet grains were being produced and consumed by pigs in Yaoundé and its surroundings.

Among the advantages of livestock production in urban areas is its important contribution as a source of employment. Our study shows that 10–20 000 persons, including producers, retailers, processors, animal input and feed traders, are employed in the industry. Because pigs and poultry both grow very rapidly, their production provides a high level of protein and a quick return on investment, thus generating income for the farmers and contributing to economic growth in general. In addition, urban livestock production benefits the diet of low-income households by providing city dwellers with an important local source of protein and minerals.

Livestock production within a fenced area permits easy collection of manure to fertilize farms or fish ponds, thus minimizing damage to the environment through a controlled discharge of pollutants. Small livestock production, especially poultry, is an important source of manure for use in the cultivation of leafy vegetables and other perishable products which can be consumed by city-dwellers. In this way, urban livestock production constitutes an important element of food security also offers great potential for economic development through effective use of urban and by-products.

Conclusion

This chapter has characterized the main livestock farming systems in Yaoundé as they interact with crop systems for a mutual increased productivity, and has pointed out their benefit to environmental management through effective re-use of livestock manure. The picture that emerges is that of a vibrant sub-sector of Cameroon's agriculture, based on small scale mainly mixed crop-livestock family farms, yet one that is insufficiently recognized for its specifically urban characteristics. Yaoundé is a city that produces over 20 000 tonnes of organic fertilizer per annum and supports a market for it to a major provincial urban capital, but is not cognizant of this fact in its official agricultural planning.

Farmers have been taking measures to improve crop-livestock system efficiency but lack information and support from the policy makers and the public sector in general. Research can play a crucial role in providing the much needed information necessary for making informed policy decisions. For instance, it seems clear that the 31 percent of organic fertilizer currently being wasted and causing environmental pollution could easily become a market commodity and be put to good use.

Urban livestock production is an important activity that leads to increased food security and to reduction of poverty and hunger in our cities. It is a practice that represents not only a cultural phenomenon but also a crucial source of revenue for the low-income urban populations of Yaoundé. While livestock activities can be a source of problems, specific actions can be taken to address these issues, with research playing a facilitating role in the process. Farmers need to be helped and encouraged to take measures that will improve livestock production and prevent it from becoming a problem. Above all, urban livestock production presents opportunities for effective environmental management and sustainable development, because it creates linkages to urban crop production by recycling nutrients to produce food in an ecologically sound manner.

At present, much of Yaoundé's livestock production takes place on mixed farms where the recycling of nutrients from livestock to crop production is relatively straightforward. This study did not look into the institutional arrangements for marketing and exchange of nutrients from the larger enterprises to crop farmers, and this aspect would need to be both researched and managed as specialization emerges with higher levels of commercialization.

Initiatives to organize support for urban and peri-urban agriculture have been taken already by research institutions and their partners, especially universities and non-governmental organizations (NGOs), several described in this book. Recommendations aimed at consolidating the outcomes of these research efforts which have been organized through cooperation between the Institut de Recherche Agricole pour le Developpement (IRAD) and the International Institute of Tropical Agriculture (IITA), and directed toward the goal of generating a body of knowledge to support and facilitate effective public policy planning, can be summarized as follows:

- Continued documentation is needed to facilitate an in-depth knowledge of urban and peri-urban agricultural production;
- Further studies of all the processing and distribution activities contributing to food security of urban populations are needed;
- Investigations and proposals of institutional arrangements that will provide support to the valorization of all urban and peri-urban agricultural production activities should be made;
- Further research on the constraints and problems of urban livestock-keeping should be undertaken in order to generate appropriate information and technologies for farmers, marketing agents and producer organizations such as cooperatives.

Farmers' efforts need to be supported and improved upon. Proper planning is necessary, so that an ecological livestock production is allowed to develop which will improve productivity and make optimum use of cultivable land, animal manure and the production of livestock feed. These aspects would constitute the elements of a sound public policy for the integration of agriculture and livestock production in Yaoundé. Support services to producers are essential for an effective implementation of such a policy. Based on this and other research – that described, for example, in the following chapter – it is clear that these goals can be achieved through collaboration among government and development organizations, including non-governmental organizations.

References

Brilleau, A 1993, *Analyse du système d'information sur l'élevage au Cameroun et propositions pour une enquête sur les effectifs du cheptel, Direction des Etudes, des Projets et de la Formation*, Rapport provisoire d'une consultation financée par le Projet CAPP et l'USAID (project number 631-0059).

Damesse, F, David, O & Zoyium, A 2003, *IIème rapport d'etape sur la recherche action sur l'agriculture peri-urbaine de Yaoundé*, intermediate report for IRAD Nkolbisson.

Dongmo, T & Hernandez, S 1999, *Identification d'un programme d'appui au développement des zones de production maraîchères et petit élevage dans la zone péri-urbaine de Yaoundé*, Direction des projets Agricoles – Ministère de l'Agriculture (MINAGRI), Yaoundé.

DRSP 2003, *Document de stratégie de réduction de la pauvreté*, République du Cameroun.

DSCN 2000, *Annuaire statistique du Cameroun 1999*, Direction de la Statistique et de la Comptabilité Nationale, Ministère de l'Economie et des Finances, Yaoundé.

Eusebio, JA 1980, *Pig production in the tropics*, Intermediate Tropical Agricultural Series, Longmans Group, Essex.

Gockowski, J, Mbazo'o, J, Mbah, G & Fouda Moulende, T 2003, 'African traditional leafy vegetables and the urban and peri-urban poor', *Food Policy*, vol. 28, no. 3, pp. 221–235.

Gockowski, J & Ndoumbé, M 1999, 'An analysis of horticulture production and marketing systems in the forest margins ecoregional benchmark of southern Cameroun', *Resource and crop management research monograph No. 27*, IITA, Ibadan.

Ministère de la Coopération 1991, *Memento de l'agronome*, République Française dans collection 'Techniques rurales en Afrique', Government, Paris.

MINREST (Ministère de la Recherche Scientifique et Technique) 1995, *Analyse et proposition de stratégies pour le long terme*, SNRA. TCP/CMR/2354, Government of Cameroun, Yaoundé.

Oluyemi, JA & Roberts, FA 1979, *Poultry production in warm wet climates*, McMillan Tropical Agriculture, Horticulture and Applied Ecology series, London.

Salichon, Y 1975, *Les déjections de volailles. Quelle est leur composition? Comment les utiliser*, INRA, No. 238 1er No. Novembre, France.

Temple, L 2003, cited by Damesse, F *in IIème rapport d'etape du fonds de recherche sur base compétitive: Equipe de recherche action sur l'agriculture périurbaine de Yaoundé*, Octobre, 2003.

World Bank 2003, *African development indicators*, World Bank, Washington DC.

Chapter 5
Institutional Development of Urban Agriculture – An Ongoing History of Yaoundé

Athanase Pr Bopda and Louis Awono

Background to the Institutional Study of UA in Yaoundé

The premise of this chapter is that urban agriculture (UA), though widespread, lacks an adequate institutional response. The study presented is based on empirical surveys carried out in 2002–2003, but it is also an urban history, presenting a rich description coming from 20 years of documentary research on Yaoundé by the lead author. We also went back to examine and make a critique of the way the Yaoundé Structure Plan of 1981 was put together and, in doing so, engaged the collaboration of representatives of a variety of institutions involved with UA. Our basic question throughout has been to what extent the reality of UA has found expression in institutions. The study was one of those executed by the inter-disciplinary team mentioned in Chapter 3, who presented their findings collectively at the Town Hall exhibition and public meeting in Yaoundé in early 2004. This meeting again brought together the main stakeholders involved in UA, as a way of moving forward the agenda of institutional development.

UA is at once a form of land-use, an economic sub-sector of agriculture and an expression of the multiple ways in which the urban and rural worlds intersect – a characteristic of many African capital cities. The practice of agriculture is intrinsic to Yaoundé, a town created by people of largely rural origin and the capital of a country where agriculture is the main activity of 70 percent of the population. The town has been in a process of rapid ongoing expansion at its periphery for about a century, as touched upon in Chapter 3, above.

Institutionally, the resulting town is a haphazard entity imposed on people who find themselves there due to the vagaries of history. It remains to be re-created as a modern African urban entity, with a citizenry that identifies itself with the place. This is essentially a task of institution building, much needed not only by the people living there but also in order to develop a really workable and recognizable African city for the 21st century (Bopda 1997). We use the term institution to also

---

A. Pr Bopda (✉)
Institut National de Cartographie (INC), 779, Avenue Mgr Vogt, BP 157, Yaounde, Cameroon
e-mail: bopda20001@yahoo.com

encompass organizations – the former meaning social norms and practices established over time and the latter meaning intentional formal structures created for a specific outcome. As we shall see, there is a problematic disconnection between both organizations and institutions of a formal nature and the norms and practices of many citizens, especially with regard to urban agriculture. Urban agriculture is but one aspect of the gradual transformation of African towns from rural to urban, and provides a useful way of examining the process of modern institution building in that it manifests many of the conflicts that arise over access to the natural and human-made environments of cities.

In order to study this transformation from rural to urban and the opportunities provided for building institutions, the complexity of human interaction with the environment involving UA must be examined. Agriculture taking place in and around urban areas must be understood in the interests of improved agricultural development in general. As the previous two chapters have shown, UA is not a deviant part of agriculture to be tolerated, but an intrinsic part of the sector's growth.

As described in Chapter 3, agriculture has been going on in Yaoundé since it began. The military station hired workers who were also farmers and the German settlement remained self-sufficient in food as a result. Since then urban agriculture has played hide-and-seek with urban management for a century, growing dramatically in the 1980s, when the economic crisis followed by Structural Adjustment Policies slowed down rural agriculture, reduced public sector employment and increased urban unemployment. The informal sector – previously considered evidence of failure and fought by officialdom with often bloody confrontation – was suddenly regarded as having the right to exist and urban agriculture was seen as part of it. It was viewed by the authorities as a means of survival as well as a source of income and social stability, while for the population it was just a spontaneous activity, part of tradition, even quasi-religious or at least an unquestioned aspect of everyday life. In this chapter we shall examine this historical evolution in much more detail using a variety of data sources.

Our major concern in this chapter is to critically examine the institutions that emerge from this history and their actual and potential relationship to urban agriculture. Currently, UA only appears in plans as "green space", while an attitude of tolerance or complicity exists between farmers and the authorities. In order to engage with UA as a sector as opposed to just tolerating it, there have to be economic, land, health and environmental policies that recognize its existence and links with many other sectors. Yet it is a real problem to integrate a phenomenon that evolves rapidly as the town develops. Since UA does not stay still it is hard to see how it can be regulated so as to protect the interests of all concerned.

Another difficulty is the current administration of Yaoundé – the four levels of hierarchy, including chiefdoms, are neither integral to, nor coherent with, the totality of the town. The division of Mfoundi into six sub-prefectures corresponding to six communes is neither an optimal way of connecting the administration with the population, nor an effective form of decentralization. This type of decentralization does not help integrate public services at local and central level in a town that is both a regional and national centre.

In every urban agglomeration, UA is the virtual frontline where the rural world of the farming plot meets the urban world of the street. It is a meeting point pertinent to food security, where citizens' needs for food challenge the capacity of rural production at the same time as the food production by rural people is stimulated to move to commercial production by the proximity of the citizens. This is not, however, an idyllic world, and there have been physical conflicts around use of land for UA in Yaoundé in recent times.

The greater socio-economic and political influence found in towns can lead some to dominate and exploit others involved in this transition. Gender exploitation, with the labour of women at the core of families being ignored in policy and planning, is one aspect of this. More generally, long-established citizens see anarchy in the rural ways of thinking and behaviour brought by more recent arrivals. The latter can feel threatened by the way urbanization works and apply means they see at their disposal to combat it, even resorting to occult practices that scare the overtly more powerful. Such a rich context of symbolic exchanges of all kinds creates confusion and uncertainty that do not auger well for a reasonable resolution to problems of urban and peri-urban land-use, whether for farming or other activities.

Yet tensions are always what animate social discourse and negotiation. In the capital of Cameroon, sporadic conflict between residents of different origins has peppered their interactions. These are often the result of different perceptions the actors have of their town as much as of each other. Whatever their cause, they indicate the points at which the invention and negotiation of a modern Yaoundé remains to be constructed. Among the items to be so renegotiated, UA appears as one where goodwill and ingenuity of all sides are needed and would be well invested.

On the positive side, the official views of UA may not be as entrenched as they seem. The same senior official who may order the cutting of maize plants or other crops laboriously planted in the valley gardens may himself raise a few chickens and tomatoes or other plants for his salad. He would, anyway, have to be completely insensitive not to notice the struggles of his own grandmother toiling to produce food or fetch fuel while he carried out such bureaucratic duties, especially if he is doing his job in order to earn money to support her.

Yet several decades after colonization, ways of thinking of the Cameroonian elites and their bureaucracies remain those of the dominant international development discourse, including thinking about the place of agriculture in urban African life. Agriculture and other policies still have the same tone as that of the international bodies, who themselves have not properly defined an agriculture policy that suits the global community they guide and dominate. Similarly, the environmental discourse has yet to clearly articulate its position on the relationship between agriculture and an urbanization that is proceeding rapidly.

In political terms, the effort to reduce social segmentation is what drives the power of local leaders and elites, who then proceed to construct the public domain. Elites emerge as they try to create the public realm in both feeling and substance, without the realization of which they would cease to be elites. In Yaoundé as elsewhere, their role is to create local public awareness and sense of identity. If much of the population is engaged in small-scale trade and farming, then that is part of the identity.

Based on research and a collective effort by stakeholders, this chapter aims to throw some light on the complexities of this discourse and to propose realistic interventions. Representatives from a number of research, government and non-governmental institutions constituted the research team and took part in the structured dialogue on the results of all the studies. It is the message of this chapter that only this type of inter-disciplinary research and multi-stakeholder dialogue can lead to institutions that will satisfy the concerns of all and, from them, to sustainable outcomes.

After explaining the methods used, the chapter presents a detailed history of Yaoundé based on documentary research. Following this, the confused institutional arrangements that currently govern UA are described and then examined in the context of the results of the 2003 research. Finally, the recommendations that emerged from the final stakeholder workshop in 2004 are presented, with their implications for institution building.

Methods

Against the background of the town's history – necessary to establish the context – key questions for our empirical enquiry were to discover which current institutions have a potential impact on UA, whether and how these impacts are felt on the ground, and what specific reasons there are for promoting UA through institutions in Yaoundé. Implicit in this enquiry is the uncovering of differences between institutional norms and rules and peoples' behaviours. Observations and interviews were carried out at five major places where food is marketed, to annotate statistics derived from secondary sources. A total of 1089 traders found operating there were interviewed in 2003, taking every fifth trader encountered.

To answer the first question effectively, the various ways in which UA affects urban life first had to be set out. Since it touches on food supply, employing a large number of different actors in production, processing and marketing, as well as on land-use planning and land management, waste management, health and environment, a very wide range of organizations (both public and private, formal and informal, governmental and in civil society) were necessarily involved. Although not yet properly defined in Cameroon, it is reasonable to assert that UA comes within the institutional purview of the agriculture sector. Indeed at the urban periphery in particular, the Ministry of Agriculture engages actively with farmers.

To cover the entire range of institutions involved and their views, we undertook a survey as follows:

- Identification of all administrative structures whose mandate touches on UA;
- An inventory of all organizations identifiable on the ground;
- An assessment of the importance of UA to traders inside and outside of Yaoundé's markets.

The first task was carried out November 2002, based on a literature search as well as asking various people in governmental and non-governmental organizations what the concerned bodies are, and then asking these bodies what level of interest and activity they have in UA. We were particularly interested in persons and organizations contributing to the plan for Yaoundé of 1981. A series of meetings on subjects ranging from spatial and land-use planning to transport and housing were held from 1980 to1984, and we were able to establish the level of participation by different organizations in these meetings based on our interviews.

The results of our enquiry were presented, along with those of other surveys described in the previous two chapters, at the public exhibition and stakeholder meeting in early 2004, again involving the same broad range of organizations. We have adopted the recommendations of that collective of interests as the conclusion to this chapter.

The History and Development of Urban Yaoundé

The site of modern Yaoundé was described by German military officers reconnoitring the South of Cameroon in 1888 on their way to Chad in the North as being occupied by the numerous Ewondo people, living in dispersed rural homesteads (Laburthe-Tolra 1981). Migratory people constantly on the move, the Ewondo searched for fertile land, good hunting or escape from inter-clan conflict. Like other Beti groups, they have a creation myth describing a Northern origin with a migration South, probably in the 18th century, but thought to have taken place historically over time (Franqueville 1984).

German occupation stabilized the pattern of settlement by inhibiting indigenous peoples' movements (Laburthe-Tolra 1970) but under later French administration some populations were displaced and others moved due to social changes, with people in Yaoundé being drawn from further afield than the Ewondo. In the 1950s, the growth of Yaoundé accelerated with increased influx of the Ewondo and other nearby Beti people in particular.

In 1988 Yaoundé consisted of 51 quarters or villages, with further additions later. Some of them have contested names, such as Awae, which Yaoundé residents call Mvog-Mbi, after the lineage resident there, Djoungolo III, again called after the resident lineage, Mvog-Ada, and a large part of the Briqueterie, which is always called by its original name Ekoudou. In Nsimeyong, the name Biyemassi is used to designate the parts of the quarter along the River Biyeme. A large part of Ngoa-Ekélé is called Obili, in memory of the resettlement there of many residents after their forced eviction by the French colonial administration from a previous location. This contestation of names signifies the divergence of residents' perceptions of urban space from those of the authorities (Table 5.1).

At the turn of the 21st century Yaoundé had become the second largest urban area of Cameroon after Douala, growing from a few thousand in 1950 to more than 1.5 million. This growth was little related to natural increase (Bopda & Grasland 1994) but mostly to the influx of migrants from the whole of Cameroon (MINPAT 1993a).

Table 5.1 Evolution of Yaoundé's population over 50 years

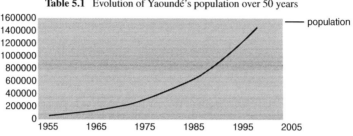

The low-density town centre is surrounded by a belt of first generation indigenous quarters with densities over 800 persons per km^2 since 1957. A second ring comprises quarters with densities over 800 persons per km^2 since 1969; these are larger, less integrated into the town and less split up into chiefdoms. The most recent part of the capital is a third ring of peripheral villages with densities of less than 1000 persons per km^2 in 1987. Here populations are increasing, mostly as a result of migration from the denser parts of town. Since the end of World War II a redistribution of population has taken place, settlements of over 800 persons per km^2 getting closer to the centre as well as stretching farther away toward the periphery (Fig. 5.1).

In villages more than 3.5 km from the town centre, the arrival of new residents – generally those leaving the first or more often the second ring of dense areas – increased rapidly between 1976 and 1987, small villages of a few hundred often doubling or tripling in size. The increase was less dramatic to the South, but some high- and middle-income central urban areas experienced similar rates of increase.

There is nowhere in the town that is not cultivated. Even in the town centre the sides of the streets are cultivated. The old indigenous quarters near the centre have abundant fruit trees and palms in what appears to be a completely saturated urban space. In the first ring of settlement, 2.5–4.5 km from the centre, maize and peanuts are planted alongside every house while in the next ring, 4.5–8 km from the centre, every plot acquired for development is first cultivated. In the peri-urban area 5.5–25 km from the centre, traditional farms are interspersed with large commercial plantations. Beyond the urban built space, at 20–100 km, crop and livestock production prevail, stimulated by demand from 1.5 million urban people. The same applies to surrounding small towns and settlements that are essentially rural in character but dominated by urban demand.

The poor urban slums, over-populated and subject to forced evictions, such as La Briqueterie and Mokolo, have suffered stagnating or declining growth. The demographic picture masks a turbulent reality of movement of people from one area to another, especially within the second ring of settlement (République Fédérale du Cameroun 1967). It is there that new immigrants are received and from where most of them depart for different areas of the periphery, following migratory trajectories linked to their places of origin and ethnicity.

In 1990, 60 percent of Yaoundé's population were migrants (MINPAT 1993b, p. 55) those from West, South and East of Cameroon predominating, the most

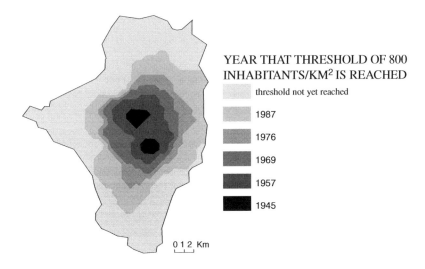

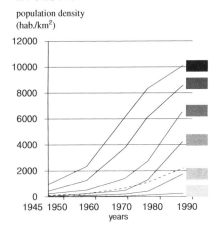

Fig. 5.1 Map showing the timing of population densities reaching 800 ppkm^2

intense in-migration being from the immediate surrounding areas. There was constant movement of people and goods between the town and its surrounding rural areas as the populations of South and Central Cameroon circulated between town and countryside (Franqueville 1987). Apart from this dominant pattern, large numbers of migrants from the East and West of Cameroon, and even from the coast, departed for Yaoundé every year (Bopda & Grasland 1994). Most migrants to the capital were adolescents aged 10–20 years, young adults of both sexes (20–30 years) and men aged 30–40 years (Bopda & Grasland 1996). They proceeded as if being drawn, step by step, the final one being within the capital, from centre to periphery.

They did this either seeking education to improve their lives (40 percent) or seeking employment (60 percent) (MINPAT 1993b, p. 213).

On arrival in town, barely 6 percent actually found employment, while 23 percent accessed an education system full to overflowing. The struggle to pay for basic essentials intensified during the economic crisis of the 1980s while during the period 1985–1995 unemployment rose from 10 to 23 percent for those aged 30–34 and from 12 to 33 percent for those aged 25–29 years (Kouame et al. 2001) (Fig. 5.2).

Many unemployed farm in town or at the periphery in order to meet their needs. This takes place in a cosmopolitan context, people from all parts of Cameroon farming side by side. The complex migratory patterns have produced an increasingly diverse population in Yaoundé, the proportion of local Beti residents falling from 55 to 41 percent between 1957 and 1993. We could discern this from inspection of documentary sources even though ethnicity was a variable dropped from the national census. In fact, province of birth has become a new indicator of ethnicity, which in fact governs spatial distribution in the town.

This is shown by the following map using the Piaseki index to examine ethnic homogeneity in Yaoundé. The index examines the probability of two residents taken at random in one quarter of the town coming from the same place of origin. We took the ten provinces of Cameroon and then further distinguished those born in Mfoundi (urban and peri-urban Yaoundé) from those born in are the rest of Central Province, as well as those born outside Cameroon who are classed as foreigners, giving a total of twelve categories of people.

The map clearly shows a heterogeneous town in the midst of a more homogeneous rural hinterland, with a gradient from urban to peri-urban. Pockets of weak heterogeneity near the centre correspond to residual indigenous villages such as Awae (Mvog-Mbi) or historic settlements designated for particular groups by the colonial authorities. For example an area near the catholic mission in Messa Mezala was designated "from Mokolo" for natives of Central Province especially the Lékié, and Ekoudou (in La Briqueterie) for Moslem populations mostly from Northern Cameroon (Fig. 5.3).

Rather than ethnic ghettos, the map of Yaoundé portrays some areas where one ethnicity predominates over others. The probability of finding people from Central Province is higher at the periphery than elsewhere, while the probability of finding people from Southern Province is overall low but much higher in the settlements along Mbalmayo road by which they enter and leave town. People from the North, including the extreme North of Cameroon, are found mostly in the mosque area of La Briqueterie and around a new abattoir right at the north-west of the town at Nkolmbong, where they enter Yaoundé from Northern Cameroon with their cattle. More dispersed throughout the town, people from West of Cameroon are found in greatest concentration in quarters like Nkomkana, Messa-Carrière and Oyomabang. The homes of people from the two Anglophone provinces of the north-west and south-west are few and mostly found in Biyemassi and Nsimeyong and, above all, the University of Ngoa-Ekélé which many early residents from this group attended on arrival in their capital.

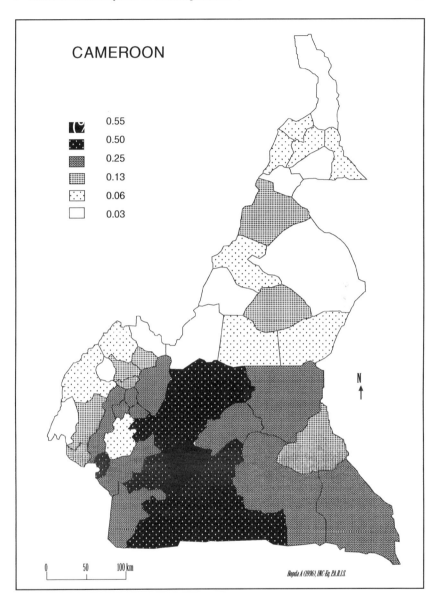

Fig. 5.2 Percentage of emigrants departing for Yaoundé from different parts of Cameroon

Everywhere, the waves of migrants from the ten provinces settled themselves, disturbing and eroding the villages of the indigenous communities that gradually reduced in proportion. The strong religious and traditional connection of the indigenous people to the land was weakened by the eruption of urbanism and commercial life, as they transformed themselves into sellers of land to accommodate the

Fig. 5.3 Homogeneity of the population according to birth place in 1987 and UA landscape

migrants. Their strategies of speculation in land and housing created the urban pattern, which in turn exerts a profound influence on the organization and functioning of the town.

The advantage of being there first meant the indigenous Ewono predominated as chiefs and leaders in the neighbourhoods, and it was their village ambiance that animated neighbourhood and community life initially. Their perceptions and responses to changing conditions, including their interactions with others, were what made the social life of Yaoundé what it has been for more than a century: socially diverse, lively, rich and, above all, peaceful. At the same time, urban life brought new economic realities that fundamentally affected day-to-day activities and the way things were being done.

Not counting the informal sector, most employed persons in Yaoundé work for the central and local government bureaucracies. This administrative town has few private sector employers, although domestic service is a major source of jobs. In 1990, 17 percent of Yaoundé residents were unemployed – 30 percent of women and 16 percent of men (MINPAT 1993b). Officially, the economically active population is predominantly male (72 percent), women being generally classed as housewives and economically inactive. These statistics tend to overlook the fact that women are the main workers in agriculture in Cameroon as well as in the informal sector in Yaoundé, sometimes being the main household income earners. In fact, women and children in school form a significant part of the urban population, such that, with the economic crisis in the 1980s aggravating the normal exit from school of large numbers of 15–25 year olds, the numbers of those seeking work increased dramatically.

Housing is an indicator of economic level and income. According to the 1987 census, 31 percent of households lived in detached houses and 54 percent in multi-family housing. However, only 8 percent of households lived in properly constructed modern houses, whether villas or flats, meaning 92 percent lived in inadequate housing. Sixty percent of housing was made up of rental units and these were generally the single story shacks that housed 46 percent of the population. Only 10 percent of housing units had title deeds, meaning those who occupied the rest lived precariously, lacking secure tenure.

Social infrastructure was concentrated in the town centre, requiring the majority to travel long distances to access services. Informal sector employment – basically self-employment in commerce – was also concentrated in areas close to the centre in Mvog-Mbi (19 percent) and La Briqueterie (58 percent) (MINUH 1981). Thus, formal and informal sectors appear to function in parallel with a similar spatial distribution, the difference being that informal sector employment is risky and precarious, being outside the realm of regulation. In other respects it reflects all aspects of the modern sector, but in a form more easily accessible to the bulk of the population, filling the gaps in state functions especially after these were cut back in the 1980s.

The marked disparity between a downtown concentration of employment and services and a rapidly moving periphery has led to a daily mass movement of thousands of people from their peripheral homes to the town centre, with a return movement at the end of the day. Congestion of vehicles and people is obviously greatest at the centre, while the movement itself creates an industry of small vehicles negotiating the inadequate or non-existent roads. It is clear that employment and services could be better located in relation to population.

Several plans have been made since colonial times, starting in 1963, but all of these have foundered on the impossibility of reconciling the divergent interests of the ruling elite and the population at large. For example, the official strategy of eviction and demolition to relocate poorly housed people at the urban periphery has been unsuccessful (Fig. 5.4).

We conclude this section of the chapter with some data on food production and marketing. Official statistics on UA are few – as with the informal sector in general – but agriculture is indeed recorded as an urban occupation. As shown in Table 5.2, the population of farmers increased even as they decreased as a percentage of the urban population. But the low percentage of farmers is a poor indicator of the fact that agriculture covers about 60 percent of the 150 km^2 urban area, and that more people than are officially recorded work in agriculture as part of household livelihood strategies. Many women, in particular, supplement household food and income with farming.

Inadequate though they are, official statistics nevertheless show the diversity of UA production in Yaoundé (Table 5.3). While cocoa and palm oil plantations (which

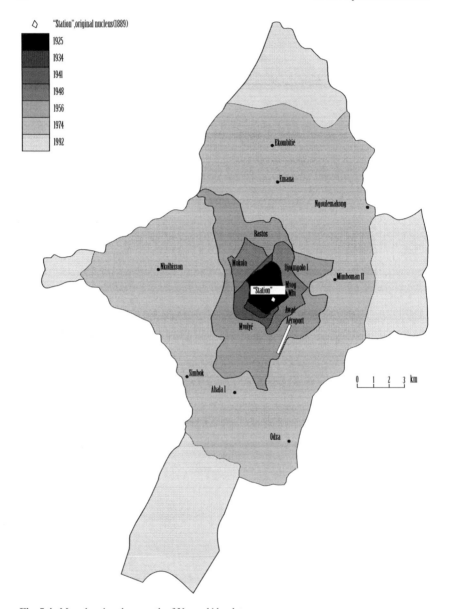

Fig. 5.4 Map showing the growth of Yaoundé by date

dominate rural production) occupy over half the space, they provide only 5 percent of UA revenue. Horticulture, staples and fruit provide about equal amounts, although staples score highest.

The markets found all over Yaoundé are large, crowded spaces with complex and conflicting needs for sales, storage space, parking, traffic movement and recreation.

5 Institutional Development of Urban Agriculture

Table 5.2 Farmers as a percentage of economically active population in Yaoundé

Year	No. of farmers	No. economically active	Farmers as % of economically active	Modal age
1957	653	3721	18	35–39 yrs
1964	3571	31 077	11	
1976	1985	86 909	2	20–29 & 30–39
1983	2500	106 000	2	
1987	5500	275 000	2	

Table 5.3 Composition of officially recorded UA production in Yaoundé

		Hectare (ha)	Tonne (t)	Price/ kilo[a]	% area	FCFA million	% FCFA
Plantation sector		760	112		52.6		
Cocoa		733	72	300	50.7	22	
Oil palm		27	40	150	1.9	6	5.4
Market gardening		94	196		6.5	145	28.4
	tomato	13	59	480		28	
	peppers	12	28	900			25
	celery parsley	11	4	1100		5	
	other	58	105	827		87	
Staples		472	1495		32.7	195	38.2
	maize	119	142	200	8.2	28	
	peanut	31	10	309	2.1	3	
	cassava	139	463	190	9.6	88	
	yam	92	156	80	6.4	12	
	plantain	77	724	87	5.3	63	
	other	14			1.0		
Fruit		120	1236		8.3	143	28.0
	mango	49	369	150		27	
	pineapple	22	511	200		39	
	Safou	15	48	400		10	
	guava	13	120	300		18	
	other		188	263		49	
Total		1445	3039		100	511	100

[a]Consumer price (Dury et al. 2000, pp. 271–275)

While hard to plan and looking disordered, these spaces and their functions are essential to the town's food supply, especially for the lower income groups, as shown in the section on food supply flows in Chapter 2 above. As places of exchange and commerce, markets contain numerous formal and informal activities and are the sites of conflict and negotiation of the multiple agendas of different stakeholders. They represent huge management problems.

Table 5.4 Distribution of different types of markets in Yaoundé by distance from the town centre, at the end of the 1980s

Location on rural–urban continuum	Radial distance from town centre (km)	Number of markets		Type of market	
		Formal	Informal	% points of sale for agricultural produce	% points of sale for non-agricultural goods
Central urban	0–1	3	0	20–50	50–80
Urban	1–3	11	8	40	60
Peri-urban	3–7	5	3	90	10
Near rural	7–40	7	0	90	10
Rural uninfluenced by urban area	40–100	16	0	70	30

N.B.: With the exception of the areas not influenced, this typology has been developed by the Yaoundé Urban Observatory of the Yaoundé Urban Community (CUY)

Rural, urban and peri-urban producers alike come to sell at these markets. As shown in Table 5.4, the newer and more central ones focus more on non-agricultural products while those at the periphery focus more on food.

Having described the reality within which UA has developed and operates within Yaoundé, we next proceed to examine the range of institutions that address it, based on our research.

The Institutional Study and Its Results

Description of the Formal Structures of Government

The administrative structure of Yaoundé is complicated and rather ineffective due to its dualistic nature. Figure 5.5 shows the parallel and overlapping rural and urban structures of Mfoundi and Yaoundé, both of which fall under the Provincial Governor. We shall now examine how each structure deals with UA.

The *Communaute de Yaoundé* (CUY) was created in 1987 with four subdivisions: Yaoundé I, II, III and IV. Later, Yaoundé V and VI were established in 1993. Those parts of CUY concerned with UA are as follows:

- Cabinet, in relation to matters of public relations as well as inspection and administration;
- Directorate of Finance and Economic Affairs, Department of Economic Development;
- Directorate of Legal and Administrative Affairs, Departments of Adjudication of Disputes and Social Affairs;
- Directorate of Technical Services, Departments of Architecture, Public Health and Hygiene, Parks and Gardens, Urbanism, Highways, and the section on Urban Development.

5 Institutional Development of Urban Agriculture

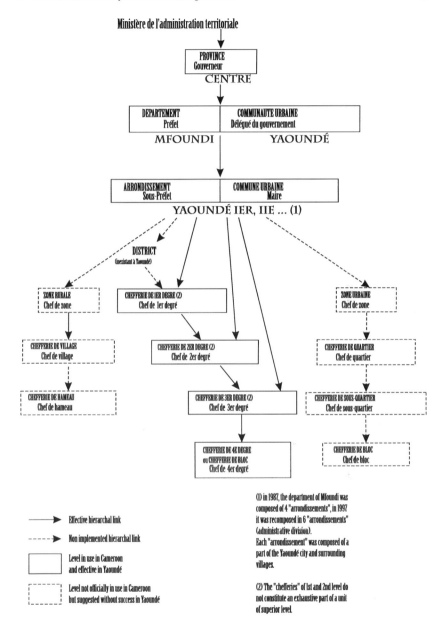

Fig. 5.5 Administrative levels and powers within Mfoundi Division

Meanwhile, as in the rest of Cameroon, each central government ministry has its local representatives and these operate through Mfoundi Division (*Departement*). Several ministries are organized to deliver services or to delegate them at provincial, divisional and sub-divisional (*Arrondissement*) level and even through the local traditional chiefs (*Chefferie de 2E Degre*). While the Ministries of Agriculture and Livestock (MINAGRI and MINEPIA) are of most relevance to UA, others such as environment, forests, public health, education, women's affairs, land administration, urban affairs, finance, economic planning, science and technology, social affairs, employment and social security also have a stake.

Agriculture services are provided at sub-division level within the urban area as follows:

- Research, Statistics and Projects;
- Agricultural Production;
- Rural Life and Community Development;
- Phyto-sanitary Control;
- National Project on Agricultural Research and Public Awareness.

One sub-division, in the Nkolondom area, even has an Education and Extension Centre that is used by urban farmers.

Mfoundi Division also administers the forces of law and order and supervises the formation of associations, many of which are traditional or indigenous social structures that have become surrounded by urbanization. The police force is significant for UA as many people think areas under cultivation are a security risk. The force, guided by a decree placing it under National Security, consists of four central police stations in different parts of Yaoundé, each of which collaborates with the municipal authorities in maintaining law and order.

In practice, many tasks of government need to be acted upon at sub-division level, especially sanitation, waste management, and public health and hygiene. However, it is through the vertical links between the levels of the system of command that action on UA is usually managed, from the top down to chiefs at the level of village or neighbourhood. Official responses to agricultural, especially livestock raising, activities are in fact sporadic. Sometimes, due to the sensitivity of such actions as destroying crops, a concerted action on behalf of the forces of law and order is undertaken, involving collaboration of all parts of the administration. When it comes to revenue collection, there is no proper definition of the nature and amount of taxes to be levied in relation to urban farming, as there is in the case of markets. This is where demanding of bribes and private accumulation creeps in.

Civil Society

Alongside these government structures are parastatal organizations with similar concerns, as well as research and development organizations, including universities and other educational institutions. The recent expansion of organizations in civil society includes several dealing with UA. In the private sector, there are organizations such as those working in waste management and recycling, including manure for re-use in urban farming.

Local grassroots associations form a vibrant element of Cameroonian society both in rural areas and in urban centres where they spring up as an adaptation of those in their adherents' place of origin. These traditional associations, being dedicated to pursuing members' common interests, play a strong role in UA. Several non-governmental organizations (NGOs) work with them on both food production and trade. For example, the Centre for Accompaniment of Initiatives for Development (CAID) – only legalized after 10 years of operation in 1997 – helps farmers deal with flooding, ditch digging and building of culverts, protecting springs to improve safe water supply, and promoting rabbit keeping for protein and income. The Association of Volunteers for Development (AVD), legalized in 1999, was actually formed 2 years earlier to help traders. Working with local associations to install sanitation and electricity in markets, it inspired other NGOs who addressed issues such as composting of domestic waste to clean up neighbourhoods and benefit UA production. Like AVD, CAID's members include many women and unemployed youths, and its resources are the contributions of members and the income from the water supply and other services provided, with an annual budget of about 4–6 million FCFA. They have had some outside assistance, including from CUY (Madiot 2001).

Chiefdoms – an essential base in traditional Cameroonian society – have also sprung up in urban centres. As they are also an adaptation of those from where their adherents came, there is an issue regarding their ethnic nature and relationship to government. This issue is unresolved. The authorities watch out for these informally created "traditional" chiefdoms, which they like to recognise and build into the central administration. The truly traditional chiefdom, Efoulan, covers the whole of Mfoundi, but others have proliferated. Some quarters of Yaoundé have second-level chiefdoms, usually with the same name as the quarter, whereas others do not. On the other hand, third-level chiefdoms (see Fig. 5.5) cover the whole of Yaoundé. In fact these chiefs are not part of civil society as they are officers of government who execute its policies. In 2002 there were 24 such official chiefdoms in Yaoundé VI, 33 in Yaoundé I, 36 in Yaoundé III while Yaoundé V had 40.

While the chiefdoms tend to be co-opted by government and are undoubtedly used as a means of control and coercion, the chiefs themselves are also concerned about income and employment for the populations in their areas and therefore tend to defend and articulate the problems and concerns of urban farmers.

Alongside these officially sanctioned so-called traditional chiefdoms, there are numerous other institutional structures also labelled "traditional". Sometimes spatial and sometimes socio-political, they co-exist in a complex array and are often used to good purpose as a means of conflict resolution in communities. They also assist their adherents to farm crops or to keep livestock on small plots, especially toward the town periphery, and this form of institutional support helps farmers to resist when others come to buy land for development.

Thus civil society can be seen as a contested area, with government attempting to incorporate it through co-optation into its hierarchies as well as through controlling legal registration. It is also an arena of opportunity for new institution building that responds to local concerns and initiatives coming from the population.

Institutional Mobilization for Yaoundé's 1981 Structure Plan

Our investigations showed considerable mobilization of organizations for UA during preparation of Yaoundé's structure plan of 1981. We measured levels of participation in UA and in the production of the plan in a variety of ways. As shown in Table 5.5 below, the scientific organization responsible for public sector agricultural research and development, IRAD, had most persons involved in UA, followed by the University of Yaoundé I and the National Institute of Cartography (INC), followed by the Ministry of Agriculture (MINAGRI), the International Institute of Tropical Agriculture (IITA) and the Higher National Polytechnic.

Table 5.5 Ranking of institutions involved in UA in Yaoundé by publications

Institution	Acronym	No. of publications	Average publications per researcher
Institute for Agricultural Research and Development	IRAD	32	3.6
University of Yaoundé 1		16	4.0
National Institute of Cartography	INC	13	3.3
Ministry of Agriculture	MINAGRI	9	3.0
International Institute of Tropical Agriculture	IITA	8	2.7
Higher National Polytechnic School	ENSP	8	4.0
International Centre for Promotion of Economic Recovery	CIPRE	5	5.0
Ministry of Livestock, Fish and Animal Industries	MINEPIA	5	5.0
African Research Centre on Bananas and Plantains	CARBAP	4	4.0
University of Ngaoundere		4	4.0
International Centre for Research on Agriculture for Development	CIRAD	3	3.0
University of Yaoundé II		3	3.0
Food and Agriculture Organization of the United Nations	FAO	2	2.0
Urban Community of Yaoundé	CUY	1	1.0
The Research Group on Social and Environmental issues	GRES	1	1.0
International Network for the Improvement of Banana and Plantain	INIBAP	1	1.0
Ministry of Scientific and Technological Research	MINREST	1	1.0

Regarding engagement in developing the 1981 structure plan, 123 persons attended meetings for this purpose, over one-third being from the Directorate of Urban Affairs. There were a few each from transport, the urban local authority, the Office of the President, the armed forces, and one or two from 31 other organizations. Analyzing them by sector, 47 percent were urban specialists, 10 percent from transport, 7 percent from energy, 6 percent from infrastructure, 4 percent from tourism and lesser amounts from education, Office of the President, armed forces, health, local government, agriculture, livestock (2 percent each), finance, and planning.

Since this simple enumeration did not indicate actual input to the plan, we developed an index for this purpose. The index measured participation in the different meetings and whether actions proposed were included in the plan. A system of weighting was used, meeting attendance being weighted less than actions proposed being included in the plan. The highest rating on the index was given if actions proposed were included in the plan without attending the meeting. This allowed us to determine each organization's capacity to influence decisions. The results were very clear. The only organizations influencing the plan were the Municipality itself (now CUY), the Office of the President and the armed forces. Local sub-divisions had no influence whatever.

In reality, the plan of 1981 was never implemented, and a new one was supposed to be in preparation in 2006. As the institution operating at the interface of administrative and elected powers and primarily responsible for what appears in the 1981 plan, the CUY (which emerged from the Municipality) will obviously be concerned that any new plan will prove more effective. This institution has undergone some important changes since 1982, notably the creation of many new sub-divisions and the introduction of multi-partyism in Cameroon's political life, potentially offering opportunities for a different type of engagement in future.

As a complement to this analysis, we searched the media for activities or interventions concerning UA in 2002. Although aspects such as waste management, information, control and regulation were included in our search, little was found. This is not to say nothing was going on, only that it was not recorded anywhere as of interest.

The Views of Market Traders About Institutions Dealing with UA

Because the many markets of Yaoundé are most definitely concerned with UA, we also conducted a study of these institutions. We found 30 and divided them into two types, those selling mainly maize and dried beans and those selling tomatoes and a profusion of local and indigenous vegetables. Through a rapid appraisal of the point of origin of the vehicles bringing produce to these markets we were able to determine that more than two-thirds of the produce in the town's markets is from urban and peri-urban production. Out of 177 vehicles counted, 119 were coming from within Yaoundé and its peri-urban area (namely Mfoundi and its close environs), 38 from the surrounding areas of Central Province, and only very few from elsewhere.

The most common points of origin of produce were Nkolondom, Nkozoa, Okolo and Nkolbisson.

Despite the fact that men are supposed to predominate among the economically active population of Yaoundé, our survey of traders revealed three times as many women as men, consistent with the findings of the survey of cassava root traders described in Chapter 3. The youth seemed to be less involved, most traders being older, married women and men. The traders' level of education varied, a little under a third having no education, another third having only primary education and just short of the last third having secondary education. In proportion to their overall numbers, women predominated among those with primary or no education. Sizes of traders' households varied greatly, with 50 percent of the women traders having large households of six persons or more, compared to 40 percent of the men.

Trading was not always the main occupation of those we found doing it. As shown in Table 5.6, although nearly a quarter of those interviewed did not respond to the question, of those that did, 82 percent said they were traders, 7 percent said they were farmers, and smaller percentages said they were artisans of some kind as well as traders, some housewives, students and even a couple of teachers.

Table 5.6 Main occupation stated by traders in five Yaoundé markets

Occupation	No.	%
Trader	685	81.6
Farmer	56	6.7
Artisan/trader	44	5.2
Housewife	32	3.8
Student	13	1.5
Tailor	4	0.6
Teacher	2	0.2
Driver	1	0.1
Night watchman	1	0.1
Artist/comedian	1	0.1
Total responding	839	100.0

Source: 2003 authors' study carried out in five markets in Yaoundé

Most women and men traders thought that the produce they sold from UA was not as good as that from elsewhere even though they recognized its economic advantage, being fresh and accessible. It was said that people in and around the town grow products like maize because they like to have their own and sell the surplus. Those who saw the UA produce as inferior quality attributed this to the poorer soil making it less tasty. As much of it is grown in the marshy valleys (which are also polluted) it is often too watery. Traders think that growing food using wastewater makes it spoil more quickly, and that frequent infestations of pests such as caterpillars mean that there is over-use of pesticides. They thought that rural produce tastes better but is more expensive and scarce due to the state of the roads, lack of transport, and lack of local demand. Still, many traders did think that UA maize was good when well cultivated, and this corresponds with the findings on UA livelihoods in Chapter 3 above, as more maize comes from the household producers higher up the slopes and vegetables from the commercial producers in the polluted valley bottoms, where more pesticides are also used.

More than half the traders said they had no contact at all with any kind of institution, while a few mentioned more than one. As shown in Table 5.7, institutions most frequently mentioned were the town hall followed by the police. Comparing these data with the list of institutions derived from the literature search, it can be seen there is not much connection.

Table 5.7 Institutions dealt with by traders in five Yaoundé markets

Institution	No. of mentions	% of mentions
Town hall	418	67.6
Police	151	24.4
Ethnic association	16	2.6
Peasant association	9	1.5
Savings association	4	0.7
IRAD	11	1.8
Railways	4	0.7
MINAGRI	3	0.5
MINEFI	2	0.3
Total mentions	618	100.0

Source: Survey of every fifth trader encountered in Yaoundé markets, 2003

When asked about the nature of the contacts, traders made it clear they were repressive and predatory. Collection of taxes was the main contact any trader had with the authorities. Payment of 100 FCFA was the most common occurrence, clearing blocked pathways was second, followed by other cash payments, evictions and payment of rent. Less often there were points of contact around clearing of waste, clearing of badly built stalls, or advice on conservation or other matters, including the planting of eucalyptus in the marshes, which has been an official policy in order to reclaim these lands for development. Collective activities were another category of infrequent institutional contact. For example, groups organized security watches and group work on fields as well as meeting to collect membership dues.

Traders' engagement with these institutions varied according to the type of market. The local authorities and the police were most occupied at the larger central markets of Mokolo, Mvog-Mbi and Mfoundi, whereas the indigenous institutions were found to be more active in the markets at the periphery. The local authority and the police were the institutions most involved with coercive actions such as evictions and the clearing of traders from pathways.

Indigenous or traditional bodies were identified as important institutions in terms of getting access to a place in the market, since they have power to allocate such space. This is apparently most prevalent in the peripheral markets where UA produce predominates. The survey also underlines the considerable disconnection between the traders dealing daily with agricultural products and ministries with explicit responsibilities for food and agriculture.

Conclusions

The widespread presence of UA in African cities and towns like Yaoundé demands an institutional response. Our data have shown that urban and peri-urban production and distribution of food play a large role in social life and the economy. Currently, formal public institutions are only weakly engaged, but the ample range of government, traditional and civil society organizations involved in the issue offers opportunities for more effective structures to be built through public consultation.

The Urban Harvest-supported research described in this and the previous two chapters was the subject of a public meeting in February 2004 that began just such a process, involving all the institutions identified in this study. This meeting, held at the Town Hall in Yaoundé on 19 February 2004, was attended by about 40–50 participants from international and national research institutions, government and local government officers – including those concerned with health and agriculture – donor agencies, NGOs and the public. Being a public forum involving the key stakeholders identified in this study and most of those that took part in the earlier deliberations on the 1981 structure plan, it can be seen as a necessary milestone in establishing future policy and action on UA.

The late Honourable Nicolas Amougou Noma, the *Delegue du Gouvernment* to the Urban Community of Yaoundé, opened the workshop by outlining the problems of urban growth and stating that UA is a source of revenue, nutrition, employment and a valorization of urban land but also presents problems of food quality, pollution and contamination of food. He stressed the need to identify all stakeholders, and the types of operations in the city, and urged action in the short term. He set the long-term goals as environmental protection, improved quality of life, and employment creation. His thoughts were echoed and endorsed by other speakers including the Prefect of Mfoundi Division, the Scientific Director of IRAD, the Regional Coordinator of Urban Harvest and the representative of IITA.

There was strong support for the stakeholder-based model of inter-disciplinary team research that generated the results described in Chapters 3, 4, and 5 and involved twelve organizations and seven research studies. The main findings were summarized in posters prepared specifically for the workshop and mounted in a public exhibition. To increase public awareness, the media conducted numerous interviews and there was good coverage over the next few days in both print and audio–visual media.

Despite the wealth of information presented, it remained clear that Yaoundé has a huge number of farmers about whom little is known and who lack organization, consumers who are only poorly aware of the issues involved, a negative official perception of UA, an insufficient involvement of the state and local government, a very limited involvement of civil society, and an environment that both enables and constrains the activity. UA continues to be a little explored area of opportunity for stabilizing the rural–urban transition in a way that satisfies the various actors involved.

The public debate highlighted the current situation where fields and plots in and around the capital are generally being farmed through the intervention of chiefs, acting as managers of this world in transition from customary to modern rules. However, chiefs are renting land, whether or not it is technically vacant, on an informal basis – rather than in their official capacity – from which they also make a profit. Clearly, all institutional processes at work to facilitate the pursuit of UA, not to mention the informal sector at large, need to be reconciled and brought into focus with formal institutions in the interests of good governance and addressing the needs of the population.

At the workshop the following steps were recommended:

- Seminars to raise awareness of UA in the large towns of Cameroon;
- Preliminary assignment of institutional responsibility for UA in Cameroon;
- Better recognition of UA to enable urban farmers to access small-scale credit programs;
- Establishing a forum for UA stakeholders to develop a strategy that addresses the concerns of farmers, food security and the need for health and environmental protection;
- Including promotion of UA in the health and security strategies of local authorities;
- Better integration of UA in agriculture and livestock extension and promotion of the agriculture sector in general;
- Integration of UA in plans for protection of culture, tradition and green space in and around towns and cities;
- Better waste management through integration with UA, building on the activities of NGOs;
- Build the capacities of all involved in the production chain of UA in order to promote a better quality of life in towns as well as in the countryside.

Participants called for the creation of a forum to continue the debate and explore options. Specifically, the creation of a UA program in Yaoundé was endorsed, with links to other institutions. It was said that, while doing nothing about UA avoids conflict, it is not sustainable in the long run. It was proposed that a debate platform involving different ministries and agencies be set up, and that the lead should be taken by CUY, which would set the priorities for follow-up action. The workshop itself proposed the following priorities to this group to be convened by CUY:

- Follow-up to the institutional analysis by creation of a steering group or other mechanism to address UA;
- The identification of areas for action, including waste management in the city, linked to crop–livestock interaction and the re-use of city organic wastes;
- Identification of actions needed on public health issues concerning UA, including further research, public awareness campaigns, policy and regulatory review, perhaps in collaboration with other cities involved in this, such as Kampala.

Urban agriculture is important to food security, employment creation and poverty alleviation, as well as being a tool of environmental management by re-using wastes, creating green space and cleaner air. The needed institutional responses to UA are both complex and urgent. However, it is apparent that a process such as that used to develop the unrealized 1981 structure plan needs to be replaced by a process of institution building that engages the many stakeholders affected. Only a transparent and public engagement with all aspects of urban food production and distribution based on evidence derived from sound research is likely to lead to the building of effective institutions. Whether that means, for example, a new organization dealing with UA or a new regulatory framework for how UA ought to function, is something that needs to emerge from a process of consultation. Institutions – as social norms and practices that emerge over time – have to be based on a common understanding.

References

Bopda, A 1997, *Yaoundé dans la construction nationale au Cameroun: territoire urbain et intégration*, Thèse de Doctorat, Université de Paris I, France.
BOPDA (A.) et GRASLAND (C.), 1994 – Migrations, régionalisation et régionalismes au Cameroun. In *Espaces, population et société*, 1994-1. Lille-Flandres-Artois, U.F.R. de géographie, pp 109–129.
Bopda, A & Grasland, C 1996, 'Noyaux régionaux et limites territoriales au Cameroun. Migrations et structures par âge de la population en 1987', in Bocquet-Appel, JP, Courgeau, D & Pumain, D (eds) *Spatial analysis of biodemographic Data*, John Libey Eurotext/INED, Paris, pp. 131–156.
Dury, S, Gautier, N, Jazet, E, Mba, M, Tchamda, C & Tsafack, G 2000, *La consommation alimentaire au Cameroun en 1996; données de l'enquête camerounaise auprès des ménages*, DSCN, CIRAD & IITA, Yaoundé.
Franqueville, A 1984, 'Yaoundé, construire une capitale', *Coll. Mémoire ORSTOM, Etudes urbaines 104*, ORSTOM, Paris.
Franqueville, A 1987, 'Une Afrique entre le village et la ville: les migrations dans le sud du Cameroun', *Collection mémoires, no.109*, ORSTOM, Paris.
Kouame, A, Kishimba N, Kuepie M & Tameko D 2001, 'Crise, reformes des politiques economiques et emploi a Yaoundé', *Les dossiers du CEDEP no. 64*, CEDEP, Paris.
Laburthe-Tolra, PH 1970, 'Yaoundé d'après Zenker (1895)', *Annales de la faculté des lettres et sciences humaines, no. 2,* Université fédérale du Cameroun, Yaoundé.
Laburthe-Tolra, PH 1981, *Minlaaba I: Les seigneurs de la forêt*, Publications de la Sorbonne, Paris.
Madiot, C 2001, *Des associations de quartiers et des ONGs pour l'entretien urbain à Yaoundé (Cameroun): Des clés de lecture de la crise urbaine*, Mémoire de DEA, Université de Paris I et de Paris IV, France.
MINPAT 1993a, *Indicateurs démographiques sur le Cameroun*, DRDI/République du Cameroun, Yaoundé.
MINPAT 1993b, *Migrations et urbanisation: Le cas de Yaoundé et de Douala*, DRDI/République du Cameroun, Yaoundé.
MINUH 1981, *Les emplois dans l'espace urbain de Yaoundé et de Douala: situation et perspectives*, SEDES/République Unie du Cameroun, Yaoundé.
République Fédérale du Cameroun 1967, *Enquête sur le niveau de vie à Yaoundé 1964–1965*, fasc. I, Yaoundé.

Part II
Uganda

Uganda Overview

The first three chapters in this section dealing with Kampala present research led by the International Centre for Tropical Agriculture (CIAT) while the fourth describes a study of health and urban agriculture that ran in parallel, led by a research-policy platform. New data in Chapter 6 show not only that the extent of farming in the city is even higher than previously measured in the 1980s, at around 49 percent of all households, but also that there are differences between urban and peri-urban farming systems. It also presents a typology of urban farming systems to allow for more effective policy and programs. Confirming and updating findings from surveys in the 1980s data show that farmers fall in four categories:

- Commercial
- Sufficiency
- Food security
- Survival

The first two types have fewer farmers who primarily need help with marketing, while the latter two types are much larger, have more women farmers and mainly need help with support and social safety nets. However, all four need help both with a supportive policy framework and encouragement with marketing.

This is followed by Chapter 7 on an experiment working with schools on urban agriculture, including using schools as centres for seed production. While the seed-production experiment was not successful, extension was more so and a great deal was learned that has since had significant follow-up in the city. Chapter 8 describes research on markets for urban and peri-urban produce, conducted rigorously but in consultation with farmers. Poultry, pigs, mushrooms, fruit and vegetables were products of interest, but with different conditions for urban and peri-urban farmers and specific needs for marketing support structures. Finally in this section, Chapter 9 describes an extensive study on the health impacts of urban agriculture. Treated more extensively in a companion book (Cole et al. 2008), the chapter presents findings of scientific studies and distills these as straightforward public health messages. These cover food security and nutrition, contaminant risks – both biological and chemical – and livestock-related benefits and risks.

Chapter 6
Changing Trends in Urban Agriculture in Kampala

Sonii David, Diana Lee-Smith, Julius Kyaligonza, Wasike Mangeni,
Sarah Kimeze, Lucy Aliguma, Abdelrahman Lubowa,
and George W. Nasinyama

Introduction

Kampala in the 21st century is the showcase of Uganda's economic, political and social transformation following the economic decay and civil war of the 1970s and 1980s. A capital city that reflects the country's diversity, it is also the site of the historic Buganda Kingdom and its traditions, including agriculture. This and other socio-economic factors have contributed to agriculture being a visible part of the city's life.

With its tropical climate and ample rainfall Kampala is fertile, like the rest of Uganda, 75 percent of the country being suitable for agriculture, which forms 42 percent of the national economy. The agriculture sector accounts for 90 percent of Uganda's exports, 80 percent of employment and most of the raw materials that go to the mainly agro-based industrial sector, much of which is located in Kampala. About 64 percent of Uganda's agricultural gross domestic product (GDP) is in the form of food crops, mainly produced by around three million farm households, of which 80 percent have less than 4 ha of land and for whom the hand-hoe is the predominant technology (MFPED 2002, 2003).

In the early 1990s urban agriculture (UA) was widespread both within the built-up areas of Kampala City and in peri-urban areas. A 1993 survey of three neighbourhoods found that 35 percent of households engaged in agriculture, mainly crop cultivation. In 1992, 56 percent of land within municipal boundaries was used for agriculture, while an estimated 70 percent of poultry products consumed in Kampala were produced in the city (Maxwell 1995b).

Maxwell identified three significant periods between 1970 and the early 1990s when farming expanded in Kampala, beginning in the 1970s when households farmed to cope with the harsh economic circumstances under the dictatorship of Idi Amin. Then, after 1986 when the Museveni "Movement" government came to

---

S. David (✉)
Sustainable Tree Crops Program, International Institute of Tropical Agriculture (IITA),
P.O. Box 135, Accra, Ghana
e-mail: s.david@cgiar.org

power, people felt secure enough to engage in agricultural production away from gardens close to their homes. The third period was the 1990s when the impacts of structural adjustment policies combined with greater security to increase urban farming (Maxwell 1995b).

This chapter uses 2002–2004 data to describe and analyze Kampala's urban agriculture since then, as a baseline for future interventions and policy, on the assumption that urban farmers could be important players in the sector for various reasons, not least their proximity to the market and their role in the alleviation of hunger. The country's current Plan for the Modernization of Agriculture (PMA) aims at increased and sustainable food production by helping farmers improve production and commercialize, but does not yet incorporate urban agriculture. This chapter and the three that follow are intended to help the PMA address this gap.

This chapter first provides a background to the research and methodology and introduces the four areas along the urban to peri-urban agriculture continuum where research was carried out. It then describes Kampala's urban farming production and livelihood systems annotated with new statistical data suggesting that the phenomenon is even more extensive than previously understood. The chapter concludes by revisiting and updating Maxwell's typology of urban farming livelihood systems to derive a policy direction for urban agriculture. Variations along the urban to peri-urban continuum give an important new perspective to such policy direction.

Background

In 2002–2003, Urban Harvest, a global program of the Consultative Group on International Agricultural Research (CGIAR), supported the project "Strengthening Urban Agriculture in Kampala" involving three CGIAR centres along with local agricultural research institutions including university departments and civil society organizations. The various studies – aimed at characterizing urban farming systems and developing technical interventions to support them – form the subject of the four chapters in the Kampala section of this book. This chapter draws on multiple surveys to give an overview, and is followed by chapters on market opportunities for urban farmers, dissemination of UA technical interventions through schools, and the health impacts of UA.

An important outcome of the Urban Harvest project was that Kampala City Council undertook a review of its legislation governing urban agriculture and food handling, and set out to list all enterprises involved in these activities in the city. The project research team, which subsequently formed itself into a policy-research institution, helped start that list with a pilot study of two parishes, one urban and one peri-urban, and the results are included in this chapter as new statistics for the city. The new legislation governing UA in Kampala, approved by City Council in 2004 and gazetted as law in 2007, constitutes a substantial change in the policy environment, making it possible for urban farmers and food handlers to operate within a new and more supportive framework (Hooton et al. 2007, 2006).

In 2002 Kampala had a population of 1.2 million, comprising about 300 000 four-person households living at a density of about 7400 persons per square kilometre (Aliguma 2004). Situated in the Lake Victoria Basin with an average annual rainfall of 1180 mm, today's city reflects its establishment in the 19th century by the British in agreement with the kingdom of Buganda, with dual centres for colonial administration and the kingdom, respectively. The systems of land administration and use, including farming, also embody this history (Calas 1998; van Nostrand 1994).

Administratively, Kampala City is also a district, currently having five divisions: Central, Makindye, Nakawa, Kawempe and Rubaga, further subdivided into parishes and zones, each administered by the relevant level of Local Council (LC). The city is built on a number of hills interspersed by wetland valleys into which run sewage, domestic and industrial waste, and which are commonly exploited for farming. Kampala City Council (KCC) – more advanced than most in addressing urban agriculture – uses a classification of farming styles according to four categories: urban old, urban new, peri-urban in transition to urban and peri-urban proper (Table 6.1).

Table 6.1 KCC urban agriculture classification system

	Urban, old	Urban, new (dense slum)	Peri-urban in transition	Peri-urban (peripheral)
Population density	Very high	High	Medium	Low
Prevalence of crop production	Low	Low	Medium	High
Prevalence of local livestock	Low	Low	Low	High
Prevalence of improved livestock	Medium	High	High	Low
Land availability	Limited	Limited	Moderate	Very good

Methods

To get a broad overview of UA activities in Kampala, a site was selected from each category for the initial participatory appraisal (David 2003). The unit of analysis was one or more villages from selected parishes typifying each category: Bukesa in Central Division (urban old), Banda in Nakawa Division (urban new), Buziga in Makindye Division (peri-urban transition) and Komamboga in Kawempe Division (peri-urban peripheral). In each parish, half-day Participatory Urban Appraisals (PUAs) were conducted – to introduce communities and local authorities to the project as well as collecting preliminary data on UA and local wealth classification. These data are used for the portrayal of each area below. Consistently, more women than men participated in the PUA, reflecting their predominance as urban farmers.

Table 6.2 Summary of data collection methods used

	Participatory urban appraisal	Farming household survey	Livelihood study	Production systems study	Listing survey
Urban locations	Bukesa, Banda,	Bukesa, Banda,	Banda,	Bukesa, Banda,	Banda,
Peri-urban locations	Buziga, Komamboga	Buziga, Komamboga	Komamboga	Buziga, Komamboga	Buziga
Sample size and sampling method	Purposive, based on calls by local leaders	175 farmers randomly selected from lists provided by local leaders	40 farmers purposively selected based on PUA and transect observation	80 farmers purposively selected based on PUA and transect observation	Complete household census of selected zones
Methods used	Group exercises and discussions	Formal questionnaire survey	Key informant interviews	Transect walk, key informant interviews	Formal questionnaire survey
Year	2002	2003	2003	2003	2004

Table 6.2 summarizes all data collection methods used between 2002 and 2004 (see Plate 6.1).

The City Council's effort to list all UA enterprises with the help of the project team also forms part of our data. The pilot listing was done in selected zones of two representative parishes, one urban and one peri-urban. The data, which reveal new information about the pattern of urban and peri-urban farming, are based on complete (100 percent) sampling in the selected zones. However, it must be pointed

Plate 6.1 Bringing grass to stall-fed dairy cows creates small business opportunities in many cities, including Nakuru, Nairobi and Kampala

out that these zones constituted only 8 percent of the population in the urban parish (Zones B2 and B10 in Banda Parish) and 69 percent in the peri-urban parish (Zones A, B and C Kiruddu and Katuso zones of Buziga parish), and that the data derived are not based on a random sample of the city's population from which statistically accurate projections can be made.

Urban agriculture is defined for the purposes of this chapter as the growing of crops and raising of livestock for food within the built-up urban area. Peri-urban agriculture is defined as the growing of crops and raising of livestock for food on the fringes of the urban area. Our purpose being to explore characteristics of and differences between urban and peri-urban agriculture in Kampala, the categories of urban and peri-urban farming areas used by the City Council were adopted, all such areas being within the municipal boundary.

A Description of Urban Agriculture Across the Urban to Peri-Urban Continuum in 2004

Bukesa: An Old Urban Area

A heavily populated parish near the city centre, Bukesa is close to Mengo, the centre of the Buganda Kingdom and the royal tombs at Kasubi. Although Bukesa was originally part of the Kingdom's lands, its farmers were diverse, about a quarter coming from 15 ethnic groups other than the Baganda. Some original colonial buildings and old structures characterized the uphill areas, many of them affected by disputes and rapid changes of ownership during the Idi Amin dictatorship of 1971–1986, especially after he expelled the Asian community from Uganda in 1973 (Calas 1998). Signs of war and conflict were still evident, foundations of knocked-down buildings being cultivated with crops. The steep slopes were eroded while lower in the valley, small spaces were used for gardening in between densely packed structures used for informal dwellings, shops, roadside kiosks and garages.

Cooking bananas (*matoke*) were intermixed with beans, yams, maize, cassava and sometimes local vegetables, with goats and cattle tethered or grazing. The plot sizes were very small, less than 0.1 ha (quarter acre). Gullies cut through the road and gardens and a mixture of soil, decaying matter and refuse was strewn down the hill, by the roadside or in the gardens below. Some gardens were abandoned, as it was said KCC officials and absentee landlords strongly restrict use of the land for farming. Though densely settled, Bukesa retains the residential and farming land use pattern of the Buganda Kingdom, uphill being the place of status and downhill the place of the poor, with farming on the slopes and in the valleys (Calas 1998, pp. 273–289).

Banda: A New Urban Slum Area

Banda covers a huge area either side of the Kampala–Jinja highway, with densely populated housing and informal businesses in makeshift structures. Poor people, including those evicted when an up-market housing area was built, settled on a wetland near the Kinawataka drainage channel, which discharges water from the city down to Lake Victoria. Residents grew cocoyams for income as well as food, and kept poultry and cattle for income. They also grew green vegetables, beans, maize and cooking bananas for food.

Nearly half (44 percent) of Banda's farmers were from groups other than the Baganda, who are indigenous to Kampala. Formal employment was the main means of livelihood for men in farming households, though generally both husband and wife worked. The self-employed kept retail shops or dealt in hardware and building materials, while casual labourers mainly worked as stone crushers in a nearby quarry.

There was little sanitation or water supply and several large industries discharged waste into the channel, which floods during rains, spreading industrial and other pollutants into the area used for farming. The wetland was also used as a dumping site for animal waste (especially from cattle) as well as waste from small industries.

Middle-class and poor farming households lived in the unplanned slum, while the few rich farmers lived in the up-market neighbourhood, keeping livestock (for income) and growing crops (for food), sometimes using hired labour. The big permanent houses have spacious compounds where farming can be done in an organized fashion.

The middle class comprised mainly couples engaged in small-scale trade as well as mixed crop–livestock farming on land near the channel where they produced food for home consumption and sales, sometimes hiring labour. Their houses were mostly of temporary construction and they often rented out rooms to the poor for income purposes, and sold crops from their gardens at local stalls or in Banda market. Some men also brewed local gin or worked as *boda boda* (taxi) cyclists while some women made handicrafts or worked as maids or housekeepers. A middle-class farmer said:

> I operate a small hardware shop as head of household and my wife is a full-time farmer. We keep livestock and grow crops like cocoyams, sweetpotatoes, beans, vegetables and sugarcanes. There are two things I benefit from farming activities, which are the very reasons why we engage in them. I get food for my people and some money from what I sell. Also my people are able to have milk and vegetables any time without spending. From my earnings from the hardware business, I am able to treat the livestock, buy feed, farm equipments and hire farm labour. For example, I have two labourers each of whom I pay 40 000 shillings a month.

A woman described the family livestock enterprise:

> I started cattle keeping in 1994 with one Friesian cow purchased from Mukono at a cost of two hundred and fifty thousand Uganda shillings and it has since produced six times. Presently there are four cows; I have so far sold off three at nine hundred and fifty thousand Uganda shillings. The three cows were sold off during times when my husband could not raise all the school fees for the children. I am planning to slaughter one bull to get some money for school fees again. (see Plate 6.2)

Another middle-class farmer explained his enterprise in economic terms as follows:

Plate 6.2 Fish farming is a common source of income and nutrition, as for the members of this group in Kampala that cares for AIDS orphans

> I started thinking about this activity having realized that the bank was going to retrench me in the year 2000. Yet I had just closed my maize factory. Initially, I was planning for poultry, and then I got advised by some friends who discouraged me from keeping poultry and advised me to try cattle or goats other than chicken, which are so involving and yet their prices are so low. Therefore, toward the end of the year 2001, I started on the goat project. Then in May 2002, I bought the cows too.

The poor constituted the majority of farmers in the parish. Mainly women, they lacked basic needs like food, shelter, education and health care and many were illiterate. Most live in rented rooms in the densely packed dwellings with small, cultivated plots squeezed in between. They also planted cocoyams in the low-lying swamps and wetlands. Predominantly unemployed single mothers and elderly widows or women abandoned by their husbands, poor farmers produced crops for their own consumption on borrowed land, some also doing casual labour in the stone quarry. A few poor men farmers also did casual work, mainly brick making, while some women also looked after orphaned children with support from charitable organizations. On average they only ate one meal of maize and beans a day. One said:

> The major problem my family and I encounter is lack of money to take care of our entire needs. And to be specific, the major difficulty has been that of losing the job and I am just gambling life now. Actually, school fees are my acute problem, especially whenever my children are sent back home as school fees defaulters. My coping strategy has been to engage in farming but I lack the land and the money to expand my agricultural activities to be able to raise some good income from it.

Buziga: A Peri-Urban Area in Transition to Urban

Located on the shores of Lake Victoria South of Kampala toward Entebbe, Buziga has a lower density and less-diverse population than inner areas of Kampala. Virtually all available space was occupied by housing and farming with no idle land except near the swamps down hill and at the community field, where a herd of zebu cattle grazed, while indigenous goats were tethered and chickens scavenged along roadsides. Farmers kept poultry and cattle and sold fruit for income, growing sweetpotatoes, cassava and cooking bananas for food. Poultry keeping was the highest income earner, followed by dairy and fruit. Food security was the farmers' priority and few households farmed commercially, although the potential of income from pig keeping, mushrooms and sugar cane was also identified by some.

Three classes of farmers, rich, middle class and poor, were identified, as in the other areas. Indigenous Baganda headed 90 percent of farming households, the remainder coming from 11 other ethnic groups including foreigners – mainly better-off newcomers – thus exacerbating the gap between rich and poor. The rich owned tile-roofed houses, mainly located up-hill in walled or fenced plots with lawns, flower gardens, fruit and improved varieties of banana. More empty plots were fenced off, suggesting further development.

Middle-class and poor farmers had iron-roofed houses, with or without fenced plots, and in various states of repair, located downhill. Rainwater from the steep slopes had excavated deep gullies which farmers managed by digging trenches to channel water and control erosion. Their gardens of cassava, sweetpotatoes, banana, beans and maize grew vigorously, with small neat plots of cabbage, onions, and indigenous vegetables well managed using trenches filled with compost and manure. Extension advice provided by the non-governmental organization (NGO) Environmental Alert accounts for the good practices.

Komamboga: A Peripheral Peri-Urban Area

Farming was the commonest occupation in Komamboga on the northern boundary of Kampala. There were no industries and with its low density and plots varying in size from 0.1 to 1 ha it looked similar to a typical rural landscape. Ninety-six percent of the farmers were Baganda. Those who only farmed, as opposed to having a range of income sources, were usually poor, except for a few large commercial farmers. Those formally employed worked for government, NGOs or international aid agencies. Some business people were retail traders or operated food kiosks, while others among the self-employed were machine operators, special hire drivers, and technicians involved in vehicle and bicycle repair.

Poultry keeping was the major commercial farming activity in Komamboga while sweetpotato and cassava were the two major crops, mainly produced for household food consumption. Besides these, there were pig farms, abundant fruit trees, mainly jackfruit, avocado, mangoes, and guava, as well as pineapples. Vendors on bicycles purchased fruit from individual households and took them to the major markets. Jackfruit – perennial and easy to grow – was important for food security, children preferring it to a meal at lunch time and adults commonly eating it as a snack with tea.

Milk, poultry and eggs were sold at local markets and in neighbouring communities, with cattle being sold only at times of financial need. Piglets were sold to other farmers while mature pigs were sold to nearby pork dealers. Cattle were assembled every day from different homesteads and herded in search of pasture at school compounds, churches or alongside roads. Herders supplemented grazing with residues when pasture was scarce, while a few farmers grew Napier grass for fodder.

The rich farmed for their own consumption on plots of a hectare or more, mostly uphill, inland from the lakeshore. Generally employed and well educated, some also kept livestock and employed workers. Pig and dairy production were the most lucrative enterprises, though they also sold poultry, fruit and cassava.

Middle-class farmers mostly did mixed crop–livestock commercial faming, as well as having jobs or being self-employed traders and shopkeepers. Having enough land to produce for home consumption as well as sales, many women in these households engaged in small-scale trading, market vending or running food kiosks selling their own produce. Some raised poultry, pigs or zero-grazed cattle. They ate well, except during food crises due to poor rainfall, when some took only one meal a day. A woman commercial poultry farmer described her success at farming:

> Poultry keeping is the only farming activity of this household. I started a commercial enterprise with two hundred broiler chicks in 1993. Today the stock ranges between two hundred and four hundred birds. The birds are raised for sale to retailers. The selling price is between three thousand and three thousand three hundred shillings per bird. There are three sales in a year, yielding between one million and three million shillings annually.

Despite the advantage of the nearby urban market, poor farmers were the majority in the parish, their small plots of land (0.1–0.4 ha) permitting only small-scale production. They mostly kept local livestock breeds and many would sell off their produce to provide for other needs such as school fees and medical care, even when they did not produce enough for household consumption. These households mostly ate only one meal a day. The men often did casual work in brick making or construction or ran small bars and groceries while the women farmed. A poor farmer in a polygamous household (a man and two wives) said:

> Personally, I do engage in farming mainly to raise some money and to contribute some food for the household since I am a full-time housewife. So when I sell off some crops and milk, I am able to have my own money so that I can meet my needs without waiting for the man's money. Actually, farming is now my job since it is my only source. Our husband never provides everything to us because he earns little money. So it is up to you to find what to do. I am therefore able to buy school requirements for my children and also take care of myself.

An unemployed poor farmer with little education described his situation thus:

> Actually farming here is a full-time job for us. Our three children above the age of eight help us during holidays. This is the major source of income for our household and it is also the only way we get our daily food. What my wife sells off in the evening is from our gardens. I

am doing all this to be able to get some income to feed our children and pay school fees for the bigger ones. As you see, we have many children, yet their father is no longer employed.

But another poor woman emphasized the positive aspect of farming:

My husband is a special hire [taxi] driver and we have always depended on farming to sustain our household by getting food for both consumption and some income. There are even times when my husband is not employed and we entirely survive on farming, that is, sale of crops and livestock for all the household requirements. So presently, although my husband is working, farming still plays a major role as far as food and the household needs are concerned. Actually, it is the job I do and I am sure I am better off than some people who go to Kampala every day to work, yet at the end of the day they come back with no money for food and end up borrowing money, which never happens in my household.

Urban Agriculture Production Systems

Broadly speaking, three crop and livestock production systems were observed in Kampala in 2002–2004: commercial (livestock and/or crops), semi-commercial (mixed or crop farming) and subsistence (mixed or crop farming). This section describes these systems in some detail while the following section provides statistics on the livelihood systems already described along the continuum. We then go on to synthesize our review of urban farming typologies.

Farming households in Kampala engage in over 20 types of agricultural enterprise, the combination of these being linked to location, available land and household farming objectives. The urban end of the continuum is characterized by lack of space, higher intensity of production and a higher proportion of commercial, albeit smaller, enterprises. The peri-urban areas are characterized by larger enterprises, a lower proportion of which are commercial, higher numbers of livestock per farm and larger areas under crops, with several plots of land per farm household. In all areas, livestock and crops both play an important role.

The KCC listing exercise showed that in urban Banda the proportion of farmers engaged in mixed crop–livestock farming and crop production only was about equal, whereas in peri-urban Komamboga, the proportion engaged solely in crop production was a bit higher. The proportion raising livestock only was small in both urban and peri-urban locations (9 percent in Banda and 3 percent in Buziga).

Cropping Systems

Farmers in all areas grew multiple crops as described. The most common were cooking bananas, common bean (*Phaseolus vulgaris* L.), cassava, sweetpotato, maize and cocoyam (*Colocasia esculenta* and *Xanthosoma* spp.), all of which constitute staples in the Ugandan diet.

We found no clear association between location and type of crops grown, although all the major crops, with the exception of bananas, were less widely grown in densely populated Bukesa compared with the other three locations. Cocoyam production was particularly important in Banda due to its proximity to a large wetland.

Intercropping was common for most crops, particularly banana, maize, cassava and beans, as a strategy for maximizing land use and reducing risk, but sweetpotato and cocoyam were usually mono-cropped. Fruit trees were mainly planted among other crops or in home compounds. Annual crops were typically planted during both the first (March to June) and second season (September to December), but some farmers did not follow a seasonal pattern, planting maize, beans and vegetables continuously.

Farm-plot locations varied with almost 50 percent of urban farming plots located away from homesteads due to land scarcity, compared with only a quarter of peri-urban and peri-urban transition plots (Table 6.3). Nevertheless, there was considerable difference between the inner urban and urban new parishes. In the new urban slum, Banda, farmers used public land – in this case the wetland – more than in other areas, but even here land next to their homes was the most common place for farming, just as in the other areas studied.

Table 6.3 Location of plots by paris (percentage of plots)

Parish	Bukesa		Banda		Buziga		Komamboga		All areas	
	N°.	%	N°.	%	N°.	%	N°.	%	N°.	%
Next to homestead	39	75.0	35	38.9	52	82.5	63	68.5	189	63.6
Road side	4	7.7	14	15.6	8	12.7	6	6.5	32	10.8
Wetlands	4	7.7	29	32.2	3	4.8	18	19.6	54	18.2
Other	5	9.6	12	13.3	0	0.0	5	5.4	22	7.4
Total plots	52	100	90	100	63	100	92	100	297	100

Source: Farming household survey

Many people farmed more than one plot, between one and six depending on the degree of urbanization, urban Bukesa having the lowest mean number of plots cultivated. Farmers elsewhere cultivated at least two plots, those at the peri-urban periphery having most. Plot sizes varied similarly, being as small as one by two metres in urban areas and as big as 4 ha or more in the peri-urban parishes. However, the average peri-urban plot size was only 0.8 ha while in urban Bukesa plots of more than 0.1 ha (quarter acre) were scarce. Male-headed households farmed larger plots than female-headed households, but there was no gender difference in the number of cultivated plots.

Kampala farmers mostly practiced low-input crop production, relying largely on household labour and hand tools, with children as young as 8 years working during weekends and school holidays. It was generally the commercial or semi-commercial farmers who hired labour, usually for clearing and preparation of land and for weeding. More farmers hired labour in Komamboga because of the larger farm sizes there.

Farmers typically obtained seed or planting materials from other farmers or purchased them from shops or markets. An important reason why farmers join self-help initiatives is to access seed or planting materials and new varieties. The following quotation from Banda provides an example of farmers' annual expenditures on seeds:

> Each planting season my household buys one kilogram of maize and a kilogram of beans at a cost of about six hundred Uganda shillings each totalling to one thousand two hundred shillings per season and two thousand four hundred shillings per year. The cabbages are grown three times a year and for each season the household buys one sachet of seeds at a cost of five hundred shillings, totalling to one thousand five hundred shillings for a whole year. One sachet of tomato seeds is also purchased each season at a cost of five hundred shillings, totalling to one thousand shillings for the two seasons. One kilogram of pea seeds is also purchased at a cost of one thousand shillings per season, totalling to two thousand shillings in a year.

Most households (75 percent) that grew crops used organic fertilizers (chicken dung, cattle urine, banana peelings and compost), while only a few applied chemical fertilizer, mainly to vegetables (*amaranthus* spp., eggplant, cabbage, tomato or bitter tomato). Sixteen percent of farmers used pesticides, mainly botanicals such as ash, urine and pepper, primarily on bananas and vegetables (tomato and cabbage). Most households applied mulch to bananas and crops, typically crop residues, and a few dug trenches to control erosion.

While it was difficult to get reliable crop yield information, farmers in all locations reported that yields had decreased in the past few years due to pests and diseases, unreliable rainfall, reduced space for farming due to land being sold for construction and declining soil fertility. Other key crop production constraints identified included lack of technical knowledge, lack of clean seed or planting materials, soil, water and air pollution, insufficient labour, and theft. In addition, Table 6.4 lists some crop-specific production constraints.

Table 6.4 Crop production constraints identified by urban farmers

Crop	Constraint
Sweetpotatoes	Foliar pests, weevils, sweetpotato virus infection, rotting, low yields, limited land with low soil fertility, theft
Cassava	Cassava mosaic virus, poor quality varieties that give bitter or large fibrous roots, limited land
Cocoyams	Rotting, millipedes, bad taste of tubers grown near sewage and erosion depositions from toilets
Beans	Pests, diseases, lack of knowledge on proper spacing and planting methods, limited land with poor soil fertility, theft
Maize	Pests, lack of knowledge about crop management, low soil fertility, lodging
Bananas	Banana weevil and other pests, diseases, polythene bags, limited land for expansion, wind, soil erosion, theft, air pollution
Fruits	Insects, pests (monkeys and birds), theft

Source: Participatory Urban Appraisal

Livestock Systems

Poultry, cattle and pigs, in that order, were the most commonly kept livestock in both urban and peri-urban areas of Kampala. Other types of animals included goats, rabbits, ducks and sheep. Among all farming households surveyed, 37 percent raised poultry, 19 percent kept cattle and 15 percent kept pigs. Poultry production predominated as the most common agricultural income-generating enterprise in all areas of Kampala, but most households that kept livestock did so for semi-commercial purposes (Table 6.5).

Table 6.5 Types of livestock by paris (as a percentage of all farming households)

Parish	Bukesa		Banda		Buziga		Komamboga		All areas	
	No.	%	No.	%	No.	%	No.	%	No.	%
Total households with livestock	18		49		40		46		175	
Chickens	10	25.0	9	18.4	20	50.0	26	56.5	65	37.1
Cattle	4	10.0	4	8.2	8	20.0	18	39.1	34	19.4
Pigs	–	0.0	4	8.2	8	20.0	14	30.4	26	14.9
Goats	3	7.5	2	4.1	13	32.5	5	10.9	23	13.1
Sheep	–	0.0	0	0.0	1	2.5	2	4.5	3	1.7
Rabbits	1	2.5	3	6.1	2	5.0	0	0.0	6	3.4
Ducks	–	0.0	1	2.0	2	5.0	1	2.2	4	2.3

Source: Farming household survey

It was mostly women who kept chickens, the local ones for household consumption and exotic chickens for commercial purposes, with high demand for chicken and eggs being an important incentive. Except for chickens, more male-headed than female-headed households raised livestock, while for all types of livestock the numbers kept were higher for male- than female-headed households. Cattle-rearing yielded multiple products (milk, manure, beef) for both household and commercial use, while the advantages of raising pigs were high demand for pork, quick maturity of the animal and low feeding costs. Pigs were not found in the old urban area.

Most livestock owners kept improved breeds: 65, 46 and 42 percent of cattle, poultry and pig owners respectively. The average number of improved poultry and pigs kept was higher than the numbers for local breeds except for cattle, since improved cattle have higher initial cost as well as higher feeding and treatment expenses (Table 6.6).

In fact livestock management systems were closely associated with the type of breed and whether farmers could meet the animals' feeding requirements. Most households keeping improved poultry did so under zero-grazing and semi-intensive systems. By contrast, 70 percent of households raising local poultry used the free-range system and farmers said that they preferred free-range chickens for their own consumption. Most households with improved cattle (60 percent) practiced zero

Table 6.6 Numbers of improved and local livestock kept

Livestock type	Improved breeds		Local breeds	
	Maximum	Average	Maximum	Average
Cattle	50	4	50	6
Goats	20	4	2	2
Pigs	14	6	10	5
Poultry	1180	164	200	20
Rabbits	13	7	–	–
Ducks	–	–	8	6

Source: Farming household survey

grazing, 23 percent tethered them, 4 percent kept them in fenced paddocks and the rest used a combination of methods. For local cattle, tethering (50 percent) and communal grazing (25 percent) were the most common methods employed. Pigs were mainly kept in fenced paddocks or were tethered.

Broiler mash and maize bran were the usual chicken feeds; others included layers' mash, broilers' grain, legumes or grasses and chick mash. Most farmers obtained these from farm shops, while other sources were feed factories, gardens and neighbours. Very few cattle and goat keepers bought feed from shops however. They fed their animals mainly on crop residues or peelings plus Napier grass, all obtained from their own or neighbours' farms. The purchased cattle feed included molasses, maize bran and concentrates. Pigs were fed on maize bran, sweetpotato vines, fish and food residues, only about 40 percent of pig keepers buying any feed from shops and 36 percent sourcing them from neighbours. Family labour predominated, only 27 percent of livestock keepers, mostly commercial farmers, employing labour for milking and feeding.

The major problems associated with livestock production were diseases and pests (for example, *coccicidiosis,* Newcastle disease, infectious bursitis and influenza in poultry, and worms, ectoparasites such as lice and swine fever in pigs), difficulties in breeding, high cost of poultry feed, shortage of feed for pigs and dairy cows, high cost of medicine and veterinary services and lack of adequate space for pig rearing.

Social Capital and Institutional Support

Most urban farmers participate in social networks to facilitate access to land, funding, farm inputs (seed, new varieties) and information. As one farmer in Banda observed:

> As a farmer I have to relate well to our neighbours. This enables us to share planting materials. And in case I am not around, my neighbour helps to take care of my cow. I also think the same is true of other farmers in the village. For example, good relations can help you get seeds, new information, as well as animal feeds.

Farmers formed self-help groups, in some cases specifically to gain access to services provided by the growing number of institutions supporting UA. The situation has changed since the early 1990s, when NGOs in Kampala shied away from supporting UA activities due to their illegality (Maxwell 1994). Thirteen organizations providing services such as agricultural extension, training, improved varieties and micro-credit were identified in the study parishes. They included community-based organizations, local and international NGOs, the national farmers' association and the KCC.

However, most farmers said their main source of agricultural information was the radio or other farmers. Most of our respondents receiving training in agriculture during the past 2 years were from Buziga, where the NGO Environmental Alert operates, and evidence of this training was clearly visible there in the form of more organized plots, with crops planted in rows, trenches and composting for soil management.

Agriculture as a Means of Livelihood Across the Urban to Peri-Urban Continuum

Although there has never been a reliable statistical survey of urban farming in Kampala – in fact there are few such surveys anywhere in the world – Table 6.7 provides a good indication of the proportion of households engaged in urban and peri-urban agriculture in 2003. Percentages for the urban zones are notably close to those documented by Maxwell in the early 1990s, when figures ranged from 36 percent (for a non-representative sample of households in different parts of the city) to 25 percent (for a sample of low-income Kampala households with children) (Maxwell 1994). The marked difference between urban and peri-urban areas shown by our data is an important new finding indicating much higher but also fluctuating proportions of farmers among peri-urban households.

Table 6.7 Number of households and percent engaged in farming, selected zones of urban Banda and peri-urban Buziga, based on a census of all households

Parish	Zone	HHs engaged in farming		HHs not engaged in farming		Total households (2002 census)	
		N°.	%	N°.	%	N°.	%
Urban Banda	Zone B2	32	27.8	83	72.2	115	100.0
	Zone B10	33	25.4	97	74.6	130	100.0
Subtotal		65	26.5	180	73.5	245	100.0
Peri-urban Buziga	Kiruddu	209	38.3	336	61.7	545	100.0
	Katuso	235	95.9	10	4.1	245	100.0
Subtotal		444	56.2	346	43.8	790	100.0
Total (convenience sample)		509	49.2	526	50.8	1035	100.0

Source: Listing survey

As well as higher (and highly variable) proportions of households engaged in urban farming in the peri-urban area, other demographics also vary. Peri-urban households in general were larger (4.5–4.6 persons) than urban ones (3.1–3.3 persons), yet our farming households survey in four parishes found – as did Maxwell (1995b) in the 1990s – that urban farming households were larger (average 7 persons per household in all four areas) than non-farming households. Furthermore, most members of these large farming households were part of nuclear families, extended family members accounting for only 17 percent, 6 percent being orphans and 1 percent being workers.

One-third (33 percent) of farming households were headed by women, half of them widows and the others divorced, separated or single. Farming household heads were older than the norm, the average age being 49 years (51 for women and 47 for men), with 19 years being the minimum and 86 years the maximum reported. Most men and women heads of farming households however belonged to the active age group (25–50 years). Literacy levels were high, 88 percent of all farming household members having some form of education. However, while most farming household heads had primary or secondary education, men's education levels were higher and most household heads with no education were women.

Farming households in peri-urban Buziga and Komamboga had better structures and sometimes more assets than those in Bukesa and Banda urban parishes (Table 6.8). A majority (66 percent) of respondents lived in their own houses while 34 percent were tenants and more peri-urban then urban houses were owner-occupied. Fifty-eight percent of farming households had electricity, with little variation across parishes. Almost every farming household (97 percent) had a radio, 44 percent had television and 40 percent owned mobile phones, again with little

Table 6.8 Assets and structures owned by parish

Parish	Bukesa		Banda		Buziga		Komamboga		All areas	
Asset/Structure	N°.	%	N°.	%	N°.	%	N°.	%	N°.	%
Own house	24	60.0	21	42.9	30	75.0	40	87.0	115	65.7
Rented house/room	16	40.0	28	57.1	10	25.0	6	13.0	60	34.3
Electricity	22	55.0	26	53.1	23	57.0	30	65.2	101	57.7
Car	8	20.0	7	14.3	3	7.5	6	13.0	24	13.7
Motor cycle	–	–	1	2.0	–	–	3	6.5	4	2.3
Bicycle	3	7.5	1	2.0	8	20.0	23	50.0	35	20.0
Radio	39	97.5	45	91.8	39	97.5	46	100	169	96.6
Television	17	42.5	24	49.0	18	45.0	18	39.1	77	44.0
Refrigerator	13	32.5	13	26.5	13	32.5	8	17.4	47	26.9
Cooker	6	15.0	3	6.1	1	2.5	4	8.7	14	8.0
Phone (Land line)	4	10.0	1	2.0	4	10.0	2	4.3	11	6.3
Phone (Mobile)	19	47.5	17	34.7	19	47.5	15	32.6	70	40.0
Total Households	40		49		40		46		175	

Source: Farming household survey

variation from urban to peri-urban areas. Bicycles were more common in the peri-urban areas, while there were more refrigerators in the peri-urban transition and old urban areas.

Farming, though important to livelihoods, was seldom the primary economic mainstay. While it was reported to be the main activity for between about 25 and 40 percent of farming household heads across the urban to peri-urban continuum, less – about one-sixth to a third – said it was their main source of income (Tables 6.9 and 6.10). Trading was the main economic activity in the old urban area and peri-urban transition area. Formal employment was the main economic activity and source of income in the new urban (slum) area, with farming the biggest source of income only in the peri-urban area proper.

Table 6.9 Primary activities of household head by parish

Parish primary activity	Bukesa N°.	%	Banda N°.	%	Buziga N°.	%	Komamboga N°.	%	All areas N°.	%
Farming (crops & livestock)	9	22.5	12	24.4	11	27.5	18	39.1	50	28.6
Business / trading	21	52.5	10	20.4	14	35	14	30.4	59	33.7
Casual employment	4	10	8	16.3	7	17.5	11	24.0	30	17.1
Formal employment	5	12.5	19	38.7	5	12.5	3	6.5	32	18.3
Missing	1	2.5	–	–	3	7.5	–	–	4	2.3
Total	40	100	49	100	40	100	46	100	175	100

Source: Farming household survey

Table 6.10 Main sources of income of household head by parish

Parish	Bukesa No.	%	Banda No.	%	Buziga No.	%	Komamboga No.	%	All areas No.	%
Farming (crops & livestock)	6	15.0	7	14.2	10	25.0	16	34.8	39	22.3
Business/trading	18	45.0	11	22.4	17	42.5	14	30.4	60	34.3
Casual employment	8	20.0	9	18.4	5	12.5	3	6.5	25	14.3
Formal employment	8	20.0	22	44.9	8	20.0	10	21.7	48	27.4
Missing	–	–	–	–	–	–	3	6.5	3	1.7
Total	40	100	49	100	40	100	46	100	175	100

Source: Farming household survey

Most farming households (85 percent) had more than one income source, 28 percent having three. Sources were ranked in order of importance as business, mixed farming, livestock keeping, salary or wages, rent and selling crops. Secondary sources of income included casual labour, brewing, traditional healing, pensions and fishing, while the most frequent business categories mentioned were petty trading, retail shops, tailoring, handicrafts, brick-making, charcoal selling, carpentry and catering (see Plate 6.3).

Plate 6.3 Women participate in the urban appraisal study in Kampala

Consistent with findings from other African urban areas such as Nakuru in Kenya and Yaoundé, Cameroon, (Foeken 2006; Chapter 3 above) a large proportion (about 70 percent of farming households in all four areas) earned more than the national income per capita, equivalent to US $330 (approximately 590 000 Ugandan Shillings (USh)) while 10 percent earned five times as much, more than USh 3 million annually (US$1680). Highest income levels were reported in peri-urban Buziga, followed by peri-urban Komamboga, then Bukesa. Despite its having the most formal employment, the lowest average incomes were recorded in urban Banda.

School fees were the highest expenditure item for over half the households, with food ranked second or third and other major expenses being rent, transport, medical needs, utilities and remittances. The major item of expenditure on agricultural inputs was livestock feed – for poultry – on which three times more was spent (average 863 000 USh/year) than on the next item, labour, with transport costs (wheelbarrows) costing a sixth of what was spent on labour. Lesser amounts were spent on seed, fertilizers, manure, herbicides and tools such as hoes, *pangas,* slashers, rakes, spades and knives.

Having enough to eat was said to be the main contribution of UA to the households, followed by income, with others of less importance. Table 6.11 shows agriculture activities across the urban to peri-urban continuum, again emphasizing the relationship of livestock to income and crops to consumption.

Table 6.12 suggests a possible positive relationship between commercialization and urbanization (one urban zone having a high level of commercialization

Table 6.11 Ranking of agricultural activities by prevalence, food consumption and income value by parish

Most important activities by:	Bukesa		Banda		Buziga		Komamboga	
	1st	2nd	1st	2nd	1st	2nd	1st	2nd
Prevalence	Poultry	Bananas	Cocoyam	Poultry	Sweet-potatoes	Cassava	Sweet-potatoes	Cassava
Consumption value	Fruits	Bananas	Indigenous vegetables	Cocoyam	Sweet-potatoes	Cassava	Sweet-potatoes	Cassava
Income value	Poultry	Dairy	Cocoyam	Poultry	Poultry	Dairy	Pigs	Dairy

Source: Participatory Urban Appraisal

Table 6.12 Main farming objectives in selected urban and peri-urban zones

	Food consumption		Commercial		All farms	
Urban (Banda)	N°.	%	N°.	%	N°.	%
Zone 2	26	81.2	6	18.8	32	100.0
Zone 10	16	48.5	17	51.5	33	100.0
Peri-urban (Buziga)	N°.	%	N°.	%	N°.	%
Kiruddu	160	76.5	49	23.5	209	100.0
Katuso	205	87.2	30	12.8	235	100.0

Source: Listing survey (KUFSALCC & Urban Harvest 2004) (row percentages)

of farming), although this requires further study. The commercially oriented farm households in Banda were largely middle class, nearly all involved in mixed crop–livestock farming, with a few keeping livestock exclusively. Almost no farmers growing crops only did so for commercial purposes, whether in urban or peri-urban areas, yet about half the farmers keeping livestock (solely or as part of a mixed crop–livestock system) did so commercially.

In general, livestock production (poultry, dairy and pigs) was the key commercial enterprise across all sites, cocoyams in urban Banda being the exception. Fruit was widely grown for both food and income but compared to a city like Yaoundé commercial vegetable production was not visible in Kampala, cocoyams being a staple root crop.

While the proportion of food consumed coming from farming was not investigated, as Maxwell (1995a, p. 1676) notes, "farming is a major source of fungible income in terms of saving on food expenditure" especially for low-income households. Food production as a form of savings was still clearly the main purpose for which households farmed in 2003.

Table 6.13 shows the numbers and proportions of urban and peri-urban households (not just farming households) growing crops, keeping livestock and trading in animal-based foods, based on listing all UA activities in selected zones. While around a quarter of urban households and one-third to over 90 percent of peri-urban households grew crops, only around a sixth of urban and up to one-third of

Table 6.13 Farming and food trading households (HHs) as percentage of all households in selected urban and peri-urban zones

	HHs growing crops		HHs with livestock		HHs in food trade		Total HHs (2002 census)	Crops/ livestock/ trade ratios
Urban (Banda)	N°.	%	N°.	%	N°.	%	N°.	
Zone 2	30	26.1	15	13.0	5	4.3	115	2:1
Zone 10	29	22.3	22	17.0	10	7.7	130	1.3:1
Subtotal	59	24.1	37	15.1	15	6.1	245	1.6:1
Peri-urban (Buziga)	N°.	%	N°.	%	N°.	%		
Kiruddu	204	37.4	109	20.0	5	1.0	545	1.9:1
Katuso	226	92.2	85	34.7	3	1.2	245	2.6:1
Subtotal	430	54.4	194	24.6	8	1.0	790	2.2:1
Total	**489**	**47.3**	**231**	**22.3**	**23**	**2.2**	**1035**	

Source: Listing survey

Note: The numbers of households growing crops, keeping livestock and trading in food do not total the number of households farming (see Table 6.7 above) as the categories are not mutually exclusive.

peri-urban households kept livestock[1]. Food traders (butchers, milk vendors, fishmongers and cooked food sellers) were very few – 4–8 percent in the urban and 1 percent in the peri-urban areas. Despite public concern about the mess and chaos caused by food trading and livestock keeping in Kampala, our data suggest they need to be expanded and managed if commercialization of small-scale agriculture is to take place and the potential of peri-urban farming for supplying urban markets realized.

As with home ownership, there was higher farm-plot ownership (inherited or purchased) in peri-urban Komamboga (64 percent) and Buziga (54 percent). More farm plots were owned in the old urban area associated with the historic kingdom of Buganda, (46 percent) than in the urban slum (22 percent), though it is interesting that more were bought than inherited. More plots were "borrowed" in urban than peri-urban areas, over 70 percent in Banda, where farmers were basically squatting on public land. Little farmland was rented, except at the peri-urban periphery (Table 6.14).

The three wealth categories identified by our informants (rich, middle and poor) have some correspondence with Maxwell's (1994) farming household typology:

---

[1] There is inconsistency in the data from our two samples. The farming household survey indicates a ratio of crop: livestock keepers of around 2:1 in urban and 1:1 in peri-urban case study areas, while our listing survey gives ratios of around 2:1 or less in urban and 2:1 or more in peri-urban case study areas measured. This indicates the high variability of urban farming patterns since the samples are not from exactly the same areas. The farming household survey drew farmers from across the whole parish, based on random sampling from leaders' lists. The listing survey counted every farmer in selected zones of the same parish. We conclude that our case study data are only indicative of trends although the listing data are accurate for the limited areas studied.

Table 6.14 Ownership of plots by parish

Category	Bukesa		Banda		Buziga		Komamboga		All areas	
	No.	%	No.	%	No.	%	No.	%	No.	%
Rented	3	5.8	5	5.6	5	7.9	25	27.2	38	12.8
Borrowed	21	40.4	64	71.1	23	36.5	7	7.6	115	38.7
Own inherited	8	15.4	3	3.3	11	17.5	33	35.9	55	18.5
Own bought	16	30.8	17	18.9	23	36.5	26	28.3	82	27.6
Other/not stated	4	7.7	1	1.1	1	1.6	1	1.1	7	2.4
Total plots	52	100	90	100	63	100	92	100	297	100

Source: Farming household survey

- *Commercial* – well-off households who started farming for home consumption but mobilize credit from elsewhere to develop their enterprise;
- *Food self-sufficiency* – households with several means of livelihood who have enough land to produce for their own needs and a small surplus that helps with other expenses;
- *Food security* – households where some land is farmed, usually by the woman, to produce a contribution to household consumption, but food is rarely sold;
- *"No other means" (survival)* – food and land-insecure households, often headed by widowed or abandoned women, who grow food anywhere they can to survive.

The following section revisits Maxwell's farming typology.

Policy Implications of the Urban to Peri-Urban Continuum

The large amount of complementary quantitative and qualitative data on Kampala's urban and peri-urban farming presented in this chapter permits an analysis of changes in the city-wide agricultural system between the early 1990s and 2004 that can be used to draw policy conclusions. Though lacking a citywide census on agriculture, our studies suggest the proportion of households farming, and by implication the amounts of food produced, are not diminishing. Logically, since Kampala's population is increasing at a rate of 3.8 percent per annum according to the national census of 2002, the numbers of people involved and the quantities of food produced must have increased substantially. UA contributes to food security through household auto-consumption, as well as in terms of the large group of farmers that can be reached with strategies for the modernization of agriculture, which in turn can improve the quality and amount of food available on the market for urban consumers.

As to the percentages of households engaged in urban agriculture, the 1990s figures of 25 or 36 percent, depending on the sample taken, are matched by the 2004 figures of 25 and 28 percent in two urban areas sampled and 38 percent in one peri-urban area sampled. However, the figure of 96 percent for the second peri-urban location far exceeds 1990 figures. It is possible to draw five conclusions from our data on Kampala, despite its limitations:

1. Agriculture is practiced in all areas even close to the city centre;
2. There is uneven distribution of urban agriculture, with pockets of higher and lower intensity;
3. An urban agriculture gradient exists with farming households constituting a lower proportion of all households in urban areas and a higher proportion toward the periphery;
4. Overall, percentages of urban farming households (roughly 49 percent, but possibly even more) may be higher than previously measured because more space is occupied by peri-urban than urban areas due to the concentric spatial pattern of the city (although densities are lower);
5. Commercialization of agriculture is mainly based on livestock components of mixed farming systems, and it is the urban rather than the peri-urban livestock keepers who seem to be commercializing despite their lesser access to land.

Further, we find, as did Maxwell in the 1990s, that urban farming households are larger than other households, and endorse his statement:

> Given that farming households are considerably larger than the overall mean for the city, this means that the diet and/or livelihoods of about half the city's population is directly affected by urban farming (Maxwell 1995b).

Given the higher proportion of farmers among peri-urban populations we surmise that the overall percentage of population affected will be higher than Maxwell stated. However, this could only be confirmed by the planned urban farming household census or by a statistically valid sample for Kampala.

While Maxwell's categories remain relevant they need to be adapted to changing socio-economic conditions in Kampala. The categories, based on the logic household members use that UA is an "important spoke in the wheel of economic life" (Maxwell 1994, p. 53), can now be reworked in the light of our 2003 in-depth data on livelihood and production systems, and annotate them in relation to the urban to peri-urban continuum.

At the urban periphery there was less evidence of large commercial producers farming tracts of land they did not own using credit obtained through social influence (Maxwell 1994). In 2003, people with access to both land and credit appeared to have smaller areas of land at their disposal while more middle-income households were engaged in commercial livestock production than observed earlier. The commercial category must therefore be redefined to include the many emerging small and medium-sized mixed crop–livestock farmers. While there were still well-off commercial farmers at the peri-urban periphery, the shift to commerce is mostly among urbanites farming for self-sufficiency and relying on income from livestock production (Table 6.15).

According to the farmers themselves in 2003, there were food self-sufficient producers everywhere except the old inner urban area, and they were rich or middle class. But this category was more commercially oriented than 20 years earlier,

6 Changing Trends in Urban Agriculture in Kampala

Table 6.15 Perceived wealth categories of farmers across the urban to peri-urban continuum

	Urban old	Urban slum	Peri-urban transition	Peri-urban periphery
Commercial	–	–	–	Rich
Self-sufficiency		Rich, middle-class	Rich, middle-class	Rich, middle-class
Food security	Rich, middle-class	Middle-class	Poor	Poor
Survival	Poor	Poor	–	–

Source: Participatory Urban Appraisal

mostly through livestock production, but also with a wider range of crops, as they sought other "spokes in the wheel of economic life".

By contrast, food security producers were found in all areas, but they were rich or middle class in the old urban area, middle class in the urban slum and poor in the peri-urban areas. Mostly women, they farmed to supplement household income, seldom for commerce, although again a shift was observed as a few such households generated income from livestock, especially cattle keeping, a livelihood strategy supported by NGOs and international charities.

Producers with "no other means" than UA for survival were found in the two urban areas but hardly in the peri-urban areas where, though poor, people have enough land to achieve a level of food security. The note of desperation found in our livelihoods data from the Banda urban slum was not found in peripheral peri-urban Komamboga.

With growing interest in, and support for, UA at the level of city and national government institutions, a more useful set of categories of households pursuing UA as a livelihood strategy would be:

- *Commercial* – UA enterprise is the main source of household income;
- *Sufficiency / commercial* – Livestock keeping mixed UA farms provide household food supply and/or a major secondary source of income;
- *Food security* – Small-scale mixed or crop UA farms provide some household food, with occasional sales;
- *Survival* – Low-income households engage in UA to prevent starvation.

These are more or less the same categories as Maxwell's, but up-dated and related to urban and peri-urban farmers' conditions in the 21st century. There is also a shift from emphasis on household logic to social-economic planning, so that the categories can be incorporated into the Plan for the Modernization of Agriculture and other policy instruments. The support systems and regulation needed by farmers in the different categories vary, with emphasis more on marketing for the first two, and on safety nets for the last two. It should also be remembered that in all areas of Kampala, the food security and survival groups form the majority of farmers.

Our data confirm that urban farming is a major reality in Kampala's social and economic life, suggesting review of the policy context. The Kampala Structure Plan of 1994 encourages UA, suggesting that agriculture on residential plots should be permitted and that land set aside for environmental protection should, where appropriate, be set aside for small-plot agricultural purposes. The new Urban Agriculture Ordinances introduced by KCC in 2006 allow for such activities to be regulated (Hooton et al. 2006, 2007; Lee-Smith et al. 2008). But while the Structure Plan emphasizes the environmental protection advantages of growing certain crops, fish, reeds and grasses in wetlands (van Nostrand 1994), the Ordinances limit farming of wetlands except with prior permission; this could affect low-income groups' food and incomes unless such permissions are routinely granted.

Public concern about UA food safety is addressed by the health research described below in Chapter 9, which investigated both benefits and risks of UA, including pathogenic and toxic contamination from farming in Kampala's wetlands. While there are risks from traffic pollution next to roads, heavy metal uptake by cocoyams grown in wetlands was found to be limited to those parts not usually eaten, namely outer skin and roots. There was also minimal or no transmission of pathogens to the inner part of tubers, indicating they are safe to eat if peeled and cooked (Cole et al. 2008; Nabulo 2006).

Farming, as part of the livelihood strategy for all wealth groups in all areas of Kampala, needs policy support. While farming systems vary across the urban to peri-urban continuum, in all areas livestock and crops both play a role, subsistence production predominates and most farming is done next to the house. The typical Baganda livelihood and farming system of Kampala also prevails across the urban to peri-urban continuum and among all ethnic groups, with richer dwellings on hilltops and farming on the slopes and in the valleys where poorer people live. However, the urban areas are most densely settled with farming squeezed in between, while the pattern is more spacious in peri-urban areas.

The following chapters elucidate some of these aspects further, and explore possibilities of enhancing and improving UA livelihood and production systems. Meanwhile the data presented in this chapter should complement the new KCC Urban Agriculture Ordinances and help to further incorporate urban agriculture in Uganda's policies. It is clear that opportunities for agricultural intensification provided by proximity to the urban market need to be seized, especially the dynamism of urban farm enterprises that could become successful commercially. Livestock production in particular needs to be encouraged within a regulatory system that responds to farmers' real situations. At the same time, weak commercialization of urban crop production needs to be addressed.

A bi-focal perspective on food security and commercializing urban agriculture should be maintained. While peri-urban land and farming systems need to be encouraged to meet market demand, it must not be forgotten that the people who produce food currently are mainly doing it to feed themselves and secondarily to save money. In order to meet the Millennium Development Goals of addressing hunger and poverty and improving the lives of slum dwellers, most urban farmers need policy and program support to feed themselves better, to increase savings by producing more and to sell more of their produce.

The two largest categories of urban farm enterprise, those farming to achieve food security and survival, require an emphasis on support and social safety net planning, while the smaller categories of commercial and sufficiency require help primarily with marketing. However, the many farmers engaged in food security and survival farming also need help with marketing and must not be excluded from it or restricted to remain within their category. Rather, like national economies, they need assistance as they strive to meet their needs and generate income as well.

References

Aliguma, L 2004, *Strengthening urban and peri-urban agriculture in Kampala, formal survey*, Unpublished Final Draft Report, CIAT, Kampala.

Calas, B 1998, *Kampala: la ville et la violence*, Karthala & IFRA, Paris & Nairobi.

Cole, DC, Lee-Smith, D & Nasinyama, GW (eds) 2008, *Healthy city harvests: Generating evidence to guide policy on urban agriculture*, CIP/Urban Harvest and Makerere University Press, Lima, Peru.

David, S (ed) 2003, *Farming in the city: Participatory appraisal of urban and peri-urban agriculture in Kampala, Uganda*, CIAT Africa Occasional Publications Series no 42, CIAT, Kampala.

Foeken, D 2006, *"To Subsidise My Income": Urban farming in an East-African town*, Brill Academic, Leiden.

Hooton, N, Lee-Smith, D, Nasinyama, G & Romney, D 2007, 'Championing urban farmers in Kampala; influences on local policy change in Uganda', ILRI/ODI/KUFSALCC/Urban Harvest Working Paper, *ILRI Research Report 2*, ILRI, Nairobi.

Hooton, N, Nasinyama, G, Lee-Smith, D, Njenga, M, Azuba, M, Kaweesa, M, Lubowa, A, Muwanga, J & Romney, D 2006, 'Innovative policy change to support urban farmers in Kampala: what influenced development of the new City Ordinances on urban agriculture?' Paper presented at *Innovation Africa Symposium*, Kampala, 20–23 November, 2006.

KUFSALCC & Urban Harvest 2004, *Report of a participatory consultation process on the Kampala City draft bills for ordinances on food production and distribution*, Kampala Urban Food Security, Agriculture and Livestock Coordinating Committee (KUFSALCC), Kampala & Urban Harvest, Nairobi.

Lee-Smith, D, Azuba, SM, Kaweesa, M, Musisi, JM & Nasinyama, G 2008, 'The story of the health coordinating committee, KUFSALCC and the urban agriculture ordinances', in Cole D, Lee-Smith, D & Nasinyama, GW (eds) *Healthy city harvests: Generating evidence to guide policy on urban agriculture*, CIP/Urban Harvest and Makerere University Press, Lima, Peru, pp. 219–229.

Maxwell, DG 1994, 'The household logic of urban farming in Kampala', in Egziabher, AG, Lee-Smith, D, Maxwell, DG, Memon, PA, Mougeot, LJA & Sawio, CJ (eds) *Cities feeding people: an examination of urban agriculture in East Africa*, International Development Research Centre, Ottawa, ON, pp. 45–62.

Maxwell, D 1995a, *Labour, land, food and farming: a household analysis of urban agriculture in Kampala, Uganda*, unpublished PhD Dissertation, University of Wisconsin-Madison, United States of America.

Maxwell, DG 1995b, 'Alternative food security strategy: a household analysis of urban agriculture in Kampala', *World Development*, vol. 23, no. 10, pp. 1669–1681.

MFPED (Ministry of Finance, Planning and Economic Development) & Government of Uganda 2002, *Uganda poverty reduction strategy paper: progress report*, Government of Uganda, Kampala.

MFPED (Ministry of Finance, Planning and Economic Development) & Government of Uganda 2003, *Uganda Bureau of Statistics, Statistical Abstracts*, Government of Uganda, Kampala.

Nabulo, G 2006, *Assessment of heavy metal contamination of food crops and vegetables grown in and around Kampala city, Uganda*, Ph.D. Thesis, Makerere University, Uganda.

van Nostrand, J 1994, *Kampala urban study, three-volume report*, John van Nostrand Associates Ltd., Ministry of Lands, Housing and Urban Development, Ministry of Local Government, Ministry of Justice & Kampala City Council, Kampala & Toronto, articles 6E.05, 7C.09, 9B04-08.

Chapter 7
Can Schools be Agents of Urban Agriculture Extension and Seed Production?

Fred Baseke, Richard Miiro, Margaret Azuba, Maria Kaweesa, Moses Kalyebara, and Peter King'ori

Introduction

With increasing recognition of the role of agriculture in the livelihoods of urban and peri-urban communities all over the world, attention is focusing on it as one possible mechanism toward achievement of the Millennium Development Goals on hunger, poverty and the alleviation of urban slum conditions. This chapter describes an experiment aimed at improving urban and peri-urban agriculture outputs in the city of Kampala by working with schools.

It is estimated that, if figures from East and Southern Africa are projected to the region, 200 million people could be directly depending on food (both crop and livestock products) produced in urban and peri-urban areas of Africa by 2020 (Urban Harvest 2003; Denninger et al. 1998). In the 1990s about 35 percent of residents in the capital city Kampala were reported to be engaged in farming with up to 70 percent of the city's poultry products being produced in the urban area (Maxwell 1994; KUFSALCC & Urban Harvest 2004). This local production has been recognized for its role in supporting the food needs of the urban poor, who otherwise have difficulty in buying food. Kaweesa (2000) found that up to 72 percent of the inhabitants in one part of the city experienced food shortages, with a prevalence of stunting, underweight and wasting in children at 19, 12 and 4 percent, respectively. She pointed out that increased urban production could help fill this gap.

Urban Harvest, a program of the Consultative Group on International Agricultural Research (CGIAR), first established linkages with Kampala organizations in 2000, with the aim of improving food security and living standards of the urban poor through urban agriculture research and development. In 2002 it supported the project "Strengthening Urban Agriculture in Kampala", in which a number of institutions collaborated, led by the International Center for Research on Tropical Agriculture (CIAT). This chapter concerns the component investigating technical interventions to support urban agriculture (UA).

F. Baseke (✉)
Department of Agricultural Extension, Makerere University, P.O. Box 7062, Kampala, Uganda

The questions addressed by this research on technical interventions were as follows:

- Can schools function as effective extension channels for UA?
- What are the most feasible extension approaches and methods to be used by urban schools?
- Is it feasible for urban schools to produce seed commercially for planting purposes?
- How can schools engage farmers in the process?
- What are the constraints and the outcomes of this process?

The action research project was carried out in two phases: the first involved dissemination and extension of crop technologies demanded by schools, while the second tested schools' commercial seed production for urban and peri-urban communities. The first phase was carried out in the second growing season of 2002 (September to December) and the second phase during the first growing season of 2003 (March to June).

By the end of the project in 2004, although findings on commercial viability of seed production were less than positive, the questions raised and the preliminary findings had generated a lot of interest leading to follow-up interventions. Based on several favourable conditions – the farmers' demands, the lessons learnt in the process and an already existing institutional collaborative network – there was an immediate follow-up project that took advantage of the schools connection. Supported by technology transfer funds of Farm Africa, this project, which ran from 2004 to 2006, successfully promoted the growing of pro-vitamin A orange-fleshed sweetpotato at 11 schools working in farming communities in two divisions of the city. This chapter focuses on results from the research between 2002 and 2003.

Background and Rationale for the Study

Uganda has a national policy that schools should teach and practice agriculture as a subject at primary and secondary level, the purpose being to develop a strong public awareness of the importance of agriculture to the national economy, as well as providing school pupils with the basic knowledge and skills to practice it. In the 21st century the teaching of agriculture in schools was projected to move toward vocational education, whereas before it was seen more as an extra-curricular activity for school agricultural clubs. Yet there has been little research and development to explore how the policy goals could be achieved through effective programs in the schools.

In particular, practical ways of integrating agricultural activities in urban and peri-urban communities to meet food security needs have not been either promoted or tested through applied research. This is mainly because urban agriculture has been a policy blind-spot for Uganda, much as it has been in the whole Sub-Saharan Africa region. There have been important studies on urban agriculture in Uganda, including those by Maxwell in the 1980s (Maxwell 1994), David (2003) on bean production, the recent work of Urban Harvest (e.g. Cole et al. 2008), and numerous non-governmental initiatives, such as that of Environmental Alert in the 1990s which helped build the capacity of Kampala City Council's Agriculture Department. Despite all these, there is as yet no official urban agriculture policy in Uganda.

Most urban farmers lack extension services, especially information on the food items they produce. For example, David's study revealed that only 13 percent of the bean farmers in Kampala were growing modern bush and climbing bean varieties in the year 2000. Since beans are an essential part of the diet in Kampala and are becoming an important low-value cash crop in Uganda in general, urban and peri-urban farmers should be able to access information on them (David 2003; Hoogendijk & David 1997).

There has been no interaction between schools and communities concerning farming activities. Urban farmers in Uganda generally aim to satisfy their food and income needs at home with insufficient extension support, while urban schools only teach basic learning and skills development for agriculture without actively linking them to the surrounding situation.

The project was developed as a response to the national policies on agriculture and education. With the education policy incorporating agriculture in schools and the Plan for the Modernization of Agriculture focusing on commercial and productive agriculture, it was apparent that the two stakeholders, namely the schools and the farmers, needed to be brought together. Schools can play a useful role in their communities not only by educating the children but by actively reaching out to their parents as well, becoming community change agents.

It is suggested that schools can conduct agricultural extension by acting as sources of useful agricultural knowledge and skills to the pupils, who can then transfer them to their homes. Institutions such as parent–teacher associations have the potential to sustain a more or less permanent but dynamic interaction between schools and parents, particularly those involved in farming. Schools with organized agricultural activities could also provide extension services through school demonstrations and mini agricultural shows.

Schools in Africa in particular embody resources of economic and social capital that should not be underestimated as focal points of development. It is the essence of education that the time and attention of staff and students are devoted to issues of public concern, while schools have skilled personnel as well as the more concrete physical capital of land, buildings and (sometimes) water that can be applied to such purposes as agriculture. This has been demonstrated in a number of projects and programs directed toward environmentally sound and sustainable development (Lee-Smith & Chaudhry 1990; Vandenbosch 2003).

In addition, schools may be motivated to invest in an activity that can bring some financial benefit, such as the sale of seed and planting material. Several studies have cited the lack of inputs, especially seed, as one of the constraints to urban agriculture

(Maxwell 1994; Lee-Smith et al. 1987), suggesting that schools could perhaps take on the role of seed producers for those crop enterprises demanded by the community.

The Urban Harvest project provided a means of exploring ways of applying these ideas, and the Urban Schools Agricultural Initiatives Project (USAIP), involving several institutions, was established as a result. Its aim was to support urban and peri-urban schools in the provision of agricultural services demanded by their communities. The communities in this case included farm families living in the school surroundings or those who could come to the school for agriculture-related learning activities. While the educational benefits for school students were considered an important aspect of the intervention, they were not however the subject of the research described here.

Methods Used for the Research

The USAIP took an innovative direction for research in its structure at all levels and in its approach to gathering and applying knowledge. The method used was action research. Two understandings, both related to challenges of African research at the start of the 21st century, informed this direction. First, there was recognition of schools as key institutions of knowledge and resources for development in communities. Second, there was an understanding of the potential of committed teams of individuals in different institutions collaborating to accomplish research and development goals. Both of these are based on the lack of resources of all types to carry out research and development, and on making the most of some very real and positive resources that are available (Lee-Smith & Chaudhry 1990).

Led by the Department of Agriculture Extension and Education at Makerere University, the interdisciplinary team from different local government, civil society and academic institutions set out to:

- Provide farming skills to pupils of the schools;
- Provide agricultural extension services to the community;
- Encourage the schools to generate income through agriculture; and
- Monitor and assess the effectiveness of these interventions.

In addition, USAIP had an implicit fifth objective, to create a platform in schools for the dissemination of health and other important urban agriculture messages, and to gather information from communities about these same issues as part of a research method. This objective was implicit in its commitment to the goals of the Urban Harvest-funded project of which it formed an integral part.

USAIP therefore took part in the Participatory Urban Appraisal (PUA) conducted in 2002 and helped other project components select schools as study sites as well as helping in their execution through contacts with the schools. In particular the Health Impact Analysis of UA in Kampala described in Chapter 9 was closely integrated with USAIP activities (see also Cole et al. 2008).

Phase One: Technology Dissemination and Extension

The Feasibility Study and Selection of Schools

The feasibility study to select participating primary schools in Kampala assessed seven schools from four of Kampala's five divisions, leaving out Central where agricultural activity was limited. The schools were designated as being in urban or peri-urban areas following Kampala City Council's classification system of agricultural areas within its boundaries.

A checklist was used to collect pertinent information about the schools and their agricultural activities. Three teachers, including the head teacher, were interviewed in each school, for a total of 21 respondents. The following criteria guided school selection:

- Ability to demonstrate accountability, transparency and control;
- Being as far as possible government-aided schools;
- Having land up to at least half a hectare for farming;
- Evidence of school farming practices;
- Willing to collaborate and commit two staff members to oversee the activities;
- Having farming in the surroundings;
- Being in areas where non-governmental organizations (NGOs) in the team operate;
- Approximately equal distribution of schools in urban and peri-urban areas (Table 7.1).

The three schools selected on this basis were as follows: Lubiri Nabagereka Primary School in Rubaga (urban), Valley St. Mary's Primary School in Kawempe (peri-urban), and Reach-out Primary School in Makindye (peri-urban).

Despite being in an urban setting Lubiri Nabagereka Primary School had about 1.5 ha of arable land and was surrounded by a farming community because both the school's and surrounding land belong to the Kabaka – the King of Buganda. The land has a special status because farming is integral to the kingdom's traditions and this also meant that the school had an active farming program even though it had no outside support from the City or NGOs.

Table 7.1 Location of the schools surveyed during the feasibility study

Division	School	Location: urban or peri-urban
Rubaga	Lubiri Nabagereka Primary School	Urban
Kawempe	Kawempe Church of Uganda Primary School	Urban
Makindye	St. Kizito Senior School	Urban
Nakawa	Kyanja Moslem Primary School	Peri-urban
Kawempe	Valley St. Mary's Primary School	Peri-urban
Makindye	Reach-out Primary School	Peri-urban
Makindye	Munyonyo Primary School	Peri-urban

Valley St. Mary Primary School had about 0.6 ha of land available for farming, as well as farming in its surrounding community and extensive school farming activities. It was also linked to Plan International, one of the NGOs involved in USAIP. Reach-out Primary School, with almost 1 ha of arable land in a peri-urban setting, was actively collaborating with Environmental Alert, the other NGO involved in the project, and already had ongoing school farming activities involving the community. Schools not selected had very little land, inactive farming activities, or no contact with a support NGO doing urban farming.

Participatory Planning for Community Entry and Selection of Enterprises

In early September 2002 the USAIP team organized a workshop for school representatives and community leaders to establish a work plan and guidelines for field activities. A memorandum of understanding between the schools and the USAIP team was drawn up and debated and preparations made to introduce the project to the communities. Mobilization was planned to be led by community leaders with teachers assisting.

The mobilization included announcements on local community radios and village notice boards, as well as local leaders moving to individual homes to invite people to the schools for the first project meetings. These were held in late September for the communities of Valley St. Mary's and Reach-out Primary Schools respectively, and in October for the community around Lubiri Nabagereka Primary School.

The meetings were used to conduct a group appraisal of community farming activities, assess demand for seed and select the crops to be worked with. Two checklists were used, one to guide the participatory appraisal of the farming activities in the area and assessment of seed demand and the other to guide monitoring the implementation of the selected enterprises after they were established. Participants identified a number of crops for which they needed technical support and these were then prioritised as the critical ones for which USAIP support would be provided through the schools.

Technical support was offered in the form of training for schools and communities on the agronomy, post-harvest management and marketing of the selected crops. The trainers, from Namulonge Agricultural and Animal Research Institute (NAARI) and Kawanda Seed Project, conducted "lecturettes", group discussions, and method demonstrations including the setting up of demonstration plots. The first of these were established at the schools in October, in time for the second rains. There was a second round of planting in February, in time for the first rains of 2003.

The training of teachers was passed on to school students, who were involved throughout, the demonstration plots and the process of community involvement providing subject matter for agriculture classes. However, monitoring and evaluation only covered crop performance and not educational or community engagement objectives.

Monitoring and Evaluation of the Crops' Performance

Participatory monitoring and evaluation involved collaboration among the USAIP team, the schoolteachers and the community. Two evaluation meetings were held each season, one in the middle and the other at the end of the season. Additionally, two community meetings per school were held to review the progress of the project. The following issues were given priority in monitoring:

- Level of field preparation;
- Planting and spacing;
- Seed rate (kg/unit area) and germination percentages;
- Weed control (weed types, method of control and weeding dates);
- Pests and diseases identified on each crop (incidence and severity, cause, and control strategy);
- Date of flowering or tasseling (in the case of maize);
- Date of cob, beans or tuber development;
- Yields of beans, maize or sweetpotatoes.

Assessment of Crop Enterprises and Land Availability

Crops for which technical advice was needed included sweetpotatoes, maize, beans, onions, ground nuts, cassava, cocoyam, *matoke* (cooking banana), tomatoes, sugar cane, eggplants and vegetables. The farmers in two of the communities around Valley St. Mary's Primary School to the North of Kampala and Reach-out Primary School to the South prioritized sweetpotato, maize and beans as the ones they wanted advice on. The farmers close to Lubiri Nabagereka Primary School chose to be advised on maize, beans and tomatoes.

With the exception of tomatoes – commonly grown for sale – farmers wanted advice on their annual food crops, indicating their concerns about household food security. Nevertheless, the potential to sell off crops in excess of what is produced is high in urban and peri-urban areas, as is the need to promote small-scale enterprises for such crops. Other factors affecting the selection of these annual food crops were farmers' access to enough arable land and the type of land tenure. Seventy to eighty percent of the participating households near the three schools were small farmers with around 500 m^2 arable land. In all three communities those with more than half a hectare of arable land were few, about 5 percent in each location.

Seed-Demand Assessment in Schools

The farmers near the urban-located Lubiri Nabagereka Primary School bought 95 percent of their seed, compared with the two peri-urban locations where 60 and 55 percent respectively of seed was bought, the rest being home produced. The fact that over half the peri-urban farmers bought their seed implies a high demand for good quality, reliable and affordable seed, but it also reflects farmers' consumption habits – eating most of what they grew with nothing reserved as seed for the next season Fig. 7.1.

Farmers bought about the same amount of seeds for maize and beans, about 3 kg of seed per household for each crop. The average price per kilo was higher for beans (600–800 Ugandan shillings (USh)/kg or 35–46 US cents) than for maize (about 300 USh/kg or 17 US cents).

Peri-urban farmers always bought seed from retailers, such as grain merchants or farm supply stores, while those in the urban area frequently bought seed wholesale from a factory. Apart from the farm supply stores, these sources were not selling seed specifically for farming but rather for other purposes such as milling or livestock feed. Farmers said that some of the seed bought in such places failed to germinate and expressed concern that they were buying seed without knowing the type or variety. Some feared they had bought uncertified seed merely dusted with purple or green colour to look like the real certified ones.

Farmers in all three areas were willing to buy seed from the school as long as it was of good quality, indicating their willingness to pay about 600 USh/kg (34 US cents) for maize or bean seed. They were interested in new varieties, wanting the Longe 5 variety of maize, an open-pollinated, quality protein strain, locally known as Nnalongo. In addition, they requested planting material for the yellow or orange-fleshed sweetpotato varieties with high amounts of beta-carotene, a precursor of Vitamin A, as well as fruits such as oranges, guavas and papayas. They said their sources of agricultural information were NGOs and community-based organizations.

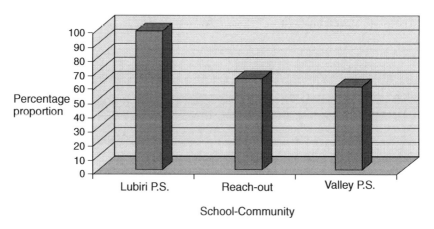

Fig. 7.1 Proportions of households in the school's communities buying seed

Using Schools for Agricultural Extension

Between 15 and 22 community members, two-thirds of them women, attended the initial field activity where demonstration plots were set up and farmers received training on the basic agronomic practices of the selected crops. Both farmers and school staff participated actively by way of asking questions related to agronomy, such as spacing, varieties and control of pests and diseases.

On 14 October 2002 farmers and teachers planted the varieties Longe 5 maize and K132 beans on the demonstration plot at Lubiri Nabagereka School. Farmers subsequently visited the school-based demonstrations and teachers working with the project took them around and answered their queries. At the end of the first season 11 of the 15 original participants who started off at the beginning were present to share their observations. The teachers said they found what the agronomists had shared with them very helpful and they passed this on to farmers who were encouraged to manage their crops better.

Farmers and teachers at Valley St. Mary's School planted the maize variety Longe 5 and four varieties of sweetpotatoes, including Tanzania, Kakamega SPK 004, Ejumula and Sowola (398A) on 23 October 2002. Teachers mobilized farmers for meetings and encouraged regular visits to the demonstration plot where they explained developments. Some farmers took planting materials to try out and compare with what was being done at school. They were keen to apply the practices demonstrated, especially in sweetpotato production, and tried out skills learned at school.

Longe 5 Maize Variety at Lubiri Nabagereka Demonstration Site

Although poor rainfall affected crop growth, the new maize variety was observed to be fast-growing. At the time of weeding the plants were strong and had a dark green colour, except in one area with poor rocky soil and a thin layer of topsoil deficient in nutrients. Although it took a long time to germinate, the maize had a 100 percent germination rate and the maize stalk borer, which attacks local varieties, was not observed because the maize seed was treated (seed dressed). Farmers observed that rodents did not eat any of the maize, possibly because the risk was spread as many people in Lubiri had planted maize. Farmers said Longe 5 performed better under low rainfall conditions than the local varieties, with most plants putting on two to three grain-filled cobs, while local varieties had only one or two, with only the top cob having good grain-fill and the lower one being smaller with a lot of gaps. The maize matured in 3 months and was also seen to dry faster than local varieties.

The harvested maize had an average fresh weight yield of 0.5 kg/m^2. Since a wet ear of maize contains about 57 percent of dry grain (Basalan et al. 1995), this gives a dry weight grain yield of 0.285 kg/m^2 or 2.9 tonnes of grain yield per hectare. This output was lower than the expected five tonnes per hectare yield predicted by the National Agricultural Research Organization (NARO 2003), the cause of this low harvest being the delay in onset of the second rains and shorter than usual rainy season (FEWSNET 2002).

K132 Bean Variety at Lubiri Nabagereka Demonstration Site

Participants observed that the K132 bean variety introduced by the project was similar to K20, locally known as *Nambale*, a variety released in 1968, which is widely grown by farmers in Uganda (David 2003). Germination was observed to be good, due to heavy rains in the early days of crop establishment, although the leaves became yellow due to too much water and inadequate aeration in the root zone. Later, when the rain reduced, the plants regained their green colour.

Weeding was hard for some of the farmers because of close spacing within the line. To facilitate the weeding of beans, one farmer suggested that a spacing of 15 × 60 cm be used instead of 10 × 75 cm as advised by agronomists. Other community members advised their colleagues to use the recommended spacing but try to prepare land well and practice early weeding to prevent competition for nutrients. At this school the beans were harvested after about 3 months, had an average of 27 pods per plant (vine) and a fresh weight yield of 600 g/m^2, equivalent to about 1.24 tonnes of dried beans per hectare. The normal yield for this K132 bush bean variety is 2 tonnes per hectare.

Longe 5 Maize Variety at Valley St. Mary's Demonstration Site

Maize was harvested 3 months after planting, with an average fresh weight yield of 0.35 kg/m^2 equivalent to a dry weight grain yield of 0.12 kg/m^2 or 1.2 tonnes per hectare. Community participants observed that late planting affected crop growth and yield, that vermin and termites ate up the maize after sowing so that about 40 percent of the maize crop had to be gap-filled, that maize grains were small and there was a high incidence of maize stalk borer (locally known as *Ndiwulira*). One farmer observed that the attack by the maize stalk borer could have been a result of late planting and the dry spell that hit the crop soon after planting. The soils in the fields were also of low fertility and roaming livestock frequently ate the crops, all of which contributed to the low yields.

Despite those problems, the community members acknowledged that the new maize variety had a good vegetative growth at the start and was fast growing. In addition, it gave a high yield with about 2–3 well-filled cobs compared to the local varieties, so that overall the farmers were very much interested in adopting it.

Sweetpotato at Valley St. Mary's Demonstration Site

Three sweetpotato varieties (SPK 004, Ejumula and Sowola) had no serious shortcomings, but the Tanzania variety was attacked by wilt and was not resistant to the drought. Yellowing of leaves seen on the same variety was identified as a lack of nutrients, particularly nitrogen, in the soils at the school garden. Generally the new sweetpotato varieties were observed to be high yielding and good for consumption. No yield measurements were kept for sweetpotato due to logistical limitations. The community requested the team to continue with the extension aspect of the project during the second season.

Further, they requested that new farmers be given some or planting materials to try at their homes while continuing with the school-based demonstration. Although funds were insufficient to give out planting materials to all the farmers who requested them, a follow-up project was developed and funded based on the demand for the crop; meanwhile farmers were advised to purchase planting materials themselves. The schoolteachers involved, and some participating farmers, were requested to share information about the new varieties and management practices with other farmers. It was agreed that the new farmers be contacted through village leaders, or through invitation letters written by the school head teachers to churches, friends and neighbours.

Phase Two: Schools as Commercial Seed Producers

Phase two started in the first season of 2003 with sensitization on commercial seed production followed by training, setting up of fields, follow-up activities and field monitoring. During this phase, technology-dissemination activities were also continued.

Participatory Planning of School-Based Commercial Seed Production

The process started with community meetings at the schools where farmers discussed the role of schools in growing and selling seed and planting materials to farmers. This phase involved only Lubiri Nabagereka and Valley St. Mary's Primary Schools. Neighbouring farmers expressed interest in buying seed produced by the schools provided that it was of good quality. In Lubiri Nabagereka, farmers even collected funds and gave them to the school to buy seed for them from reliable sources. Crops selected for commercial seed production were beans, maize and sweetpotatoes, but due to limited resources the whole exercise of planting and cost estimation was only carried through for beans.

Three training workshops conducted by staff from NAARI were organized for teachers, the first, held on 14 February 2003, focusing on the agronomy of commercial crop enterprises. The second and third workshops consisted of training on the profit implications and on commercial seed production practices, led by an agroenterprise specialist from CIAT. The following stages of setting up a commercial seed business were examined:

1. Assessing the need for a seed business in the area;
2. Assessing one's ability to produce a good quality seed;
3. Assessing the knowledge and skills possessed for successfully running a seed business;
4. Finding out what one needs to invest;
5. Analyzing and reflecting on information collected and making conclusions about estimated profitability and chances for business success;
6. Planting a seed plot for at least one season;
7. Selling seed on a trial basis;
8. Identifying whether one can sell seed profitably or not;
9. Making a decision to start a seed business.

It was agreed that profit estimates be confirmed through actual production costs and yields during the trials. Participants were cautioned on the importance of proper post-harvest handling and storage of seed for commercial purposes and a handbook on Commercial Bean Seed Production (David 1998) was distributed.

Participants were asked to estimate the production cost of bush bean seed in their schools and calculate the selling price based on a 20 percent profit mark-up. It was reported that Valley St. Mary's could sell bush bean seed at 1150 USh/kg (60 cents US) while Lubiri Nabagereka could sell at 1410 USh/kg (72 cents US). Both of these were much higher than the price that farmers had indicated they were willing to pay for bean seed (about 600 USh/kg). A bean yield of 1500 kg/ha was forecast.

Teachers' training workshops on harvest and post-harvest management of beans and sweetpotatoes were followed up by technical sessions to check crop performance and guide the process of seed crop management. Teachers were advised to keep records strictly, as this was fundamental in commercial seed production and farmers were encouraged to visit the seed production fields to ensure their engagement with the process.

Results of Commercial Seed Production in Schools

Although drought affected the crops, there was a moderate yield of bush beans at both schools, as shown in Table 7.2.

The selling price of bean seed was estimated for Lubiri Nabagereka Primary School as 5430 USh/kg and that of Valley St. Mary's as 11 150 USh/kg, both very

Table 7.2 Commercial bean seed yields from the two schools

School	Size of plot	Fresh Yield (kg)	Number of pods/ plant	Yield / ha (kg fresh weight)	Yield / ha (kg dry weight)
Lubiri Nabagereka Primary School	20 x 20 m	91.0	40	2275	469.5
Valley St. Mary's Primary School	5 x 10 m	12.0	30	2400	495.2

high compared to what the farmers were ready to pay. Table 7.3 outlines the process of estimating the selling price.

The high costs of production seen in Table 7.3 are due to the high cost of urban manual labour as well as difficulty in judging its price. Lubiri Nabagereka's lower costs appear to be due to the economies of scale in its labour pricing.

Table 7.3 Estimating the selling price of the bean seeds

School	Lubiri Nabagereka Primary School	Valley St. Mary's Primary School
Activity	Cost (USh)	Cost (USh)
Seed costs	–	–
Land clearing	30 000	5000
Planting	10 000	5000
Weeding 1st and 2nd	30 000	10 000
Harvesting	15 000	3000
Total	85 000	23 000
Plot area	400 m^2 or 0.04 ha.	50 m^2 0.005 ha
Total cost per hectare	2 125 000	4 600 000
Total yield (kg) per hectare	470	495
Cost of production UShs/kg	4526	9289
Profit costs: 20% of production costs	905	1858
Selling price per kg	5430	11 150

Constraints

Logistical difficulties as well as limited funding, especially for transport, made it hard to get the schools, communities and the USAIP team all together for the initial training, which included setting up the demonstration plots. This resulted in varying planting dates for the three schools, with longest delays at Reach-out Primary School. Although planting was done eventually, there was no follow up by staff responsible for the demonstration at Reach-out, one reason being that the demonstration was put on a more extensive school farm plot about 4 km away from the school. This was not easily accessible to interested community members or the teachers. As a result, the crop at Reach-out failed, and only the other two schools' demonstrations were used in the subsequent discussions on yields and profitability of a seed enterprise. In all cases the many other demands on the time of teachers affected their ability to adequately monitor the project.

Lack of project resources also meant there was insufficient technical backstopping. Mobilizing the communities and keeping them interested required commitment on the part of the teachers who had limited incentive to be engaged in such an amount of work. This resulted in low morale. In one case a teacher who kept project records got a job in another school and left without handing the records over. Besides these constraints, there were also challenges of poor soils in the case of Valley St. Mary's, while the drought affected all the schools.

Despite these challenges, and the failure to demonstrate economic viability at the first attempt, the schools had been initially highly motivated to produce commercial seeds so long as guidance was provided, while the farmers were also interested in buying them from schools. It is possible that schools could be an important source of seeds for farmers if this experience is taken into account and future projects are well-grounded and supported.

Conclusions and the Way Forward

Although this action research was unsuccessful in demonstrating the viability of commercial seed production by schools for the selected crop, it was more successful in using schools for agriculture extension and also provided a learning experience from which lessons can be drawn. Despite the challenges of using schools as technology-dissemination centres, the farmers in the communities had a keen interest in the project and were comfortable with the schools, which are their neighbours, as a source of useful information and technology.

Numerous institutions in Kampala and elsewhere were struck by this, and by the enthusiasm with which farmers took up the crops and agronomy, especially sweetpotato. This resulted in a follow-up project on orange-fleshed sweetpotato promotion in Kampala. Apart from the Makerere University Department of Agriculture Extension, the partners included the International Potato Center (CIP), NARO, civil society, and Kampala City Council. Working with eleven schools and communities in the city, two varieties of sweetpotato were being promoted – Ejumula and SPK004 (Kakamega) – which are orange-fleshed and have high β-carotene content, a precursor to vitamin A. Research to assess the nutritional impact of these measures was also undertaken. The World Agroforestry Centre (ICRAF) "Farmers of the Future" and the related Education for Sustainable Development Program, based in Nairobi, are among other institutions that have learned from the study.

In terms of whether urban and peri-urban schools can function as agricultural extension service providers, the project experience demonstrates potential for this to happen but within certain limits. Some key elements are vital for success: commitment of the schools, training and motivation of the teachers, involvement of the community, presence of supporting institutions, and adequate funds.

Schools need not only to be willing to serve the community in a new way by promoting improved farming, but must be willing and able to provide their resources including time, land, personnel, pupils, and readiness to host community or farmers' meetings at the school. The project experience is that teachers mobilized local leaders and the community to participate in the project, managed the demonstrations and answered questions from farmers who visited the demonstrations outside of training and joint learning sessions, but they needed more appreciation of their efforts. The response of urban farmers suggested a demand for such a service aimed at improving food security in homes through learning.

The approach and methods used successfully engaged farmers in identifying their priority crops and setting the scene for the rest of the activities. The multi-institutional, interdisciplinary team was the core of this approach as it provided the different types of expertise farmers needed. Leadership of such a team requires technical expertise in extension and crop production and USAIP provided this, organizing regular meetings for learning, checking on the demonstrations and discussing progress on all aspects among participants. However, resource limitations resulted in inadequate school mentoring, monitoring and documentation and limited transport to get to the schools.

Looking at the way forward, it seems there must be an intervening agency such as USAIP that works as a mentoring institution, identifying the best trainers (researchers and extension workers) to provide the farmers, schoolteachers and pupils with the needed knowledge and skills. Innovative training techniques are required to promote participation, such as discussions, observations, question and answer sessions, practical involvement, team work, exchange visits between schools and promoting interaction among learners and their hosts. While the required people and institutions were available, the financial resources to make it all happen were not.

Involving schools in commercial seed and planting material producers was initiated but would need more intensive inputs, especially training, in order to be effective. The training of teachers and farmers would be best done together in order to ensure transparency. The buyer (farmer) needs to know about the production process and what will come out of it.

It was also the experience of this study that strong institutional linkages were built between schools and the communities that these can be used to pass on other important policy messages. The question is whether the schools can withstand the rigour of producing commercial seed given its agronomic, monitoring, record-keeping and proper data-collection needs. Results of this study, based on a small seed grant, suggest that more research and development funds need to be invested for such a system to be developed through schools.

The farmers (who it should be remembered were mostly women, and therefore concerned about feeding their families) selected and took up crops that improved their household food security. Although the emphasis of the second phase of the project in particular was more on production for a market, the lesson should be drawn that both food security and production for the market are important to farmers. Urban farmers, especially women, need more help in balancing these two aspects.

Finally, a major issue and indeed one of the project's objectives, that of benefits to school students in terms of their knowledge of agriculture, was not examined here as a research question. Our study also did not explore whether informing school pupils can have multiplier effects. There are several possibilities of multiple pathways here, involving interactions and exchanges between students, parents, family and other community members over shorter and longer terms. These processes could not be assessed within the resources and time frame of the project, since a complex evaluation research model would be needed to describe and assess different types of outcomes. This study made a limited assessment of an innovative and complex technical intervention – to enhance the benefits of urban agriculture through working with schools – by mobilizing available resources in a creative manner.

References

Basalan, M, Bayhan, R, Secrist, DS, Hill, J, Owens, FN, Witt, M & Kreikemeier, K 1995, 'Corn maturation: changes in the grain and cob', *1995 Animal Science Research Report*, pp. 92–98, http://www.ansi.okstate.edu/research/1995RR/1995RR16.PDF [Accessed 29 February 2008].

Cole, DC, Lee-Smith, D & Nasinyama, GN 2008, *Healthy city harvests: Generating evidence to guide policy on urban agriculture*, CIP/Urban Harvest and Makerere University Press, Lima, Peru.

David, S 1998, *Producing bean seed: handbooks for small-scale bean seed producers, handbook 1*, Network on Bean Research in Africa, Occasional Publications Series no. 29, CIAT, Kampala.

David, S 2003, *Growing beans in the city: a case study in Kampala, Uganda*, Network on Bean Research in Africa, Occasional Publications Series no. 39, CIAT, Kampala.

Denninger, M, Egero, B & Lee-Smith, D 1998, *Urban food production: a survival strategy of urban households; report of a workshop on East and Southern Africa*, Workshop series 1, Regional Land Management Unit (RELMA) & Mazingira Institute, Nairobi.

FEWSNET (Famine Early Warning Systems Network) 2002, *A monthly newsletter on food security and vulnerability in Uganda, no. 12*, FEWSNET, Kampala.

Hoogendijk, M & David, S 1997, *Bean production systems in Mbale District, Uganda with emphasis on varietal diversity and the adoption of new climbing beans*, Network on Bean Research in Africa, Occasional Publications Series no. 20, CIAT, Kampala.

Kaweesa, M 2000, 'Responses to urban food security: Kampala City's case, Uganda', paper presented at a regional workshop on *Food security and nutrition in urban areas*, University of Zimbabwe, Harare, 4–16 December 2000.

KUFSALCC (Kampala Urban Food Security, Agriculture and Livestock Coordination Committee) & Urban Harvest 2004, *Report of a participatory consultation process on the Kampala City draft bills for ordinances on food production and distribution*, KUFSALCC, Kampala & Urban Harvest, Nairobi.

Lee-Smith, D & Chaudhry, T 1990, 'Environmental information for and from children', *Environment and Urbanization*, vol. 2, no. 2, pp. 27–32.

Lee-Smith, D, Manundu, M, Lamba, D & Gathuru, PK 1987, *Urban food production and the cooking fuel situation in urban Kenya – national report: results of a 1985 national survey*, Mazingira Institute, Nairobi.

Maxwell, DG 1994, 'The household logic of urban farming in Kampala', in Egziabher, AG, Lee-Smith, D, Maxwell, DG, Memon, PA, Mougeot, LJA & Sawio, CJ (eds) *Cities feeding people: an examination of urban agriculture in East Africa*, International Development Research Centre, Ottawa, ON, pp. 45–62.

NARO 2003, *Addressing the challenges of poverty eradication and modernisation of agriculture: Improved technologies by NARO, 1992–2002*, National Agricultural Research Organization, Entebbe, Uganda.

Urban Harvest 2003, *Nairobi and environs: from rural and peri-urban dairy production and marketing to urban agriculture and organic waste management*, CGIAR Annual General Meeting site visits guide, Nairobi.

Vandenbosch, T 2003, *Farmers of the future: report of a meeting on agricultural education and education for sustainable development*, World Agroforestry Centre (ICRAF), Nairobi.

Chapter 8
Identifying Market Opportunities for Urban and Peri-Urban Farmers in Kampala

Robinah Nyapendi, Rupert Best, Shaun Ferris, and John Jagwe

Introduction

Rapid urbanization in many Sub-Saharan Africa (SSA) countries is an indicator of the ever-growing pace of structural transformation. Increasing urban population requires more services, the most basic being food. This increasing demand has led to recognition of the contribution of urban and peri-urban agriculture to providing food security, employment and income generation as well as productive management of idle or under-utilized resources. Statistics from the United Nations Development Program (UNDP) indicate that approximately 800 million people are engaged in urban agriculture worldwide with the majority in Asian cities. Of these, 200 million are considered to be market-oriented producers, employing 150 million people full time (Smit et al. 1997).

Urban agriculture within a rapidly growing city such as Kampala offers opportunities and, at the same time, creates constraints, such as inappropriate regulation as well as health and environmental hazards. Lack of services and low levels of maintenance exacerbate the risks and could easily undermine the sustainability of this sector as an integral part of the food system.

In the 1990s it was established that 56 percent of the land in Kampala was used for agriculture purposes and about 30 percent of the households in Kampala were farming in the city (Maxwell 1995), while according to Muwanga (2001), 75 percent of the persons engaged in urban farming were women and about 50 percent of farming households supplemented their incomes through urban and peri-urban agriculture. More recently, as outlined above in Chapter 6, it was documented that about 49 percent of Kampala households were farming in the city in 2003, and that most were the poor and middle class who farm as a survival strategy and coping mechanism.

R. Nyapendi (✉)
Programme Officer Governance and Trade Oxfam GB in Uganda, P.O. Box 6228, Kampala, Uganda
e-mail: rnyaps@hotmail.com

The earlier studies showed that about 80 percent of Kampala farmers were involved in crop production, 13 percent in poultry, 6 percent in dairy cattle, 2 percent in rabbit keeping and 1 percent in goat rearing, while 1 percent was involved in other production enterprises such as fish and pig farming (Maxwell & Zziwa 1992). More recent data indicate that 47 percent of all Kampala households were growing crops in the city in 2003 and 22 percent were keeping some type of livestock (see Chapter 6 above). Meanwhile, urban agriculture was contributing up to 40 percent of the food and about 70 percent of all the poultry products consumed in Kampala (KUFSALCC & Urban Harvest 2004).

Kampala is the most rapidly growing urban centre in Uganda. With its large population of 1.2 million inhabitants growing at a rate of 3.8 percent per year (population census quoted in Aliguma 2004), Kampala is becoming a very big market, particularly for agricultural products. Its central location in the country and relatively well-developed market infrastructure make it a favourable destination for farm produce. The main markets for agricultural produce are located in the urban areas and Kampala City is the main market. Many of these markets offer a comparative advantage to producers and processors located within or close to the city.

Despite such a situation, there are currently no comprehensive studies on the opportunities and constraints in the Kampala market that can be used for agricultural enterprise development. Most existing urban and peri-urban studies conducted in Kampala (David 2003) do not reflect the rapidly changing situation in the country. Most reports also focus on food production, environment and social issues, leaving important gaps in the economic environment of people's lives. Market opportunities for a wide range of agricultural products therefore remain largely unknown. While the data presented in Chapter 6 suggest that market opportunities for perishable and higher value agricultural products have been increasing, a more detailed analysis of these opportunities needs to be done.

There are several types of markets in Kampala including formal produce markets, informal produce markets, food processors, hotels and restaurants, fast-food outlets, institutions including schools, supermarkets, shops, kiosks, roadside vendors, etc.

Formal produce markets are the major farm-produce outlets that have traditionally developed as legal entities. The largest formal produce markets within Kampala are as follows:

- Nakasero market that offers a wide range of fresh foods and beverages; it is popular with the rich and middle classes (see Plate 8.1);
- Owino market, a chaotic market popularly known for selling used clothes, but which also offers all types of agricultural produce, especially cereals and legumes; it is a popular market for both middle- and low-income households;
- Nakawa market, a modern market, newly constructed, selling mainly fresh fruit and vegetables;
- Natete market, another new modern market providing the same services; and
- Kalerwe market, offering all types of farm produce (see Plate 8.2).

8 Identifying Market Opportunities for Urban and Peri-Urban Farmers

Plate 8.1 Typical market stall located in a produce market serving a middle-class neighbourhood and selling assorted vegetables

Plate 8.2 Typical local butchery in a middle-class neighbourhood selling chicken, beef and goat

Informal food markets abound, in addition to the formal produce markets. These are unauthorized and un-gazetted food market places. Informal food markets operate mainly along the roadsides besides established formal food markets and in open places within residential areas. These markets are very popular for poor people because they provide cheaper and more convenient food outlets. However, they are often unhygienic and operate contrary to the Market's Act of 1964 and Kampala City Council (KCC) market bylaws (Muwanga 2001).

Supermarkets are the fastest growing segment in the food retail market in rapidly growing urban areas where incomes are higher. Supplying these supermarkets presents a large opportunity, which comes with equally big challenges for small farms and processing or food manufacturing firms. The scale of procurement is typically much larger and requires both volumes and coordination among suppliers and between suppliers and retailers and their intermediaries. In Kampala there are three large multinational supermarkets (Uchumi, Shoprite and Metro), several locally owned medium to small supermarkets spread out in all the suburbs, as well as different petrol stations which run supermarkets alongside their business.

Small groceries, butcheries and kiosks, despite being under pressure from supermarkets, are numerous, with fresh-produce kiosks and roadside vendors specializing in fruits and vegetables in strategic locations. At the lower class market, small butcheries and kiosks offer an opportunity for small farms to meet required quantities and continuity of supply. The majority of these kiosks operate informally without license.

Processing industries based on agriculture dominate Uganda's industrial production. These range from the processing of fruits and staple foods such as grains, to beverage and tobacco production for both the domestic and foreign markets. Opportunities exist for urban farmers to supply these markets, particularly with the more perishable categories such as fruits and vegetables.

Catering services in Kampala are dominated by public eating-places, which are another major outlet for farm produce. These range from hotels, restaurants, multinational fast-food chains like Nandos and Steers, local takeaways, etc. For specific farm produce such as high-value products, the food-preparation outlets present a niche market for farmers.

In 2002, as part of the Urban Harvest initiative of the Consultative Group on International Agricultural Research (CGIAR), a project was initiated to strengthen and promote urban agricultural systems in a bid to improve food security and livelihoods of the urban poor in Kampala. The project had three components, each of which is dealt with in a separate chapter of this volume:

1. An investigation of livelihoods and production systems (Chapter 6);
2. Schools as extension and seed service providers (Chapter 7);
3. An assessment of market opportunities (this Chapter).

The purpose of this particular study was, firstly, to identify a portfolio of agricultural products that have market demand and whose production is technically and economically feasible for urban and peri-urban farmers; and secondly, to collect and analyse information on purchasing conditions for the range of products identified.

Methods

The method used for identifying market opportunities was adapted from Ostertag (1999), using the following steps:

1. A participatory appraisal to establish the socio-economic and institutional profile of the area under study;
2. A market study to capture opportunities for existing and potential crop and livestock products; and
3. A participatory evaluation for the most promising market options for urban and peri-urban farmers.

Participatory Urban Appraisal

This was undertaken as a common activity to provide information for the livelihoods, production systems and market opportunity identification components of the project. The KCC provided information for the selection of four parishes, according to their classification system for urban and peri-urban areas. Each represented different socio-economic conditions, as described in Chapter 6 above. One was an older urban area, one a new, rapidly urbanizing urban informal area, one an area with peri-urban characteristics but in transition to urban characteristics, and one peri-urban setting with more rural characteristics but on the urban periphery. The criteria used for selecting parishes included:

- Presence of crop and livestock enterprises;
- Presence of farm-produce markets;
- Presence of non-governmental organizations (NGOs) working for the empowerment of local communities;
- Population density;
- Presence of local food processors.

The methods used in the rapid urban appraisal were focus group discussions, interviews with key informants and observation. A total of 190 farmers (86 men and 104 women) from the four parishes participated in the exercise. Crops and livestock produced for income generation and household food security were identified. In addition, the scale of production, market outlets, value addition and constraints for the various enterprises were established.

Market Study

This included the following activities:

(a) *Defining strategies for surveying the market.* Based on the information generated from the urban appraisal, enterprises of major importance to urban and peri-urban farmers were collated and ranked. The Ansoff product–market growth matrix (Kotler 1999) was used as a tool for planning the survey in terms of growth alternatives. The following market-research strategies were adopted:

- Products for which demand is growing;
- Products that are in scarce supply;
- Products that are currently sold by urban and peri-urban farmers;
- Alternative high-value specialty products that could be grown by urban and peri-urban farmers;
- Street foods.

(b) *Developing the research plan and corresponding tools.* A checklist of market-research strategies was developed in relation to the different categories of market outlets (see Table 8.1). For primary data collection, questionnaires for each of the different market categories of outlets were designed, tested and adjusted. All five of Kampala's major produce markets and three large supermarkets were surveyed, as well as five small supermarkets and 21 small shops (kiosks) that were randomly selected from each of the five divisions of Kampala city. Top and middle-range hotels were sampled and five food processors were selected, based on the categories of products currently being sold by urban and peri-urban farmers. Where available, secondary sources of market information were also consulted.

(c) *Data collection.* The survey was conducted using semi-structured questionnaires to obtain the information required. The methods of contact were face-to-face discussions in teams of two, and telephone interviews.

(d) *Data processing and analysis.* Data were cleaned, standardized and manually coded. Analysis was done using the Statistical Package for Social Scientists (SPSS) computer software.

Participatory Evaluation of Options

This was done in two stages, starting with a more detailed economic characterization of products identified and then discussing these with farmers. Based on the results of

Table 8.1 Matrix to indentify research tools

Contact	Market research strategies				
	Products with high sales growth	Products in scarce supply	Demand for existing products sold by urban farmers	Demand for higher value, specialty products	Demand for street foods
Multinational supermarkets (3)	✓	✓	✓	✓	
Local supermarkets (5)	✓	✓	✓	✓	
Petrol stations (2)	✓	✓	✓	✓	
Small shops/kiosks (21)	✓	✓	✓	✓	
Hotels (3)	✓	✓	✓	✓	
Wholesale produce markets (5)	✓	✓			
Food processors (5)	✓	✓	✓		
Street vendors (8)	✓	✓			✓

the market survey, a first cut selection of potentially viable options was made using a set of pre-established criteria. More precise information was then gathered about this more limited list of products, which were classified using market, production and economic parameters. The information was collated within matrices to facilitate analysis.

Following a formal evaluation of the products short-listed by the research team, the Kampala City Council extension staff reconvened farmers from the four parishes to discuss and evaluate the results. Selected information from the three matrices was presented on large sheets of Manila paper (a type of strong brown paper):

- Marketing requirements;
- Production requirements;
- Profitability.

Divided into groups, the farmers rated the different options against their own criteria, using the information on the products provided by the research team. Farmers held discussions and then decided on their priority enterprises for future investment.

Throughout this chapter, the Uganda Shilling (USh) is used as the unit of currency. At the time of the study, there were about 1,750 Uganda Shillings to the US dollar.

Results

Participatory Urban Appraisal

Table 8.2 shows the existing agricultural enterprises being undertaken by farmers in the target parishes. They are divided into those for income generation and those for household food security and are ranked in order of their importance. The ranking of enterprises is based on the number of people involved.

The most important income-generating products for urban farmers, in terms of people involved, are poultry (broilers and eggs), dairy, pigs, fruit (mango, avocado, jackfruit and pawpaw), cocoyam, mushrooms and leafy vegetables. The majority of crops grown for household consumption are not considered income-generating products, with the exception of some fruit and leafy vegetables. Among the important income-generating enterprises, poultry and dairy are the two categories of enterprise common across all urban/peri-urban categories. Table 8.3 summarizes the principal market outlets, quantities sold, prices and constraints faced by farmers, and these are also discussed below.

Poultry

Local and exotic birds are reared for both meat and egg production. Despite the fact that local chickens and eggs fetch a higher price in the market, the scale of operation

Table 8.2 Existing agricultural enterprises in order of their importance for income generation and household food security

Parish	Bukesa	Banda	Buziga	Komamboga
Classification	Old urban	New urban	Urban/peri-urban transition	Peri-urban
Participation m = male f = female	74 (41 m, 33 f)	51 (23 m, 28 f)	40 (12 m, 28 f)	25 (10 m, 15 f)
Principal income generating enterprises	– Poultry – Cattle (dairy) – Mushrooms – Fruits – Cocoyam	– Cocoyam – Poultry – Cattle (dairy) – Leafy vegetables – Pigs	– Poultry – Cattle (dairy) – Fruits – Pigs – Mushrooms – Sugarcane	– Pigs – Cattle (dairy) – Poultry – Fruits – Cassava
Food security crops/livestock	– Fruits – Matooke[a] – Maize – Beans – Cocoyam	– Leafy vegetables – Cocoyam – Beans – Maize – Matooke	– Sweetpotato – Cassava – Matooke – Beans – Maize – Fruits	– Sweetpotato – Cassava – Beans – Matooke – Fruits – Leafy vegetables
Ranking of crops/livestock based on number of people involved	– Poultry – Matooke – Flowers (potted plants) – Maize – Beans – Fruits	– Cocoyam – Poultry – Cassava – Leafy vegetables – Matooke	– Sweetpotato – Cassava – Poultry – Beans – Matooke – Fruits	– Sweetpotato – Cassava – Poultry – Pigs – Matooke – Fruits

[a]Matooke is the name given to cooking banana in Uganda

for local birds is much smaller compared to exotic birds. Poultry products would have an easy market but the enterprises are not well organized. The major constraints to production include cost of feed, diseases, especially for the local birds, inadequate mixing of feeds and theft. Most poultry producers were raising broilers because layers take a much longer time to yield returns. Poor farmers with low earnings do not have the cash flow to deal with mid- to long-term investment, namely that exceeding 3 months.

Dairy Cattle

Milk production is an important income-generating activity. However the number of farmers involved is small. One major reason is the initial capital investment required to purchase a cow, which is beyond the reach of most farmers. Land in the urban setting is another major limiting factor. Dairy farmers keep both local and exotic cattle breeds on intensive (zero grazing) and semi-intensive systems, feeding mainly on peelings and forage. Occasionally, cattle are fed on commercial dairy meal but this is generally only associated with calving. Breeding is usually by bulls and rarely with artificial insemination. The major constraints faced by dairy producers are cost of feeding and veterinary drugs, and the hazard of cattle consuming polythene bags, which is often fatal.

8 Identifying Market Opportunities for Urban and Peri-Urban Farmers

Table 8.3 Supply conditions and constraints of the main income-generating products

	Poultry	Milk	Fruits	Pigs	Mushrooms	Cocoyam	Leafy vegetables
Market outlets	Shops, small local supermarkets, roadside roasters, restaurants.	Neighbours, local retailers	Neighbours, roadside vendors, wholesalers	Butcheries, restaurants, small local supermarkets	Small local supermarkets, hotels, restaurants, shops, produce markets, neighbours	Produce markets, roadside vendors	Schools, neighbours, market vendors
Average supply quantity per household	600 broilers/year 2500 trays of eggs/year	10, 950 l/year	Avocado: 300 kg/year, Mango: 300 kg/year, Papaya: 150 kg/year, Jackfruit: 200 kg/year	50 kg/year	750 kg/year	1200 kg/year	560 kg/year
Average supply price (USh)	3300/broiler Eggs 2600/tray	450/l	Avocado and Mango-200/kg, Papaya-500/kg, Jackfruit-300/kg	1650/kg	3000/kg	500/kg	1000/kg
Supply form	Live birds, dressed	Fresh	Fresh	Live pigs, pork	Fresh and dry	Fresh and often cooked	Fresh
Constraints	Limited capital, unreliable markets, expensive inputs.	Lack of storage facilities, limited capital	Lack of storage facilities	Low prices	Unreliable markets, lack of market information, lack of storage facilities for fresh mushrooms.	Lack of organized marketing	Lack of storage facilities, limited capital investment

Piggery

Pigs are profitable because they forage on waste materials. However, middlemen often exploit farmers because the market is not well structured. Farmers keep both exotic and local breeds of pigs mainly on a semi-intensive mode of production. The major constraints to production include lack of space for expansion, swine fever and stench. Pigs require high standards of cleanliness in order to reduce odour problems. Manure disposal is problematic and there are often complaints of air pollution by neighbours.

Fruit

The main fruit sold by urban farmers are avocado, mango, papaya and jackfruit. There is a ready market for fruit but production is at a low volume. Many farmers do not plant fruit trees and depend upon those that happen to grow where they reside. During harvest periods when there is an abundance of fruit, there are high levels of loss due to a lack of storage facilities and markets to absorb the levels of production.

Mushrooms

Mushroom growing is popular among women, who attach importance to their nutritive and disease-curative attributes. This enterprise is considered to be a feasible income-generating activity for poor households, as returns on investment are rapid (3–5 weeks) and the process is not labour intensive. The most limiting factor however is the unreliable market.

Cocoyam

Farmers with access to swamps grow this crop. Production is mainly for food security, especially for poor households. The use of swamps is against the government conservation law that prohibits cultivation in the wetland.[1] The commodity is mostly consumed in the evening (already cooked) at food markets commonly known as "Toninyira".[2] The main problem faced by producers is a disease that causes withering of the leaves and eventually stunts the plant. Another problem is that swampy soils, where most cocoyam is grown, tend to be highly acidic, which reduces yield. Farmers also suffer from theft when the crop is still in the garden.

Leafy Vegetables

Few farmers are involved in growing leafy vegetables for commercial purposes, although vegetables and other perishable crops are ideally suited to urban and peri-urban production. Vegetables take a shorter time to grow than other crops, but the majority of farmers only grow them for home consumption, not for sale, due to the lack of organized markets. The most commonly grown vegetables are *Amaranthus* spp. (*dodo, bunga, jobyo*), *nakati*, spinach, cabbage, eggplant and bitter berries (*ntula*).

The key demand characteristics used to evaluate portfolio options as shown in Table 8.3 included:

- Scale of operation;
- Prices;
- Market outlets;
- Constraints faced by farmers for the principal income-generating products.

Except for dressed poultry, pork meat and cooked cocoyam as a snack food, there was little evidence of value addition, and there were no sales to food industries. The scale of production restricts the type of market outlet that farmers can access for any particular product. Among the constraints mentioned by farmers, lack of appropriate

---

[1] Although not against the Kampala Structure Plan as noted in Chapter 6

[2] These food markets are chaotic, busy, crowded and unhygienic. The name Toninyira, meaning "don't step on me", developed as a result of people always requesting others not to step on them.

(cold) storage facilities and lack of capital limit the possibility of bulk production of higher value and perishable products (poultry, fruit, mushrooms, etc.). This limits these farmers' ability to capture higher volume markets.

Across the different localities (old urban, new urban, transition and peri-urban), there is a clear difference in terms of age, degree of responsibility, ownership of resources and equality of participation among men and women, and level of interest and motivation. In the old urban situation participants were younger, appeared to have time at their disposal, were alert and open to new opportunities and had high expectations of the project. There was also more equal participation from both men and women and the discussions were conducted entirely in English. In the peri-urban situation, participants were more elderly, possessed more resources and had other commitments, meaning they had little time to devote to the project. Men dominated discussions, which were conducted in the local language because few people spoke English.

Market Survey

Results showed there is an increasing demand for the crops and livestock products produced by urban and peri-urban farmers. This implies that these products present an opportunity for urban and peri-urban farmers to increase their incomes, if they can meet the demands of traders in terms of quality, quantity, continuity and price. There is also demand for a variety of fruit and vegetables that they do not currently produce.

Products in high demand across all the market outlets included broilers, eggs, avocado, mango, tomatoes, onions, leafy vegetables, carrots and green pepper (see Table 8.4).

Pork and beef were the only livestock products with a high consumer demand but limited market outlets. Consumer demand for vegetables was highest in produce markets and large supermarkets, whereas demand for fruit was highest in small shops, kiosks, roadside vendors and local supermarkets. The high consumer demand for vegetables was based on low price in produce markets and on quality in large supermarkets. The high demand for fruit from small shops, kiosks and roadside vendors was attributed to the convenience they present to consumers who prefer to buy the products on their way home after work. Mushrooms – a high-value product of interest to urban farmers – were in high demand in produce markets (in dried form) and supermarkets (in fresh form) [see Plate 8.3].

Most of the products that exhibited the highest consumer demand were also found to be in scarce supply in the various market outlets as indicated in Table 8.5.

Most notable was the limited supply of eggs, milk, pork, mango, avocado, onions and leafy vegetables. This shortage is further evidence of the market potential these products represent for urban and peri-urban farmers. The reasons for scarcity in supply are shown in Table 8.6.

Table 8.4 Products in high demand in different market outlets

Product category	Producer markets	Local supermarkets, including petrol stations	Large supermarkets	Small shops, kiosks and roadside vendors	Food processors	Hotels
Livestock			Pork	Pork		
	Fish					Fish
					Beef	Beef
						Goat
Poultry	Broilers	Broilers	Broilers	Broilers	Broilers	Broilers
	Local chicken					
	Eggs	Eggs	Eggs	Eggs		Eggs
Dairy		Pasteurized milk	Pasteurized milk	Unpasteurized milk	Unpasteurized Milk	Pasteurized and unpasteurized milk
				Yogurt	Yogurt	
					Butter	
Fruit	Avocado	Avocado	Avocado	Avocado		
	Mango		Mango	Mango	Mango	
		Watermelon	Watermelon	Watermelon		Watermelon
		Finger banana		Finger banana		Finger banana
		Pineapple		Pineapple	Pineapple	
	Jackfruit			Jackfruit		
			Apple	Apple		
	Papaya	Papaya				
				Bogoya		Bogoya
		Processed juice	Processed juice			
					Passion fruit	
						Lemon
Vegetables	Tomato		Tomato	Tomato	Tomato	Tomato
	Carrot	Carrot	Carrot			Carrot
	Green pepper	Green pepper	Green pepper			Green pepper
	Onion		Onion	Onion		Onion
	Amaranthus spp.		Amaranthus spp.	Amaranthus spp.		
	Garlic		Garlic			
	Dried mushrooms		Fresh mushrooms			
	Cucumber	Cucumber				
		Lettuce	Lettuce			
	Nakati					
	Coriander					
	Egg plant					
		Cauliflower				
				Cabbage		
						Spinach
Legumes	Beans					
	Cowpea			Cowpea		Cowpeas
	Groundnut			Groundnut		Groundnut

Plate 8.3 The market for mushrooms is growing. This high-value product represents an opportunity for urban farmers

From the total of 38 products that demonstrated market demand, eight were selected for additional study to determine their suitability for urban and peri-urban agriculture. The criteria used to select these eight products are explained below. As an illustration, Table 8.7 shows sources and purchase conditions of these selected products from the various market outlets.

Detailed analysis indicates that most of the products are supplied from outside Kampala. Purchase conditions in the different market outlets showed that large supermarkets offer significantly higher purchase prices for products compared to the other outlets. This is undoubtedly because of the contractual nature of the supply and high-quality standards expected from the suppliers (Plate 8.4). Purchase prices were lowest in produce markets, which may largely be attributed to low-quality standards or lack of value addition to the products. Food processors have the largest bulk-purchase requirements.

The survey revealed that between 65 and 100 percent of traders in all product categories reported equal or greater sales compared to the previous year (see Table 8.8). This business trend implies that there is potential growth in demand for the identified products. Of particular note, pork is the only product for which all respondents indicated equal or greater sales over the past 2 years. Poultry, followed by dairy, are the other two categories that appear to have highest growth.

Participatory Evaluation of Options

The 38 market options identified by the market study were evaluated to select a shorter list of more feasible alternatives for in-depth analysis. The evaluation criteria to select the short list were as follows:

Table 8.5 Products in scarce supply in different market outlets

Product category	Produce markets	Local supermarkets, including petrol stations	Large supermarkets	Small shops, kiosks and roadside vendors	Food processors	Hotels
Livestock	Goat				Goat	Goat
			Pork	Pork		
Poultry products	Local chicken					
	Eggs	Eggs	Eggs	Eggs	Broiler	
Dairy	Milk	Milk	Milk	Milk	Milk	
Fruit	Pineapple	Pineapple	Pineapple		Pineapple	Pineapple
	Mango	Mango		Mango	Mango	
		Orange	Orange	Orange		
			Avocado	Avocado		Avocado
	Papaya			Papaya		
			Watermelon	Watermelon		
	Tangerine	Tangerine				
		Apple				
Vegetables	Leafy vegetables		Leafy vegetables	Leafy vegetables		
			Pumpkin	Pumpkin		Pumpkin
	Dried mushrooms					Fresh Mushrooms
			Onion	Onion		
		Garlic	Garlic			
	Tomato		Tomato			
	Egg plant					
	Green pepper					
	Red pepper					
	Cauliflower					
		Carrot				

- Market demand – those in high demand and scarce supply;
- Production feasibility in an urban setting;
- Current urban and peri-urban production;
- Government policies.

Using these criteria, market options that were not feasible in an urban farmer context were discarded. From a list of 38 options, a total of eight emerged. These were poultry, dairy cattle, pigs, avocado, mango, papaya, leafy vegetables and mushrooms. Table 8.9 summarizes the reasons for inclusion or exclusion of each product. For some non-selected products, insufficient information was at hand to be able to make a definitive decision, and these products could be included in future. The

Table 8.6 Reason for scarcity of products

Product category	Product	Reasons for scarcity
Meat and fish	Goat	Sources of supply are distant from Kampala and scattered, which increases procurement costs. Uneconomical to transport goats from source of supply, as they cannot be transported with cattle, which increases costs
Poultry products	Local chicken	Production is scattered and the sources of supply are distant from Kampala
Dairy	Milk	Seasonality of production, volumes decrease in the dry season (March to May and August to October). Sources of production are distant from Kampala, and transport is unreliable
Fruit	Pineapples	Seasonality of production
	Papaya	Seasonality of production, scarcity during the dry seasons (March to May and August to October)
	Mango	Seasonality of production, scarcity during the dry seasons (March to May and August to October)
	Tangerines	Seasonality of production, scarcity during the dry seasons (March to May and August to October)
Vegetables	Dried mushrooms	Collected from the wild, no formal production
	Red pepper	Seasonality of production, scarcity during the dry seasons (March to May and August to October)
	Green pepper	Seasonality of production, scarcity during the dry seasons (March to May and August to October)
	Cauliflower	Seasonality of production, scarcity during the dry seasons (March to May and August to October)
	Amaranthus spp.	Seasonality of production, scarcity during the dry seasons (March to May and August to October)
	Tomato	Seasonality of production, scarcity during the dry seasons (March to May and August to October)
	Egg plant	Seasonality of production, scarcity during the dry seasons (March to May and August to October)

eight options were evaluated in terms of their market, production and economic characteristics. Subsequently, these products were further subjected to a participatory evaluation with the farmers who then made a choice of the two most appropriate enterprises within their means. Table 8.10 below provides market characterization information for each of the eight products including the clientele for each and their requirements.

This comparative table illustrates that demand varies according to outlet type and product. All the products have multiple market-outlet possibilities, with the supermarkets and restaurants requiring higher quality. Local outlets, such as the produce markets, local butcheries and kiosks in general, are not so demanding, although all are looking for high quality. This may be a determining factor when deciding which outlet to target.

Continuity and guarantee of supply are priorities of supermarkets and restaurants and these are more likely to require a contract. Perishability is high to intermediate

Table 8.7 Selected products with market potential and their purchasing conditions

Product	Purchase price USh	Minimum purchase	Source	Quality standards
Fruits				
Mango	218/= per kg	250 kg/day	Masaka, Jinja	Mature, half ripe, free from external damages
Avocado	312/= per kg	42 kg/day	Kabale, Mbarara, Masaka	Mature, half ripe, free from external damage
Papaya	517/= per kg	965 kg/day	Mukono, Mpigi	Mature, half ripe, free from external damages
Livestock				
Dairy Milk	370/= per litre	205 l/day	Mbarara, Nakasongola, Mukono	Good odour Fresh and concentrated
Pork	1350/= per kg of live for carcass	3 pigs/day	Masaka, Jinja, Mpigi, Mukono	Low fat content Min. weight of live carcass (16 kg) Max. weight of live carcass (55 kg)
Vegetables				
Bugga	778/= per kg	23 kg/day	Wakiso, Mpigi	Fresh and intact green leaves
Dodo	778/= per kg	30 kg/day	Wakiso, Mpigi	Fresh and intact green leaves
Nakati	1000/= per kg	16 kg/day	Wakiso, Mpigi	Fresh and intact green leaves
Carrots	950/= per kg	37 kg/day	Kabale, Mbale, Kampala	Clean, medium size, orange colour
High value products				
Mushrooms	3000/= per kg (fresh) 4000/= per kg (dried)	11.7 kg/day	Kampala Jinja	Fresh Mature Well dried Good aroma
Poultry				
Local chicken	4500/= per bird	66 birds/day	Lira, Arua, Jinja, Iganga	Average size and in good health
Broilers	3200/= per bird	108 birds/day	Kampala, Jinja	At least 1.5 kg and in good health

/= USh

across all products, which should not be a major constraint for urban and peri-urban farmers, and in fact constitutes their comparative advantage due to location.

Production characterization is important in determining whether an agro-enterprise is technically viable in a target region. Production characterization revealed that the selected products have considerable variation in their pre-production and production cycles (see Table 8.11). Technically, the tree fruits appear not to offer any serious constraints to production. However, control of diseases and pests would be important for achieving a high-quality product. For leafy vegetables, their water requirement would be the most important factor to take into account, especially in the dry season. Access to water will therefore be an inevitable factor. Mushrooms are technically the most intensive. The production technology is widely available but possibly of limited access to small holders.

8 Identifying Market Opportunities for Urban and Peri-Urban Farmers

Plate 8.4 Supermarkets demand high quality and consistent supply

Table 8.8 Change in sales over the past two years for various product categories (percentage of respondents)

Responses/product	Fruits (n=35)	Vegetables (n=34)	Poultry (n=11)	Dairy (n=9)	Pork (n=6)	Mushrooms (n=7)
Greater than previous year	57.1	61.8	54.5	66.7	76.5	42.9
Equal to previous year	8.6	11.8	27.3	11.1	23.5	42.9
Less than previous year	17.0	20.6	18.2	22.2	0	0
No idea/no response	17.0	6.0	0.0	0.0	0.0	14.3
Total	100.0	100.0	100.0	100.0	100.0	100.0

Economic characterization is an important criterion for determining the suitability of a business for the small holder by establishing the investment requirements and probable profitability of the enterprise. The economic characterization data for the selected options identified were derived from farmer experiences in the urban setting and reliable secondary sources. Table 8.12 compares the economic parameters for the eight short-listed products.

The technological requirements for producing most of these products are only moderately demanding for small farmers and will mainly require training for the farmers to achieve more intensive production. However, the pre-production investments needed for each of the enterprises may be considered high considering the normal income of this target group.

- Of the fruit, papaya, with its shorter production cycle and relatively higher price, is the most attractive economically;

Table 8.9 Reasons for the selection of short listed products

Identified products showing demand	Selected	Rejected	Reason for inclusion or rejection
Pork	√		High demand and scarce; popular in u/p-u setting
Fish		x	Not feasible in u/p-u and govt. policies
Beef		x	Requires too much land
Goat		x	Production difficult in u/p-u setting
Local chicken		x	Requires land for free range conditions
Broilers	√		High demand and scarce; high interest by u/p-u setting
Eggs		x	High investment required for layers
Milk	√		High demand, scarce supply; high interest by p-u farmers
Yogurt		x	Corresponds to market for milk
Butter		x	Corresponds to market for milk
Avocado	√		High demand, scarce supply, available in u/p-u setting
Mango	√		High demand, scarce supply, available in u/p-u setting
Watermelon		x	Requires water and sandy soil, lacking in Kampala
Finger banana		x	Difficult to produce volume requirements
Pineapple		x	Agronomic conditions not suitable
Jackfruit		x	Market localized, could be included in future
Apple		x	Climatic and agronomic conditions not suitable
Papaya	√		High demand and scarce supply, available in u/p-u setting
Bogoya		x	Could not meet volume requirements
Processed juice		x	Not appropriate at this stage
Passion fruit		x	Difficult to grow commercially
Lemon		x	Technology not proven locally
Tomato		x	Technology difficult in u/p-u setting, could be included
Carrot		x	Climatic and agronomic conditions not suitable
Green pepper		x	Climatic and agronomic conditions not suitable
Onion		x	Not grown at present, could be included in future
Leafy vegetables	√		High demand, scarce supply, already available
Garlic		x	Climatic and agronomic conditions not suitable
Mushrooms	√		High demand, scarce supply, already available
Cucumber		x	Climatic and agronomic conditions not suitable
Lettuce		x	Climatic and agronomic conditions not suitable
Nakati		x	Ranks as another leafy vegetable
Cornmeal		x	Requires further information on production characteristics
Egg plant		x	Not produced currently, could be included in future
Cauliflower		x	Climatic and agronomic conditions not suitable
Cabbage		x	High water requirement
Spinach		x	Climatic and agronomic conditions not suitable

Table 8.10 Market characterization matrix

Product	Currently commercialized	Perishability	Client type	Type of market	Demand growth	Quality requirement	Packaging requirement	Commercial relationship
Mango	Periodic	Intermediate	Supermarkets	City	Medium	High	Boxes	Contract
			Food industries	City	High	Med-low	Sacks	Verbal agreement
			Produce markets	Local	Medium	Medium	Sacks	None
Avocado	Periodic	High	Supermarkets	City	Medium	High	Boxes	Contract
			Restaurants	City	Medium	High	Boxes	Contract
			Produce markets	Local	Medium	Medium	Sacks	None
Papaya	Periodic	Intermediate	Supermarkets	City	Medium	High	Boxes	Contract
			Restaurants	City	Medium	High	Boxes	Contract
			Produce markets	Local	Medium	Medium	Sacks	None
Milk	Yes	High	Food industries	Local	High	High	Milk churns	Contract
			Restaurants	City	Medium	High	Milk churns	Contract
			Dairy shops	Local	High	High	Milk churns	Verbal agreement
Pork	Yes	High	Supermarkets	City	High	High	To be determined	Contract
			Restaurants	City	High	High	Plastic bags	Verbal agreement
			Local butcheries	Local	High	Med-low	None	Verbal agreement
Leafy vegetables	Yes	High	Supermarkets	City	High	High	To be determined	Contract
			Restaurants	City	Medium	High	Sacks	Contract
			Kiosks	Local	Medium	High-med	Sacks	Verbal agreement
Mushroom	Yes	High (fresh) Low (dried)	Supermarkets	City	High	High	Plastic bags	Contract
			Restaurants	City	Medium	High	Plastic bags	Contract
			Kiosks	Local	High	High	Plastic bags	Verbal agreement
Broilers	Yes	High	Supermarkets	City	High	High	Dressed, in bags	Contract
			Restaurants	City	High	High	Dressed, in bags	Contract
			Produce market	Local	High	Med	Live	Verbal agreement

Table 8.11 Production characterization matrix

Alternative product	Complete cycle	Time to first harvest	Technical requirements	Soil requirements	Soil pH	Water requirement	Need for irrigation	Major pests and diseases	Planting density	Annual yield
Papaya	5 yr	1 yr	Low	Deep, well-drained fertile with organic material soils		Does not tolerate water logging	Essential in the dry season	Squirrels, bats, black spot disease, Papaya ring spot virus	444 plants per acre	26 300 kg/acre
Avocado	>20 yr	2 yr	Low	Loose, decomposed granite or sandy loamy soils	Both acidic and alkaline soils	Does not survive under poor drainage	Essential only in long dry spells	Leaf rolling, caterpillars, avocado brown mite, six spotted mite, root rots, *Dothiorella* canker	160 plants per acre	9600 kg/acre
Mango	>20 yr	4 yr	Low	Deep well drained fertile soils	Highly acidic soils	Plant thrives in rain fall between 30 and 100 inches per year		Squirrels, bats, thrips, aphids, mites, scales, anthracnose fungal disease, mango scab disease	160 plants per acre	9600 kg/acre
Mushroom	5–7 wk	2–3 wk	High	Substrate (sorghum husks) sterilized and inoculated with spawn	Not applicable		Water 2 times per day	Cobweb, mildews, nematodes, arthropods, brown mould, bacterial spot, bacterial or pit or brown blotch		200 kg of dried mushrooms per year
Leafy vegs. (dodo, nakati, bugga)	6 mth	2 mth	Low	Tolerates wide range of soil conditions	5.5–7.5		Needed in dry season	Aphids, army worm	1 kg of seed per acre	1300 kg/acre

8 Identifying Market Opportunities for Urban and Peri-Urban Farmers

Table 8.12 Economic characterization matrix

Product	Level of technology	Price stability	Pre-production investment, USh	Average number of work days/yr	Cash flow/ work day, USh	Sales per work day, USh	IRR	NPV at 22% interest, USh	NPV at 46% interest, USh
Mangoes	Medium	Fluctuate	1 958 560	126	13 683	16 609	73	2 701 700	417 900
Avocado	Medium	Fluctuate	1 725 280	126	23 771	20 487	89	5 339 400	1 141 700
Papaya	Medium	Fluctuate	975 480	126	36 108	32 207	838	10 137 500	5 087 100
Dairy	Medium	Fluctuate	1 695 350	360	2158	893	14	−181 400	−419 800
Piggery	Medium	Stable	1 394 000	360	2222	1129	29	133 300	−171 500
Leafy vegetables	Low	Stable	745 000	105	26 000	19 571	2740	6 090 000	2 025 300
Mushrooms	Medium	Stable	2 514 000	52	57 692	17 088	214	1 155 300	621 400
Broilers	Medium	Stable	2 712 000	160	24 000	11 781	233	2 414 400	1 313 600

- When feed is fully costed, pigs do not show a high return. However, for the majority of urban and peri-urban farmers, food wastes are used as feed, which increases the returns;
- Poultry is more profitable because of its shorter production cycle;
- On the other hand the profitability of dairy is very low due to the initial investment in purchasing a cow. In many situations development agencies and the government have purchased cows for the farmers;
- Mushrooms are an attractive investment because of their rapid turnover; however the start-up cost is relatively high;
- Leafy vegetables show a high return, with low initial investment compared to the other options.

Feedback to the communities was provided to groups of farmers in each of the four parishes selected for undertaking the rapid urban appraisal (Plate 8.5). The level of attendance varied between parishes, greater numbers participating in those parishes where the meeting was organised early. Following presentation and discussion of the results of the market study (Plate 8.6), the farmers discussed and selected those enterprises most attractive to them. The enterprises selected by farmers are presented in Table 8.13, together with the reasons that they gave for selection, and with some of the obstacles or constraints that they expect in developing these options as enterprises. There is a remarkable symmetry in their choices, with peri-urban farmers choosing pigs and poultry and urban farmers choosing poultry and mushrooms. It is evident that profitability is not the over-riding factor in enterprise selection. Previous experience and access to resources and appropriate inputs are important decision-making criteria. Their appreciation of the market situation also weighs heavily.

Plate 8.5 Information collected from the market was fed back to members of the target parishes

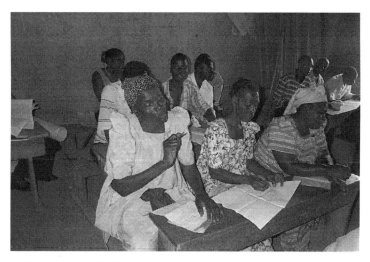

Plate 8.6 Farmers request clarification about market opportunities and decide which options are the most attractive for them

Fruit trees were really only considered an option by peri-urban farmers. The reasons given for non-selection were the limited area available for trees and the prohibition on having tall trees in urban areas, but, most importantly, the need for good quality seed, especially of pawpaw and avocado, to which they do not have access. The farmers expressed surprise at the demand for fruit and vegetables. This was something of which they were unaware. In the case of vegetables, their lack

8 Identifying Market Opportunities for Urban and Peri-Urban Farmers

Table 8.13 Evaluation and selection of enterprises with farmers

Parish	No. of farmers	Selected enterprises	Factors considered	Constraints/problems
Komamboga Peri-urban	36 (21 w & 15 m)	Piggery	High demand, high turn over, resistant to diseases, high proliferation rate, source of manure, easy to manage, feed on a wide range of products, short maturity period	Foul smell, destroy crops if on free range, theft, servicing of sows by boars predisposes them to diseases
		Poultry (broilers)	Short maturity period, requires little space, high demand	Prone to diseases and bad weather, theft
Buziga Transition	17 (10 w & 7 m)	Piggery	Profitable ("never fail to make money"), resistant to diseases, grow fast, start up capital not too high, not fussy about there feed, can use dung as fertiliser, don't cost much in drugs	Smell
		Poultry (broilers)	Profitable, requires small area, can start small and increase, use of litter as fertilizer	Many diseases, theft, prices do not increase, need a lot of feed, if you keep them longer than 8 weeks you lose money
Banda Urban new	53 (34 w & 19 m)	Poultry (broilers)	Highly profitable, requires small space, short production cycle, source of food for family, source of manure	High start up costs, high operating costs, prone to diseases, theft
		Mushrooms	Highly profitable, requires small space, short production cycle, low labour requirements, source of medicine, do not require spraying and fertilizers, nutritious, can be easily dried	Need a lot of water, require high standards of hygiene, high start up costs, difficult to market, highly perishable, seed is scarce
Bukesa Urban old	20 (10 w & 10 m)	Poultry (broilers)	Requires small space, high demand, source of food for family, source of manure, not labour intensive, short production cycle	High operating costs
		Mushrooms	Quick returns (due to short production cycle), requires small space, highly profitable, low labour requirement	

w = women; m = men

of production knowledge, access to fertilizers, and pest and disease problems were cited as obstacles to engaging in these products as an enterprise. Their experience is that they can market the little surplus they produce but problems emerge when the quantities become larger.

The milk option was seen as being attractive to only a few farmers because of management requirements and the need to find pastures on common land. Such areas, though very much a part of traditional life and farming systems, are scarce in urban areas. The production systems survey (see Chapter 6 above) showed they are still available in peri-urban but not urban areas, where grazing sometimes takes place on roadsides. Many urban and peri-urban farmers are therefore switching to stall-feeding and there is a network of small traders providing forage.

The farmers' preference for pigs and poultry raises questions about control and waste management and potential human health hazards, issues that are already foremost in the minds of city authorities.[3] If these are to be encouraged as income-generating activities, there will need to be strict adherence to a set of norms and management procedures.

The farmers expressed their appreciation in receiving the feedback, observing that on several occasions researchers had not taken the trouble to inform them of the results of their studies. They manifested their commitment to report back to other members of their community. Farmers from all the sites expressed their need for assistance in the following areas in order to improve their ability to produce for the market:

- Training in enterprise management;
- Support in establishing market linkages;
- Access to capital for expansion of their existing enterprises;
- Help in procuring improved seed.

Discussion and Conclusions

The survey revealed that quantities of food products supplied by urban and peri-urban farmers were less than the quantities demanded by traders, even the small kiosks that specialize in fruit and vegetables. This suggests that urban farmers are facing two alternatives:

1. To confine themselves to the local or niche markets that they can supply with their current production levels, or
2. To move toward collective action as a means of supplying greater volumes to those markets indicating growth possibilities.

The study shows that farmers are aware of the options (poultry, pigs and mushrooms) that represent a high value given the limits of assets (land, purchased inputs and capital), and whose production is less risky than other alternatives that might provide a higher income (fruit and vegetables). The move toward higher value options that would increase returns per unit of land area would therefore need to fit into a similar way of thinking, unless a set of services can be provided to

---

[3] However, see also Chapters 2 and 4 above and Chapters 10, 11, and 12 below, where the policy and ecosystem health benefits of nutrient cycling are discussed. (Eds.)

support farmers in overcoming these initial "barriers to entry" (with access to seeds, technical assistance, contacts with potential purchasers, etc.) (Table 8.13).

In addition, the average wholesale-purchase prices of products studied, such as milk and pork, obtained from districts outside of Kampala were found to be less than the farm gate prices of the same products produced in Kampala. This could be attributed to, *inter alia,* economies of scale enjoyed by extensive producers in rural areas (in the case of milk) or lower feed costs (in the case of pigs). There are certain commodities like poultry, mushrooms and fruit for which producers in Kampala have a comparative advantage such as closeness to market that reduces transport costs and enables the timely delivery of fresh produce.

It is possible to categorize the current products of urban and peri-urban farmers in the following manner:

- Those products for which farmers are meeting all the purchase conditions with respect to quantity, quality, continuity and price. Examples are poultry (broilers and eggs) and mushrooms;
- Those products for which farmers are meeting requirements of quality and price, but are unlikely to meet the additional requirements of quantity and continuity in order for their enterprises to grow. Examples of these are vegetables and fruit;
- Those products for which farmers are unable to compete in terms of price, quantity and continuity with producers from rural areas. These products include pork and milk. The quality of these products, especially milk, is competitive and could be used as a lever to capture a share of the market that will pay higher prices for better quality. The enthusiasm of farmers for pigs is due to the present semi-subsistence production based on the use of waste food. The upgrading of this enterprise, with the concomitant need for more intensive use of inputs, may make this activity less attractive.

Peri-urban farmers would be better placed to make use of the opportunities identified by the market study that require land as a principal input (dairy, pigs, fruit and vegetables). However, it was interesting to observe the lack of motivation by peri-urban farmers for new opportunities compared with urban farmers. This attitude might represent an obstacle to implementation. Interestingly, a much more positive response was obtained during the feedback session, when a set of concrete options was presented to the farmers.

The farmers living in those areas considered as peri-urban but moving into an urban transition have on average about 0.1 ha that provides them with the opportunity for keeping small livestock. Some of these areas will undoubtedly soon be encroached upon by urban expansion. Those that are in the more favoured areas will have the choice of selling their properties to property developers. The dynamics of land use change should be observed before developing any intervention strategy for these areas.

Urban farmers were open to new opportunities. However, the study has had limited success in identifying alternative products, beyond mushrooms and poultry, of high value and with a quick return on investment that could be produced on their reduced land areas. There are some crops and economic activities that should be investigated further, such as the production of leafy vegetables or other vegetables and ornamental (potted) plants. There could potentially be other products, such as herbs and spices that might offer attractive alternatives. But as in the case of peri-urban farmers, support services will be required to help identify and promote these alternatives.

Overall, the study has shown that demand for a wide range of food products is growing in Kampala. This is expected given the rate of growth and economic development. While the marketing of food, especially fresh food, is still predominant in the hands of small traders in produce markets, there is an increasing trend toward the "supermarketization" of the food system (Weatherspoon & Reardon 2003). Urban and peri-urban farmers could capture a share of this new market, despite very demanding purchase conditions, if supported with appropriate technical backup and help in organizing supply to meet minimum purchase requirements. The success of some mushroom producers illustrate that this can be achieved.

As mentioned in the introduction, this study was part of an exploratory and information gathering process, the results of which could form the basis of a wider proposal to support urban and peri-urban agriculture in Kampala. The preliminary results show that urban farmers in Kampala are market-oriented and that there are growth opportunities for their enterprises and their livelihoods. However, there are many obstacles for them to overcome.

To achieve the goal of improved livelihoods, farmers would need to benefit from appropriate support services. At present services either do not exist or are more oriented toward production with little attention paid to marketing or enterprise organization. The inter-institutional and multidisciplinary nature of the present project is an attempt to bring together appropriate actors from both research and development agencies, to develop such services and work with the local community to develop more efficient enterprises.

With respect to the marketing and enterprise component of this initiative, local capacity needs to be built or improved to provide the following functions:

- Market information focused on urban farming groups to assist them in deciding what crops and livestock products to produce and where to sell;
- Micro-finance schemes to motivate expansion of production and the creation of new enterprises;
- Technical and business extension and training services that are based on an analysis of demand;
- A legal framework that will encourage appropriate and environmentally sound agro- enterprise activity in urban settings, including access to suitable land resources;
- Support for farmer organizations for collective actions in marketing and the provision of other essential services (inputs etc.).

The study has led to a follow-up project that intervenes at two levels:

- Pilot projects in the parishes where contact was made, to increase the competitiveness of the selected products. These projects provide a learning ground for local service providers that could enable them to expand to other areas;
- Design and implementation of a wider intervention to strengthen key enterprise support services at the parish or division level. Appropriate enterprise strategies for each parish or division will take into account the assets of each area and the dynamics of land-use change.

References

Aliguma, L 2004, *Strengthening urban and peri-urban agriculture in Kampala, formal survey*, unpublished Final Draft Report, CIAT, Kampala.

David, S (ed) 2003, *Farming in the city: participatory appraisal of urban and peri-urban agriculture in Kampala, Uganda*, CIAT Africa Occasional Publications Series No. 42, CIAT, Kampala.

Kotler, P 1999, *Kotler on marketing. How to create, win and dominate markets*, Simon and Schuster UK, London.

KUFSALCC (Kampala Urban Food Security, Agriculture and Livestock Coordination Committee) & Urban Harvest 2004, *Report of a participatory consultation process on the Kampala City draft bills for ordinances on food production and distribution*, KUFSALCC, Kampala & Urban Harvest, Nairobi.

Maxwell, D 1995, 'Alternative food security strategy: a household analysis of urban agriculture in Kampala', *World Development*, vol. 23, no. 10, pp. 1669–1681.

Maxwell, D & Zziwa, S 1992, *Urban agriculture in Africa: the case of Kampala*, ACTS Press, Nairobi.

Muwanga, J 2001, *Informal food markets, household food provisioning and consumption pattern among the urban poor. A study of Nakawa division Kampala city*, M.A. Thesis, Makerere University, Uganda.

Ostertag, CF 1999, *Tools for decision making in natural resource management: Identifying and assessing market opportunities for small rural producers*, Centro International de Agricultura Tropical (CIAT), Cali, Colombia.

Smit, J, Ratta, A & Nasr, J 1997, *Urban agriculture: food, jobs and sustainable cities*, United Nations Development Programme, New York, NY.

Weatherspoon, DD & Reardon, T 2003, *The Rise of Supermarkets in Africa: implications for agrifood systems and the rural poor. Development Policy Review*, vol. 21, no. 3, pp. 333–355.

Chapter 9
Health Impact Assessment of Urban Agriculture in Kampala

George W. Nasinyama, Donald C. Cole, and Diana Lee-Smith

Introduction

While the Urban Harvest-supported studies described in the previous three chapters were going on in Kampala, an opportunity arose to complement them with an exploration of the health impacts associated with urban agriculture (UA) in the city. The Kampala City Council had expressed concern about the health risks associated with some forms of UA and the research team welcomed the chance to examine ways to reduce health risks and increase health benefits. A parallel research process was therefore set up, governed by a Health Coordinating Committee comprising researchers and policy-makers. Scholars from universities and research institutes in Uganda and Canada, in collaboration with local government and non-governmental organizations, were supported by the Canadian International Development Agency (CIDA) and the International Development Research Centre (IDRC). This chapter is essentially a summary of a companion book titled "Healthy City Harvests: generating evidence to guide policy on urban agriculture", which contains all these studies (Cole et al. 2008). As in that book, the research findings are translated into policy implications, not only for Kampala but also for other cities with similar conditions.

Identifying Potential Health Impacts

Urban governments have reason for their concerns about the health risks of food production in urban areas with high population density, as in parts of rapidly growing cities of the South like Kampala. Communicable (infectious) diseases from pathogens transmitted to and among people often through inadequate water and sanitation services have been compounded with exposures to potentially toxic chemicals, creating a "double health burden" among the urban poor living in informal,

G.W. Nasinyama (✉)
Department of Veterinary Public Health and Preventive Medicine, Makerere University, Kampala, Uganda
e-mail: nasinyama@vetmed.mak.ac.ug

un-serviced settlements (UN-Habitat 2001). High urban population densities are accompanied by high concentrations of energy use, resource consumption and generation of waste, even with the same levels of individual consumption, leading to a different mix of risks and benefits from urban as compared to rural agriculture (Cole et al. 2006).

With increasingly sophisticated understanding of the links between human health and environment, approaches have been developed to assess benefits and risks to human health including Health Impact Assessments (HIAs) and frameworks for managing risk and promoting healthy public policy (Cole et al. 2006). In HIAs, potential hazards are first scoped out, and those of greatest concern are identified for further investigation. A potential hazard does not always mean there is a risk to health. The pathways through which exposures occur or diseases are transmitted are examined in HIA, providing a way to work out mitigation strategies (Cole et al. 2006).

HIAs on UA involving concerned local government and community stakeholders should include four steps (adapted from Lee-Smith & Prain 2006):

1. Identifying and prioritizing key hazards and benefits associated with UA;
2. Examining hazardous exposures for particular populations and developing risk-reduction methods;
3. Identifying health benefits for particular groups and considering how to enhance these benefits;
4. Formalizing outputs into health-risk mitigation and health-benefit-promotion strategies.

To start the process in Kampala, stakeholders – including policy-makers, scientists and civil society representatives – listed the main health *benefits* of UA as:

- Improved nutrition including dietary diversity, namely protein from livestock and livestock products, energy from staples, mainly *Matoke* (green banana) but also cereal and root and tuber crops, and micronutrients from fruit and vegetables, especially traditional vegetables;
- Nutritional benefits such as mitigation of effects of contaminants and HIV-AIDS, including production of medicinal plants;
- Use of organic waste to produce crops (vegetable compost and livestock manure), and raise animals (organic residues as feed); and
- Psycho-social benefits of physical labour, greening the city, bolstering self-esteem, community organizing and social capital.

The main health *risks* associated with UA were identified as:

- Bacteriological and toxic contamination from cultivation in wetlands (Lake Victoria and its channels) due to poor sanitation and uncontrolled discharges from a variety of urban industrial and other activities;
- Bacteriological and toxic contamination from cultivation in areas where soil or well water are polluted by garbage, run-off or other sources;

- Poor handling of waste and its use for farming (mixing of organic and inorganic);
- Air pollution from industry and traffic; and
- Transmission of disease from livestock to humans (zoonoses).

A Participatory Urban Appraisal (PUA) showed urban farmers' health-risk perceptions varied with location (Chapter 6 above; Lee-Smith 2008). For example, farmers in the Banda wetland slum area identified a wider range of health risks and differentiated between them, as compared to farmers in the semi-rural peri-urban area of Komamboga who perceived fewer health risks. Awareness of risks sometimes led to mitigation actions, such as community organizations forming to manage garbage. However, the very poor, while aware of health risks from UA, could sometimes not avoid them because their survival depended on getting food and water for themselves and their families. Women's and men's perceptions and mitigation strategies also varied due to the different activities in which they are involved (Lee-Smith 2006). Women were more likely than men to grow food crops on contaminated land and were more involved in multiple UA activities, yet were least protected from its environmental hazards (Nabulo et al. 2004).

Based on the benefits and risks identified and resources available, we chose to focus on food security and nutritional benefits, contaminant risks – both biological and chemical – and livestock-related benefits and risks.

Documenting Household Food Security and Child Nutritional Benefits

Aim and Methods

Although data from Kampala in the 1990s suggested that children in farming households were better nourished, based on height-for-age (stunting), than those in non-farming households (Maxwell et al. 1998), there was limited evidence in the published literature regarding not only the mechanism by which UA could contribute to the food and nutritional security of urban populations, but also the degree to which it can. Kang'ethe and colleagues (2007) called for the inclusion of adequate controls, measurement of intermediate outcomes between urban agriculture and nutrition outputs and better control of potential confounding factors.

Hence, we sought to examine relationships among kinds of urban farming, household socio-economic characteristics, household food security (HFS), food frequency, child dietary intake and biochemical indicators among pre-school children (Sebastian 2005; Yeudall et al. 2007) as per Fig. 9.1.

Food security means that "all people, at all times, have physical, social and economic access to sufficient, safe and nutritious food that meets their dietary needs and food preferences for an active and healthy life" (FAO 2002). Confronted by food shortages the primary caregiver – the one responsible for obtaining and preparing food for the rest of the household – uses the strategies of dietary change, food-seeking, household structure changes, and rationing (Maxwell 1996; Maxwell et al. 1999). Our indicator of HFS was based on four questions regarding three of

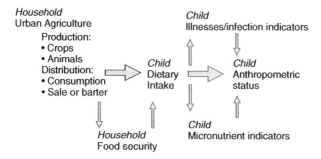

Fig. 9.1 Framework of linkages between household UA and child nutrition

these coping mechanisms, namely borrowing money or food, limiting portion size, skipping meals or eating different foods.

We interviewed 295 primary caregivers of children aged 2–5 years in randomly selected farming and non-farming households in 13 parishes of Kampala, using the Kampala City Council UA classification system of parishes on a continuum from inner-urban to peri-urban. The consent process resulted in 90–100 percent agreement, meeting the communities' need for feedback information dealing with HIV fears and providing both information on anaemia and malaria and referrals for treatment.

Anthropometrics, dietary intake and levels of vitamin A and iron were measured and underweight, stunting and wasting assessed on each child. For each household, we calculated a simple Asset Score as a wealth indicator and used Tropical Livestock Units (corresponding to approximately 250 kg animal live weight) as an indicator of livestock owned. Child-nutrition indicators included dietary quality (the percentage energy coming from animal-sourced foods and dietary diversity, the number of unique food items reported for each child in a 24 h recall), biochemical measures (haemoglobin and retinol), and the nutritional indices of weight for age Z score (WAZ) and body mass index Z score (ZBMI). Chronic infection or inflammation was assessed with C-reactive protein (CRP).

We first used multivariate linear regression analysis to model factors associated with HFS in three population subsets: all respondents (farmers and non-farmers), crop growers and livestock farmers (including those who also grow some crops). Then we used pathway analysis to explore links of these variables to child-nutrition indicators (Yeudall et al. 2007).

Findings on Household Food Security (HFS)

The main factors that improved HFS were greater wealth (assets), particularly among those with less land – land size < 0.25 acre (0.1 ha) – and livestock keeping, with more complex relationships with gender and education (Sebastian 2005; Yeudall et al. 2007). This means those cultivating more land could produce more food either for consumption or for sale, improving their HFS. Raising pigs contributed significantly to HFS, apparently related to cash obtained, since 85 percent of these farming households kept pigs only for sale.

Women were the primary caregivers of their children and generally the urban farmers as well. HFS was better in male-headed households if the woman had secondary education or higher. Her nutrition education also played a role, but it was less clear. Intriguingly, HFS was only significantly and positively associated with assets among household heads having at least a secondary education. A recent review of agriculture interventions concluded that investments in different types of human capital were needed for increased food production to improve nutrition (Berti et al. 2004).

For households raising livestock, women-headed households had greater food security than male-headed households when the size of land was a quarter acre or less. Women heading households may have fewer resources (labour time and money) to make use of land than if they receive support from the man of the household. Women farmers in Kampala have been found to grow crops requiring minimum labour time because they are also responsible for household activities such as cooking and caring for their families (Nabulo et al. 2004). Perhaps women farming larger plots of land alone may rely more on UA as a livelihood strategy than do women in male-headed households. Women's access to food for themselves and their children has been shown to be strongly associated with their control over household resources (Kabeer 1991).

Findings on Child Nutrition

In the path analyses, assets (wealth) contributed directly to children's weight-for-age status, and indirectly to nutritional security indicators such as animal-sourced-food consumption and diet diversity. We noted a significant positive relationship between HFS and subsequent animal-sourced-food consumption, which in turn was positively associated with retinol, an indicator of vitamin A status. Consumption of animal-sourced food was significantly negatively associated with CRP, which in turn was significantly negatively associated with haemoglobin. Thus infection, inflammation and anaemia were less likely in children consuming more animal-sourced foods. Haemoglobin was significantly positively associated with WAZ, highlighting the importance of efforts to increase animal-sourced foods in children's diets (Allen 2003).

Dealing with Contaminants on and in Crops

Nature of the Enquiry

To respond to Kampala City Council's (KCC) concerns about microbial and chemical contamination, it is useful to distinguish the pathogenic risks presented

Table 9.1 Differences between biological and chemical health hazards in UA

Hazard	Biological	Chemical
Type	Pathogenic[a]	Toxic
Source	Excreta, organic matter, especially in decay	Manufacturing processes & their products e.g. paints, cleaning fluids, gasoline, batteries, and combustion products e.g. diesel emissions
Main pathways of exposure	Solid waste & wastewater	Solid waste, wastewater, air
Behaviour in environment	Have limited life span but multiply	Decay slowly, if at all, but can accumulate up food chains
Disease	Infectious	Non-infectious
Conventional controls in cities of lower income countries	Composting & latrines. Sometimes sewage filtration, sedimentation, and drying of sludge for agriculture. Some water purification, local or central.	Currently limited – except for some controls at source, re-use, emission controls, and sanitary landfill burial.

[a] Some biological organisms are beneficial e.g. soil bacteria. Here we only refer to pathogenic bacteria, viruses, parasites, etc

by microbial agents that can cause infectious diseases from the risks presented by chemicals that can cause a range of non-infectious diseases (see Table 9.1). Although often lumped together as "contamination", pathogens and potentially toxic chemicals come from different sources, behave differently in different environmental media, and require different control methods to mitigate adverse health effects. A key factor is separation of the two types of contaminant wastes – analytically as well as physically.

Many UA sites in Kampala, e.g. wetlands, dumpsites and roadsides, present a heightened risk of farm household and produce contamination. Hence the first study examined pathogenic hazards, the second, metals, and the third, complex organic compounds present at these sites.

Biological Hazards Associated with Crops Grown on Untreated Sewage-Watered Soils

Aim and Methods

Our aim was to measure the presence of selected bacterial and parasitic pathogens in sewage water and soils used for UA crops, as well as on the surfaces and in the tissues of crops themselves (Serani et al. 2008).

We took samples from three contaminated sites including the Nakivubo channel, where UA uses flood irrigation in the wet season and digging of canals in the dry season, and two uncontaminated sites on hilly areas not near any drains. The Nakivubo channel, built 50 years ago to carry storm water from the city into Lake Victoria, now receives a considerable quantity of partially treated and untreated industrial and domestic wastewater as sewage volumes have grown but wastewater treatment has not. The selected crops were tomato (a fruit), Doodo (or amaranth, a leafy vegetable) and cocoyam (a root crop).

In the laboratory, we analyzed for pathogens in water or soil and on and in the plant crops. These included the broad coliform group of bacteria found in the feces of warm-blooded animals (humans and livestock) and in soil, *fecal coliform* bacteria, found only in the former, and *Salmonella*. We could not analyze for viruses. We used both the flotation technique and centrifugal sedimentation examination to test for a range of parasites including: *Entamoeba histolytica* cysts responsible for amoebic dysentery, helminth eggs and roundworm larvae.

Findings

We found coliforms and *E. coli* in appreciable numbers in most samples from contaminated sites. Concentrations of fecal coliforms in water samples were much higher than World Health Organization guidelines for irrigation water. Coliform counts obtained from the analysis of plant parts also indicated fecal contamination. Tomato samples from one contaminated site had coliform contamination on the surface of the root and stem with internal contamination of tomatoes also seen, although at low levels.

Our detection of bacterial pathogens on the inside of tomatoes is consistent with other research showing human pathogens surviving in tomatoes and tomato products, both coliform (Zhuang et al. 1995; Zhuang & Beuchat 1996; Tsai & Ingham 1997) and *Salmonella* (Jablasone et al. 2004). Fecal *E. coli* was isolated both on the surface and to a limited extent on the inside of some plant tissues, including the popular cocoyam. One sample of the surface of the leaves of *Amaranthus* from a contaminated site was positive for *Salmonella*. Control sites that did not use wastewater were less contaminated with coliforms and *E. coli* compared to the sewage-contaminated sites.

Helminth eggs, amoebal cysts or larvae were not detected in any soil or crop samples. The three positive findings (out of nine samples) were all in wastewater from different sites: parasite eggs (6/ml), *Entamoeba hystolitica* cysts (24/ml) and parasite larvae (3/ml). In contrast, in Ghana, where use of chicken manure is common in urban farming, mean parasite egg levels of 1.1, 0.4 and 2.7 eggs per gram were isolated from lettuce, cabbage and spring onion, respectively (Amoah et al. 2006).

The World Health Organization (Keraita & Dreschel 2006) states that using excreta or wastewater in agriculture can result in a public health risk ONLY if ALL the following occur:

- An infective dose reaches a field or pond, or a smaller dose multiplies there;
- The infective dose reaches a human host;
- The host becomes infected;
- The infection causes disease or further transmission.

Kampala farmers handling raw sewage and children accompanying them are likely at greater health risk than consumers, because they handle soil and water directly, exposing them to infective doses. Kampala vegetable consumers who eat crops raw, e.g. tomatoes in salads, are also at risk of infection from bacteria inside the tissue of the vegetables, even though they may be able to remove bacteria on surfaces by scrubbing and disinfection. Vegetable crops that are cooked, e.g. cocoyam and amaranth leaves, should not pose such risks because bacteria should be killed with heating.

Heavy Metal Contamination of Vegetable Food Crops

Aim and Methods

Waste dumps and roadsides are common sites where poor households grow crops. Vehicle emissions are a major source of heavy metal contamination in urban areas, exposing human beings through inhalation and through ingestion of plants grown in affected soils. Lead (Pb), a neurotoxin particularly for children, is widely used in gasoline in Kampala; cadmium (Cd), a kidney toxin and carcinogen, is found in paints and motor vehicle batteries; and zinc (Zn), which is toxic to soils and plants at high concentrations, is a component of tires that is released as they wear. This research sought to measure these persistent heavy metal levels in soil from selected sites and rank vegetables and various plant parts according to metal uptake.

Soil samples from UA sites where municipal and other wastes were disposed of, including metal workshops, industries, sewage, a hospital, garages and scrap yards, as well as wastewater-irrigated wetlands, were analyzed for heavy metal content (cadmium, lead, zinc, copper and nickel) for comparison with international guidelines. Different plant parts were analyzed for heavy metal uptake.

Eleven sites along roads radiating from Kampala were also selected, based on the presence of UA, and traffic counts taken. Surface soil was taken at each site perpendicular to the road at 5 m intervals distance, using several locations at each distance and mixing them together to form an aggregate sample. The soil was air dried and passed through a 2 mm sieve to remove gravel and superfluous matter, care being taken to avoid other sources of contamination such as industrial or other waste or compost. Three samples of each selected crop, including one edible vegetable weed *Amaranthus dubius,* were taken from the roadside gardens, and bagged for laboratory examination. The soil and plant samples were analyzed for Pb, Cd and Zn using flame atomic absorption spectrophotometry (Perkin Elmer, Model 2380).

Findings

The study found elevated levels of Pb, Zn and Cd, particularly in leafy vegetables. While not all roadside sites had heavy metals above the recommended maximum levels in their soils, contamination of soils and *Amaranthus* leaves with Pb was clearly shown to be a function of traffic density. Nevertheless, all sites, including those with less traffic, had Pb levels above those recommended in vegetables for human consumption. Accumulation of Pb in soils above background levels took place up to a distance of 30 m. Lead concentrations were found to be highest in leaves, then in roots, next in fruits and lowest in seeds. Atmospheric deposition was found to be the dominant pathway for Pb reaching leafy vegetables, more than through soil up-take. Leafy vegetables were the most effective in accumulating Pb from the atmosphere (Nabulo et al. 2006).

Certain plants, such as the *Brassica oleraceae* acephala group, commonly known as *sukuma wiki*, were found to accumulate heavy metals (such as cadmium) more than others, while the outer leaves of cabbage had higher concentrations than the inner leaves. Washing was found to reduce Pb and Cd levels in many of the vegetables studied. Previous studies have found that heavy-metal levels in soil and vegetation decrease with increasing distance from the road that is the source of vehicle emissions. This study similarly found the lowest concentration of 28 mg/kg (ppm) of Pb was at 30 m away from the road edge, allowing the growing of leafy vegetables with relatively lower health risk (Nabulo et al. 2006).

Risk Assessment for Children Exposed to Incomplete Combustion Products

Aim and Methods

Children experience higher relative exposures to contaminants because they consume more food and liquid and breathe more air in proportion to their body weight than adults. Further, their less-developed metabolic pathways hinder their ability to metabolize, detoxify and excrete toxicants. Children also have longer to develop diseases triggered by exposure to toxic contaminants early in life (Suk et al. 2003) and early life exposures are more likely to perturb sensitive ontogenetic processes.

Polycyclic aromatic hydrocarbons (PAHs) are semi-volatile, persistent organic pollutants emitted through incomplete combustion of organic matter such as gasoline, wood, coal or oil for domestic use and industrial power generation, with smaller amounts coming from printing industries and barbequed foods. Domestic cooking is a major source of PAH in dense urban areas of lower-income countries. Approximately 95 percent of Ugandans use wood or other biomass as their primary domestic energy source. Vehicle emissions are also important contributors. This research sought to measure PAHs on surfaces, to estimate children's exposure via air, soil, water and food, and to scope potential health risks attributable to PAHs among UA households (Yamamoto 2005).

We first assessed the concentrations and spatial patterns of five PAHs in surface soils and atmospherically derived surface films: phenanthrene (PHE), anthracene (ANT), fluoranthene (FLA), pyrene (PYR) and indeno[*1,2,3-cd*]pyrene (ICDP). Samples were collected from 10 households engaged in UA that had already participated in the earlier nutrition study. The households, from different areas of Kampala, with a range of high to low traffic, industrial and population densities, also provided information about agricultural activities, smoking habits, fuel used, kitchen location and the time children in the household spent indoors and outdoors.

These data and those from the literature were then used in a risk assessment to quantify the pathways of PAH exposure and to estimate likely health impacts among children aged 2–5, using the Multimedia Urban Risk Model (MUM-Risk) designed by Diamond and co-workers at the University of Toronto and modified into the MUM-FAMrisk model (Multimedia Urban Model, Family Risk) also used in Toronto by researcher Heather Jones-Otazo (2004). The most sensitive health endpoint for three PAH contaminants, anthracene (ANT), fluoranthene (FLA) and pyrene (PYR), were chosen from the five measured, based on levels detected, potential health effects, and availability of reference dose data. Hazard quotients were calculated and comparisons made to established risk levels.

Findings

PAHs were found in most surface and soil samples and ingestion was found to be the most important pathway of exposure to PAHs, consistent with findings in other cities. Overall, the most urban areas tended to have the highest total estimated daily intake (EDI) for all compounds and one peri-urban area was consistently the lowest. The dominant risk pathways for anthracene were the ingestion of below-ground vegetables and the inhalation of indoor and outdoor air for all age groups and both sexes. For pyrene the dominant pathways were ingestion of below-ground vegetables and cereals and grains and for fluoranthene, the ingestion of fish and shellfish.

We found that all the hazard indices (HI) fell below the conservative action level chosen for MUM-FAM risk for the compounds studied, as well as the levels set by the U.S. EPA and European Commission (2002). This suggests their contribution to nose, throat and lung irritation, increased carboxylesterase activity and the decreased ability to fight diseases, as well as kidney problems among children in Kampala, should be small. But if they are considered as part of complex mixtures including other PAHs and organic compounds, there could be additive effects of multiple exposures resulting in the action levels being exceeded. There are a number of things that should be noted as cautions when dealing with risk assessments in general. Risk-assessment models involve numerous assumptions that provide conservative estimates of the hazards posed by exposure to certain chemicals and the data used can be associated with a high level of uncertainty (Vostal 1994). For this study in particular, it is hard to tell if our findings are conclusive given the small subset of PAH compounds which we were able to test. Regardless of our findings, we suggest that policy makers need to consider biomass burning and vehicle emissions as important sources of pollution in Kampala.

Managing Urban Livestock for Health

Nature of Inquiry

We engaged in a multifaceted examination of livestock related practices, potential health risks and current mitigation strategies among households keeping cattle and chickens, the two most common types of livestock in Kampala. For both types we purposively selected 10 of Kampala's 98 parishes, to give a range of urban and peri-urban livestock-keeping conditions.

Dairying

Methods

Preliminary focus group discussions (FGDs) were held with, on average, seven livestock-keeping households per parish. A household sample of 150 cattle keepers was then selected based on information gathered, and 50 neighbouring households not keeping cattle were also selected for comparison. The cattle keepers reported the age, sex and breed details of 713 bovines they kept, including 357 adult cows. Biological testing for four health hazards was done using blood and milk samples from sample cows from each household:

- *Brucella abortus* in milk, the cause of undulant fever in humans;
- Total bacteria in milk, a rough indicator of milk hygiene and cow health;
- *E. coli* O157:H7, a cause of gastro-intestinal disease and sometimes of kidney disease and blood disorders;
- Anti-microbial residues in milk, which may disrupt normal gut bacterial function that acts as a barrier to infection and may occasionally cause allergic reaction in susceptible people.

Findings

In Kampala, both rich and poor urban households kept an average of four cattle, most confined in stables and brought fodder crops, household waste and concentrates, although some were tethered or graze on vacant land and roadsides. We roughly estimate there were anywhere from 50 000 to 168 000 cattle in Kampala at the time of our surveys in 2003–2004, supporting Muwanga's similarly rough estimate of about 74 000 in 2001 (Muwanga 2001), of which 73 percent were high-yielding cattle. Agriculture was the main occupation for most cattle keepers (64 percent) and over a third of participants said keeping cattle was their main occupation (consistent with findings from Nakuru in Chapters 11 and 12). Only 13 percent of our respondents were in waged employment, with employment stated as an important subsidiary benefit of urban dairying. Still, income from cattle made up less than a quarter of total household income for half the respondents, while only 15 percent obtained more than half their income from dairying, indicating a diversified portfolio of livelihood strategies.

Milk was stated to be by far the most important benefit of dairying, with manure, sale of cattle and employment occupying subsidiary roles. Dairy cows were mostly milked by hand twice a day, giving an average milk yield of 10 l per cow per day, compared to 1–2 l per day reported for indigenous African cattle. Milk and dairy products are important sources of protein and micronutrients, some of which are found only in animal products. Milk consumption in Uganda has doubled in the last decade to around 30 l (or kilograms) per capita annually (40 l in urban areas).

Milk samples from the surveyed households indicated the cows had been exposed to brucella (*Brucella abortus*) by infection or vaccination (though we did not test directly for Brucella organisms), indicating it is likely common among cattle. Some of our milk samples also had unacceptably high levels of bacterial contamination (as defined by the East African community standards), which could be due to poor milk hygiene, mastitis (inflammation of the udder) or other factors. Of the 165 samples examined for *E. coli* O157:H7, 18 out of 69 isolates were suspect when cultured, but only two of these were serologically confirmed as positive. In the absence of information about the most important exposure routes, it is difficult to estimate globally the risk of *E. coli* O157:H7 infection from dairy cattle; but given the seriousness of the disease, it certainly warrants further investigation. Other studies in Kenya and Uganda have indicated unacceptably high bacterial counts in formal sector pasteurized milk as well, with smallholder milk comparing favourably in terms of compliance with standards (Lukwago 1999; Omore et al. 2005).

A total of 165 milk samples were tested for presence of some broad-spectrum anti-microbials in milk and 14 percent tested positive for residues above the recommended maximum residue limits. Milk samples positive for residues were re-tested for beta-lactam drugs (popular antibiotics related to penicillin) and 13 percent were positive at levels above maximum residual limits.

We then developed a pathway model describing the movement of milk from the cow to the consumer and identified critical control points where interventions may prevent or eliminate a food safety hazard or reduce it to an acceptable level (FAO and WHO 2001):

1. Contact between cow and hazards in the environment;
2. Contamination of milking shed with cow excreta and secretions;
3. Contact between milk and containers or the environment;
4. Handling of milk by processors;
5. Storage and transport;
6. Practices of food processing pre-consumption.

The average of 17 risk-mitigation strategies on the pathway from stable to table showed a rich variety of farmer and consumer risk-management strategies in place (see Table 9.2). Some of these strategies are completely effective, such as boiling milk for managing risk from *B. abortus* (this is also widespread as 93 percent of consumers boil their milk) or observing milk withdrawal for managing risk from antibiotic residues, while others reduce but do not eliminate specific risks, such as keeping only one type of livestock or selling milk within six hours of milking.

We then used a linear regression model to investigate what influenced farmers to practice risk mitigation strategies. Access to services, belief that UA was legal, and a wealth and productivity orientation were associated with higher use of mitigation strategies, while those reporting cattle disease incidence had lower use of mitigation strategies. Clearly access to services (water and electricity) makes the use of

Table 9.2 Critical control points and actors carrying out mitigation strategies (%) at each step on the pathway from stable to table

1. Mitigate exposure of cow to environmental hazards		3. Mitigate environmental milk contamination		5. Mitigate growth during storage/transport	
Keep only one type livestock	29%	Hard/washable wall to shed	100%	Refrain from consuming unsold milk	10 %
Zero-graze	38	Metal/timber roof	96	Refrain from storing and selling unsold milk	90
Cut and carry from own land only	41	Store containers off the floor	29	Treatment of milk by vendor	50
Avoid common grazing	56	Keep milk bar dry	45	Sell milk quickly (<=6 hs)	82
Treat cattle often (last 3 wks)	31	Use just metal/glass containers	19		
No calves on farm	39	Use piped water	75		
Artificial insemination	44	Good hygiene of premises	51		
Vaccinate against brucellosis	1	Dispose waste >5m away from dairy	38		
Keep local breeds	27				

2. Mitigate milk shed contamination		4. Mitigate contamination of milk by personnel		6. Consumer risk mitigation	
		Use hot water in cleaning	18	Avoid drinking raw milk	93
Feed/water trough present	94	Use soap in cleaning	81	Check milk quality with visual/olfactory test	48
Concrete/stone floor	96	Protective clothing	1	Don't consume milk until withdrawal period passed	64
Waste disposal strategy	96	Wash hands with soap before handling milk	59		
Use bedding	41	Good personal hygiene	49		
Stack manure	11	No discharges or wounds in those selling milk	97		
		Hands clean	79		
		Nails clean/short	81		
		Latrine available	98		

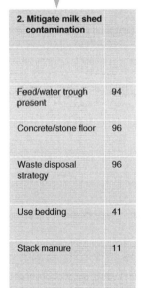

hygiene-related mitigation measures easier, while farmers may invest more in risk mitigation when they feel more secure through believing their activities are legal. Wealth, measured by a proxy index, makes it easier to invest in resources for better hygiene, suggesting special supports may be needed for lower-income producers. Farmers producing greater quantities of milk more intensively showed an increased likelihood of using mitigation strategies, probably both because these farmers were better skilled and because intensively kept animals, being more vulnerable to disease, require increased care. It may seem paradoxical that farmers reporting more cattle disease were less likely to use mitigation strategies, but the lack of resources to do anything about the risk can create a feeling of helplessness, resulting in more cattle disease.

Chicken Rearing

Methods

Focus group discussions in each parish were followed by selection of a sample of 142 households rearing chickens for sale and 50 neighbouring non-chicken-rearing households. The latter underwent questionnaire-based interviews about practices and potential health effects.

Findings

As in many other towns and cities in Sub-Saharan Africa, chicken rearing was the most common form of livestock production (Chapters 4, 6, 11, and 12). Past work found that urban poultry producers met 70 percent of the city's needs (Maxwell 1994). Women played a central role in our research, attending and contributing to FGDs and being the primary respondents in both chicken-rearing and non-chicken-rearing households. Generally they were the primary caretakers of the chickens, mostly self-taught or guided by friends or relatives. Only 14 percent were taught how to raise chickens in a rural area, and interestingly 42 percent had received some form of training or upgraded their management skills after they began raising chickens.

Chicken-rearing households were mainly high and middle income (66 percent), reflecting our focus on commercial farmers. Their non-chicken-rearing neighbours were mostly middle income as well (84 percent). It has been observed that urban livestock keeping is a response of middle-income households to growing urban demand and markets (Guendel & Richards 2003). Chicken production accounted for 38 percent of all household income, with more revenue (62 percent) coming from other business or employment. The sale of broilers brought the most income (40 percent of the chicken-rearing household income), eggs brought 34 percent and the sale of off-layers, local chickens and manure lesser amounts. Income generation was the primary reason for chicken rearing (92 percent), with half the households (52 percent) also stating it was to provide additional food and 38 percent stating that it was also to provide a source of manure. About half the chicken manure was used to fertilize household crops, 14 percent was sold and the rest thrown on waste dumps.

The typical input and marketing structures emerging from our study are shown in Fig. 9.2.

Most households raised exotic layers and broilers as well as local indigenous chickens and some raised other livestock such as cattle (24 percent), pigs (21 percent) and goats (19 percent) in addition to the chickens. Fifteen percent of rearing households said that chickens interacted with other livestock including contact with neighbouring chickens, household and neighbouring cattle, pigs, turkeys and goats. Local free-range chickens were usually kept for household consumption and exotic chickens (usually housed and fed concentrates) for commercial purposes. Almost all broilers and layers (97–100 percent) were housed (Dimoulas et al. 2008).

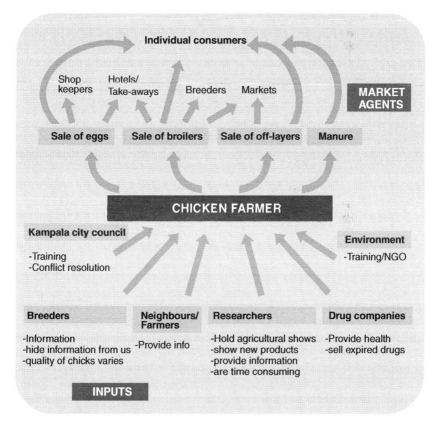

Fig. 9.2 Farm inputs and marketing pathways of chicken producers in Buziga parish, Kampala (based on focus group discussions)

Seventy percent of non-chicken-rearing households said that they benefited from farmers raising chickens in Kampala, allowing for easy access to chicken meat or eggs and manure for their gardens, although 40 percent also complained of conflict with neighbours and smell. Other benefits noted included development of the area, especially improved transport routes, and as one person said, "Friendship is formed because chickens scavenge on my land".

Nearly two-thirds of chicken keepers had bird that died within the 30 days prior to interview and 39 percent of them did not know why they died. The chicken carcasses were either disposed off in the garbage (37 percent), buried (33 percent) or their remains used as feed for other household animals (21 percent). About half of both types of household were aware that chickens could transmit influenza-like diseases directly to humans. Fewer chicken-rearing households (40 percent) and non-chicken-rearing households (28 percent) were aware that humans could acquire diseases from eating chicken meat or egg. Only 8 percent believed that chickens could be a cause of diarrhoea in people and only 4 percent of chicken-rearing households thought that consuming infected chicken meat or eggs could cause diarrhoea. One respondent did say, "When one eats raw eggs or dead birds where the cause of death is unknown, one is bound to get an infection". Another commented that her household does not share living quarters with chickens and they do not eat dead chickens or those being treated for disease, to reduce health risks. Non-chicken-rearing households expressed the following sentiments: "I try as much as possible to avoid those (chicken-rearing) houses" and "I advise farmers on proper garbage disposal".

Within chicken-rearing households, 13 percent of respondents claimed that household members had at one time or another suffered from a disease or illness due to their involvement in raising chickens. When asked if diseases acquired from chickens had any negative impact on household members' health, around a quarter of both chicken-rearing and non-chicken-rearing households thought they did, while around one-third were unsure.

Household members always washed before preparing meals in 84 percent of non-chicken-rearing households and in 96 percent of chicken-rearing households.

In the 2 weeks preceding the survey cooked eggs were eaten more often than chicken, which was eaten by over half the households. Interestingly, a number of respondents said they ate raw eggs for medicinal (67 percent) and nutritional purposes (25 percent). The practice was more common among the neighbours than the chicken keepers themselves (16 compared to 9 percent). Anecdotal information suggests that raw eggs are used in the region as a home treatment given to children with respiratory health conditions (as one respondent remarked) and are also used after excessive alcohol consumption (Nasinyama 1996).

Around half the neighbouring households bought their eggs and chicken from the local market and somewhat fewer bought directly from chicken farmers. Non-chicken-rearing households tended to consume fewer leftover eggs (5 percent) and more leftover chicken (69 percent) than chicken-rearing households. Beef and fish were more popular foods in both types of household, both having been eaten by over 70 percent in the 2 weeks prior to the interview, more often by the non-chicken keepers (90 percent).

Overall we found that enteric illness was associated with consumption of leftovers and raw eggs, as in earlier work in Kampala (Nasinyama et al. 2000), and with the interaction of chickens with other livestock. Consumption of animal-sourced foods, including chicken, was associated with less enteric illness. Zinc, an essential mineral found in beef and chicken meat, has been shown to affect human resistance to infections (Black & Sazawal 2001; Bhan et al. 1996; Bhutta et al. 1999). Eating beef and local chicken in the 2 weeks preceding the survey was consistently found to be protective for enteric illness. It can be hypothesized that local (free range) chickens consume higher levels of zinc from soils. These findings may contribute to curbing general fears of zoonotic disease acquisition relating to chicken production in Kampala but continuing epidemiological studies, combined with appropriate diagnostic assessments, are needed to further assess which production practices can reduce health risks and so be incorporated into guidance provided by those assisting urban livestock producers.

Integrating Findings into Policy and Practice

Here we summarize what has taken place so far in Kampala in the process of evidence gathering and its translation into public health and policy messages. It is important to stress that findings from elsewhere have also played a part.

In 2004 the committee overseeing the research described in this chapter transformed itself into a policy advisory body, the Kampala Urban Food Security, Agriculture and Livestock Coordinating Committee (KUFSALCC), aimed at formulating policy guidelines and public health messages based on research. At the same time, Kampala City Council (KCC), a member of KUFSALCC, led a review of legislation governing UA, for which it received inputs from the researchers. As a result, new bylaws on urban agriculture, livestock, milk, meat and fish production, processing and marketing, were passed by the City Council and gazetted in 2006. The new research and policy body continued to monitor implementation of the bylaws in collaboration with KCC (Hooton et al. 2007) and later projects in the city have also made use of the findings.

In the companion book to this one, published a little earlier (Cole et al. 2008), the findings of the studies summarized in this chapter have been reviewed and translated into policy and public health guidelines. This involved examining the research results in the light of other evidence and policy-relevant materials including the literature on urban governance and international legislation on human rights. Based on an expert panel review, the outcome sets some policy parameters and directions. It is not the last word on the subject, and directions for future research – meaning the questions still to be answered and the methods of finding those answers – are also outlined. Here we present a few of the main points emerging.

Guidelines on Promoting Food Security and Nutrition from UA

Household food security and, thus, child nutrition security, depends largely on wealth, but urban farming moderates this relationship through other household assets including the availability of land and its size, presence of livestock keeping, and household members' education, particularly that of the primary caregiver, who is almost always a woman. Because of international legislation on the Right to Food, it is wrong for national and local governments (or anyone else) to prevent people with insufficient wealth from producing food needed for their survival. This is often the case with urban agriculture, which can be an important strategy in alleviating hunger and poverty and improving the lives of slum-dwellers (three of the Millennium Development Goals). In fact, governments can help people meet these goals by supporting UA, including promoting increased consumption of animal-sourced foods through urban livestock keeping. Greater education of farming households, particularly women, may increase the benefits for household food security.

To increase the benefits of UA, our studies suggest that further research is needed on the complex interactions of gender with other variables for their impact on household food security and child nutrition, as well as on the potential for different UA practices to contribute to the alleviation of major child health challenges, including vitamin A deficiency and anaemia. Better measures of household food security and more extensive research on its relationship to UA are needed.

Guidelines on Reducing Contamination of UA Crops

The Right to Food and the Right to Health have to be balanced by policies addressing contamination of food produced by UA. Frequently both biological and chemical hazards exist and the two types have to be treated distinctly in policy, regulation and management.

Kampala is probably typical in having high levels of *biological* contamination in water and soils used for crop production, meaning improved sanitation is a high policy priority. Meanwhile, the following public health messages need to be shared and actions need to be taken around biological contaminants, especially since farmers are aware of the high value of nutrients for improved crop production using wastewater:

- Direct contact with contaminated water presents a health risk to farmers and to children accompanying them, and protective clothing needs to be worn, especially boots;
- Crops normally processed by heat or drying before human consumption (grains, yams, oilseeds, sugar beet) are recommended for growing in contaminated areas because any biological contamination would be killed by prolonged cooking;
- Growing crops that may be eaten raw, like tomatoes, lettuce, cabbages and onions, should be avoided in sewage-watered areas;
- A common pathway of food contamination with pathogenic microorganisms is the use of contaminated water to "refresh" market produce, and the provision of clean water in markets or where this process occurs may be a critical factor in preventing pathogenic contamination of food;

- Cooking of vegetables that were grown using pathogen-contaminated water should destroy most bacteria and the majority of parasite larvae, making them relatively safe to eat. Alternatively, the public should treat produce with blanching or disinfection with bleach, vinegar or sufficiently high salt concentrations to reduce bacterial loads.

Chemical contaminants present a potential health risk through various pathways associated with UA in Kampala and, again, these are not untypical. A policy priority is to regulate and prevent to the extent possible the discharge of potentially harmful quantities of heavy metals and combustion by-products into air, soil and water. Throwing batteries into pit latrines is a widespread practice that may contaminate the water table and which can be curtailed through public information. Regulation of vehicle emissions and the introduction of unleaded fuel are recommended as part of a long-term solution. Discharge of chemicals into wastewater and solid waste, where their treatment only deals with biological contaminants, must be curtailed, as these are likely to enter the food chain through re-use in agriculture. In Kampala, work on public and policy awareness has already begun, through conferences, workshops and meetings with various stakeholders including industrialists, government officials, policy-makers and farmers, and some provisions were included in the UA Ordinances and Guidelines passed by KCC in 2004.

Leafy vegetables are the most susceptible to absorption of lead (Pb) from the atmosphere and should not be grown within a 30 m distance from the edge of the road, regardless of traffic density. Some indigenous leafy vegetables are, however, safer than *brassicas,* including the popular kale or *sukuma wiki*. Crops recommended for roadside farming include those where the edible part is protected from aerial deposition, such as root crops like sweetpotato and cassava, coarse grains like corn, and legumes like pulses, beans and peas. Those vegetables that bio-accumulate heavy metals in their skins or pods, such as tubers and beans, should be peeled or the pods discarded. All vegetables should be washed thoroughly in clean water before consumption as washing was found to reduce the lead and cadmium content on the surface of many of them. Soapy water should also remove some combustion by-products from vegetable surfaces.

Further research is needed to improve and quantify risk assessments of neurotoxic and other health outcomes among children, and to assess the risks of exposure to mixtures of various organic compounds.

Guidelines for Managing Healthy Urban Livestock

Although a formal risk assessment was not carried out, our study of urban dairying and chicken raising identified current risk-mitigation practices by farm households which can be supported and improved upon. This is particularly important because of the benefits that we found to households consuming animal-sourced foods, as well as the primacy of the Right to Food. Providing infrastructure and legitimizing urban agriculture may be effective strategies for improving practices and decreasing risks. However, poorer farmers and those using less-intensive farming methods will need special support to improve their practices. Adoption of risk mitigation varied from farm to farm, suggesting a role for farmer-to-farmer extension. Livestock farmers did not appear to implement specific disease-mitigation strategies with a clear understanding of disease risk and pathways of transmission. It is suggested that public health messages be broadcast emphasizing the common pathways of disease transmission from animals to humans, and how to mitigate the risks. For example:

- Boil milk (including as tea) to destroy pathogens and protect health;
- Do not use raw eggs as a medication or as infant food because they may carry pathogens;
- Fresh human and animal feces carry microorganisms that can transmit disease. Keep hands, home and livestock sheds clean;
- Avoid cross-contamination of disease in different types of livestock by keeping them from interacting;
- Do not feed your livestock with damp or stale feed or supplements. Avoid feed being sold off cheaply from unknown sources;
- Store food in refrigerators or closed containers to protect from pathogenic organisms;
- Avoid cross-contamination from manure to milk. Avoid handling manure without protective gloves and clothing and keep these clean and away from milk, food and drinking water;
- Wash hands, cows' udders and milking utensils with detergent and warm water before milking;
- Avoid traditionally fermented milk and cheese produced from non-boiled milk;
- Keep sufficient distance between residence and livestock sheds (preferred < 10 m);
- Use protection when handling aborted foetuses (e.g. gloves).

Considering strong local preference for free-range chicken consumption and its potential association with health protection, further systematic research is needed to establish actual risks and benefits to public health.

Conclusion

Because of its history and circumstances, Kampala more than many other cities has developed a particularly strong relationship to UA. It is one of the few urban local governments with a whole department dealing with agriculture, and it is almost certainly the only one anywhere in the world that has developed a typology of urban and peri-urban farming systems to apply to what goes on within its boundaries (see Chapter 6 above). While many of its circumstances – favourable agro-climatic conditions and a surviving culture of farming-based kingdoms – differ from those of other towns and cities, there is no doubt that there are useful lessons that can be drawn from its experience in developing an evidence-based approach to managing urban agriculture.

These lessons are about both process and substance. Regarding substance, Chapter 6 has demonstrated that as Kampala expands with a high rate of urban growth, the proportion of the urban population farming (and, as a result, the absolute numbers) appears to grow also. This dynamic may have as much to do with the relationship between low levels of employment and corresponding poverty levels as it has to do with agro-climatic conditions. We suggest that growing cities in dryland climates – Khartoum is an obvious example – may also have high proportions of their populations engaging in crop and livestock farming. Given the corresponding lack of urban infrastructure, gender inequalities and divisions of labour and other similar conditions, then many of the lessons learned in Kampala may be transferable.

Regarding process, the social learning and institution-building that continues to take place in Kampala does not have to be started from scratch in every other town and city government but could usefully be copied with adaptations, as is already happening in Nakuru, Kenya, as described below in this book. Nakuru has adapted one particular aspect – legislative review – in which it was already engaged. It simply examined and modified some parts of the Kampala model.

As public health practitioners and researchers, and as the Kampala and Toronto co-leaders of the health work from 2002 to 2005, we feel strongly that UA can play an important role as a determinant of health, and that urban farmers and municipal policy-makers can make a difference in improving the health of urban populations. We suggest that some of the lessons we learned in this process may well be appropriate for use and adaptation in other places.

References

Allen, LH 2003, 'Interventions for micronutrient deficiency control in developing countries: past, present and future', *Journal of Nutrition*, vol. 133, pp. 3875S–3878SS.

Amoah, P, Dreschel, P, Abaidoo, RC & Ntow, WJ 2006, 'Pesticide and pathogen contamination of vegetables in Ghana's urban markets', *Archives of Environmental Contamination and Toxicology*, vol. 50, no. 1, pp. 1–6.

Berti, PR, Krasevec, J & FitzGerald, S 2004, 'A review of the effectiveness of agriculture interventions in improving nutrition outcomes', *Public Health Nutrition,* vol. 7, no. 5, pp. 599–609.

Bhan, MK, Bhandari, N, Bhatnagar, S & Bahl, R 1996, 'Epidemiology & management of persistent diarrhoea in children of developing countries', *Indian Journal of Medical Research*, vol. 104, pp. 103–114.

Bhutta, ZA, Black, RE, Brown, KH, Gardner, J, Gore, S, Hidayat, A, Khatun, F, Martorell, R, Ninh, NX, Penny, ME, Rosado, JL, Roy, SK, Ruel, M, Sazawal, S & Shankar, A 1999, 'Prevention of diarrhoea and pneumonia by zinc supplementation in children in developing countries: pooled analysis of randomized controlled trials', *Journal of Pediatrics,* vol. 135, pp. 689–697.

Black, RE & Sazawal, S 2001, 'Zinc and childhood infectious disease morbidity and mortality', *British Journal of Nutrition,* vol. 85, Suppl. 2, pp. S125–S129.

Cole, DC, Bassil, K, Jones-Otazo, H & Diamond, M 2006, 'Health risks and benefits associated with UA: impact assessment, risk mitigation and healthy public policy', in Boischio, A, Clegg, A & Mwagore, D (eds) *Health risks and benefits of urban and peri-urban agriculture and livestock (UA) in Sub-Saharan Africa. urban poverty and environment series report #1*, International Development Research Centre, Ottawa, ON, pp. 11–23.

Cole, DC, Lee-Smith, D & Nasinyama, GW (eds) 2008, *Healthy city harvests: Generating evidence to guide policy on urban agriculture,* CIP/Urban Harvest, and Makerere University Press, Lima, Peru.

Dimoulas, P, Waltner-Toews, D & Nasinyama, GW 2008, 'Household risk factors associated with chicken rearing and food consumption in Kampala', in Cole, DC, Lee-Smith, D & Nasinyama, GW (eds) *Healthy city harvests: generating evidence to guide policy on urban agriculture,* CIP/Urban Harvest, and Makerere University Press, Lima Peru.

European Commission 2002, *Polycyclic aromatic hydrocarbons – occurrence in foods, dietary, exposure and health effects*, Health and Consumer Protection Directorate General, Brussels.

FAO (Food and Agriculture Organization of the United Nations) 2002, *The state of food insecurity in the world 2001*, Food and Agriculture Organization, Rome.

FAO & WHO (World Health Organization) 2001, *Codex Alimentarius Commission procedural manual* 12th ed., Joint FAO/WHO Food Standards Programme, Food and Agriculture Organization of the United Nations, Rome & World Health Organization, Geneva, http://www.fao.org/DOCREP/005/Y2200E/y2200e00.htm [Accessed 30 April 2006].

Guendel, S & Richards, W 2003, 'Peri-urban and urban livestock keeping in East Africa – a coping strategy for the poor?' Paper presented at *Deutscher Tropentag 2003*, Georg-August-University, Göttingen 8–10 October 2003.

Hooton, N, Nasinyama, G, Lee-Smith, D & Romney, D in collaboration with Atukunda, G, Azuba, M, Kaweeza, M, Lubowa, A, Muwanga, J, Njenga, M & Young, J 2007, 'Championing urban farmers in Kampala: influences on local policy change in Uganda, with ILRI/ODI/KUFSALCC/Urban Harvest Working Paper', *ILRI Research Report No. 2*, ILRI, Nairobi.

Jablasone, J, Brovko, LY & Griffiths, MW 2004, 'A research note: the potential for transfer of *Salmonella* from irrigation waters to tomatoes', *Journal of the Science of Food and Agriculture*, vol. 84, no. 3, pp. 287–289.

Jones-Otazo, H 2004, *Screening-level human health risk assessment: development and application of a multimedia urban risk model*, M.Sc. Thesis, University of Toronto, Canada.

Kabeer, N 1991, 'Gender, production and well-being: rethinking the household economy', *Discussion Paper No. 288*, Institute of Development Studies, University of Sussex, Brighton.

Kang'ethe, EK, Grace, D & Randolph, TF 2007, 'Overview on urban and peri-urban agriculture: definition, impact on human health, constraints and policy issues', *East African Medical Journal*, vol. 84, no. 11, pp. S48–S56.

Keraita, B & Drechsel, P 2006, 'The use of polluted water in urban agriculture: "Livelihood realities and challenges"', in van Veenhuizen, R (ed) *Cities farming for the future: urban agriculture for green and productive cities*, RUAF Foundation, IDRC & IIRR, Ottawa, ON, pp. 261–263.

Lee-Smith, D 2006, 'Risk perceptions, communication and mitigation: community participation and gender perspectives', in Boischio, A, Clegg, A & Mwagore D (eds) *Health risks and benefits of urban and peri-urban agriculture and livestock (UA) in Sub-Saharan Africa. Urban poverty and environment series report #1*, International Development Research Centre, Ottawa, ON, pp. 103–104.

Lee-Smith, D 2008, 'Urban Food Production n Kampala: community perceptions of health impacts and how to manage them', in Cole, DC, Lee-Smith, D & Nasinyama, GW (eds) *Healthy city harvests: Generating evidence to guide policy on urban agriculture,* CIP/Urban Harvest, and Makerere University Press, Lima Peru.

Lee-Smith, D & Prain, G 2006, 'Urban agriculture and health', in Hawkes, C & Ruel, MT (eds) *Understanding the links between agriculture and health: 2020 Focus No 13,* Brief 13 of 16, IFPRI, Washington, DC.

Lukwago, M 1999, *A comparative evaluation of the microbial load of two Uganda pasteurised milk brands; Fresh Dairy and Country Taste*, M.Sc. Thesis, Makerere University, Uganda.

Maxwell, DG 1994, 'The household logic of urban farming in Kampala', in Egziabher, AG, Lee-Smith, D, Maxwell, DG, Memon, PA, Mougeot, LJA & Sawio, CJ (eds) *Cities feeding people: an examination of urban agriculture in East Africa*, International Development Research Centre, Ottawa, ON, pp. 45–62.

Maxwell, DG 1996, 'Measuring food insecurity: the frequency and severity of "coping strategies"', *Food Policy*, vol. 21, no. 3, pp. 291–303.

Maxwell, D, Ahiadeke C, Levin C, Armar-Klemesu M, Zakariah S & Lamptey GM 1999, 'Alternative food-security indicators: revisiting the frequency and severity of "coping strategies"', *Food Policy*, vol. 24, no. 4, pp. 411–429.

Maxwell, D, Levin, C & Csete, J 1998, 'Does urban agriculture help prevent malnutrition? Evidence from Kampala', *Food Policy*, vol. 23, no. 5, pp. 411–424.

Muwanga, JM 2001, *Informal food markets, household food provisioning and consumption patterns among the urban poor: a case study of Nakawa Division, Kampala City*, unpublished B.Sc. Thesis, Makerere University, Uganda.

Nabulo, G, Nasinyama, G, Lee-Smith, D & Cole, D 2004, 'Gender analysis of urban agriculture in Kampala, Uganda' in *Urban Agriculture Magazine No. 12 – Gender and Urban Agriculture*, Resource Centre on Urban Agriculture and Forestry (RUAF), http://www.ruaf.org/no12/32_33.pdf [Accessed 19 February 2008].

Nabulo, G, Oryem-Origa, H & Diamond, M 2006, 'Assessment of lead, cadmium, and zinc contamination of roadside soils, surface films, and vegetables in Kampala City, Uganda', *Environmental Research*, vol. 101, no. 1, pp. 42–52.

Nasinyama, GW 1996, *Diarrhoea and Salmonella infections in humans and animals in Kampala district, Uganda*, Ph.D. Dissertation, University of Guelph, Canada.

Nasinyama, GW, McEwen, SA, Wilson, JB, Waltener-Toews, D, Gyles, CL & Opuda-Asibo, J 2000, 'Risk factors for acute diarrhoea among inhabitants of Kampala District, Uganda', *South African Medical Journal*, vol. 90, no. 9, pp. 891–898.

Omore, A, Lore, T, Staal, S, Kutwa, J, Ouma, R, Arimi, S & Kang'ethe, E 2005, 'Addressing the public health and quality concerns towards marketed milk in Kenya', in *SDP Research and Development Report No. 3*, Smallholder Dairy (R&D) Project, International Livestock Research Institution (ILRI), Nairobi.

Sebastian, R 2005, *Associations between household food security, socio-economic characteristics, and urban farming activities in Kampala, Uganda*, M.Sc. Thesis, University of Toronto, Canada.

Serani, S, Nasinyama, GW, Lubowa, A, Makoha, M & Cole, DC 2008, 'Biological hazards associated with vegetables grown on untreated sewage-watered soils in Kampala', in Cole, DC, Lee-Smith, D & Nasinyama, GW (eds) *Healthy city harvests: Generating evidence to guide policy on urban agriculture*, CIP/Urban Harvest, and Makerere University Press, Lima Peru.

Suk, WA, Ruchirawat, KM, Balakrishnan, K, Berger, M, Carpenter, D, Damstra, T, Pronczuk de Garbino, J, Koh, D, Landrigan, PJ, Makalinao, I, Sly, PD, Xu, Y & Zheng, BS 2003, 'Environmental threats to children's health in Southeast Asia and the Western Pacific', *Environmental Health Perspectives*, vol. 111, no. 10, pp. 1340–1347.

Tsai, YW & Ingham, SC 1997, 'Survival of *Escherichia coli* O157:H7 and *Salmonella* spp. in acidic condiments', *Journal of Food Protection*, vol. 60, no. 7, pp. 751–755.

UN-Habitat (United Nations Centre for Human Settlements-HABITAT) 2001, *Cities in a globalizing world: global report on human settlements 2001*, Earthscan, London, pp. 105–127.

Vostal, J 1994, 'Physiologically based assessment of human exposure to urban air pollutants and its significance for public health risk evaluation,' *Environmental Health Perspectives*, vol. 102, Suppl. 4, pp. 101–106.

Yamamoto, SS 2005, *A screening-level risk assessment of exposure to selected PAHs among children in Kampala, Uganda*, M.Sc. Thesis, University of Toronto, Canada.

Yeudall, F, Sabastian, R, Cole, DC, Ibrahim, S, Lubowa, A & Kikafunda, J 2007, 'Food and nutritional security of children of urban farmers in Kampala, Uganda', *Food and Nutrition Bulletin*, vol. 28, Suppl. 2, pp. S237–S246.

Zhuang, RY & Beuchat, LR 1996, 'Effectiveness of trisodium phosphate for killing Salmonella montevideo on tomatoes', *Letters in Applied Microbiology*, vol. 22, no. 2, pp. 97–100.

Zhuang, RY, Beuchat, LR & Angulo, FJ 1995, 'Fate of Salmonella montevideo on and in raw tomatoes as affected by temperature and treatment with chlorine', *Applied and Environmental Microbiology*, vol. 61, no. 6, pp. 2127–2131.

Part III
Kenya

Kenya Overview

This section contains the results of research undertaken in three cities in Kenya – Nairobi, Nakuru and Kisumu – from 2002 to 2005 as part of Urban Harvest's program. There is a strong focus on integrated crop–livestock systems and how these relate to the cycling of nutrients and understanding urban ecosystem functioning. Markets are seen as a part of this and are studied in this context.

The market and nutrient flows of organic wastes that form part of urban agriculture as manure and compost are mapped for Nairobi in Chapter 10, based on a study led by the International Livestock Research Institute (ILRI). Nairobi performs quite inefficiently in this respect, recycling only a tiny proportion of available organic wastes compared to what was revealed for Yaoundé earlier in the book. One of the main suggested reasons for this is lack of market information and the failure of policy to connect with informal sector activities including farming, waste management and composting.

Nakuru, as described in a study presented in Chapter 11, does better than Nairobi but not as well as Yaoundé in terms of the recycling of nutrients through its urban farming systems. This gendered study is closely related to the following one in Chapter 12, where the health aspects of urban dairy production are explored in greater detail. Men's and women's management of livestock and understanding of health risks related to livestock-keeping are examined. The study is also complementary to the one in Chapter 9 from Kampala. It reveals the importance of getting clear messages to urban livestock keepers to increase the safety of this emergent household- based enterprise sector, critical to urban health, nutrition, employment and wealth generation.

Kisumu's agroforestry food and non-food products are mapped by the study, emerging from the World Agroforestry Centre (ICRAF), in Chapter 13. The market chains for these products are examined in relation to how the urban setting works as an ecological as well as social and political system.

Chapter 10
Recycling Nutrients from Organic Wastes in Kenya's Capital City

Mary Njenga, Dannie Romney, Nancy Karanja, Kuria Gathuru, Stephen Kimani, Sammy Carsan, and Will Frost

Introduction

Background to the Study

The question how much of the potential soil nutrients contained in urban wastes are being used and what processes are involved led to this study in the early 2000s. The issue is of central importance to understanding the potential benefits of a properly managed urban agriculture sector, since soil fertility is a major problem in Sub-Saharan Africa and urban wastes represent a large potential source of nutrients (Savala et al. 2003). Mougeot (1993, p.114) highlighted the importance of solid waste management and offered insights into the use of organic wastes by farmers as compost for their crops. When the Consultative Group on International Agricultural Research (CGIAR) was starting up its new system-wide program – Urban Harvest – in Africa in late 2000, stakeholders called for better documentation of these processes. In response, we came together from a number of institutions in Kenya to identify and map out the basic market and material flows for composts and manure in Nairobi and identify opportunities for improving the functioning of the system.

Several of us were also involved in a UN meeting at the end of 2001 on the links between waste management and urban agriculture (Kahindi et al. 2001), and the two CGIAR centres based in Nairobi both had a stake in the issue. The International Livestock Research Institute (ILRI) had done some preliminary work in 15 countries on crop–livestock system intensification in peri-urban areas (Staal 2002), and the World Agroforestry Centre (ICRAF) was interested in market chains involving urban nurseries using compost and manure. Coming as well from a local NGO and a national research organization, we formed an interdisciplinary team. Participatory methods were employed because a basic value underlying our collective approach was that research has a greater impact if the potential users of its results are engaged in the process and have a stake in the outcome.

M. Njenga (✉)
Urban Harvest, c/o CIP, P.O. Box 25171, Nairobi 00603, Kenya
e-mail: m.njenga@cgiar.org

Rapid urbanization in Sub-Saharan Africa without parallel economic and employment growth has led to urban agriculture (UA) being widely practiced as a means of supplementing household income while also contributing to food security and the alleviation of hunger (Lee-Smith et al. 1987). UA is defined as the production, processing and distribution of food and non-food items through the cultivation of plants, tree crops, aquaculture and animal husbandry within urban and peri-urban areas (Mougeot 2000). Crop–livestock integration allows small farmers to intensify production and increase ecological integrity with potential improvements in livelihoods and natural resource management. It has been found that crop–livestock interactions increase with human population densities, interactions starting at 150–160/km^2 and being optimal around 375/km^2 after which they again reduce (McIntire et al. 1992; Staal 2002). In Nairobi the population density in 1999 was 10 times higher, averaging 3079/km^2, 1284 in high- and 17 283/km^2 in low-income divisions (Republic of Kenya 2001).

Most urban areas of Sub-Saharan Africa have increasing problems of pollution and waste disposal due to rapid growth accompanied by widespread poverty, socio-cultural change and weak, resource-poor local governments. While most such governments grapple unsuccessfully with solid and liquid waste management, the urban poor are known to be involved in waste recycling and compost production on a limited scale in many towns, and a study of urban food production in Kenya identified widespread use of organic inputs by urban crop farmers (Lee-Smith & Lamba 1991). We therefore set out to investigate the operations of compost-making groups, while holding parallel stakeholders' meetings with local government and other interested organizations to lay the groundwork for producing research results that could impact the practical side of urban planning and management (Urban Harvest 2002).

The problem of solid waste recycling can no longer be treated lightly, especially in view of its many inherent advantages, which are being tapped increasingly as recycling becomes the norm in many cities of the industrialized world. However, approaches developed in rich countries are not always applicable to poor countries as conditions vary. This study is a contribution to research and development based on prevailing conditions in one African capital. Since it relates to the agricultural use of the product in nutrient cycling, both livestock manure and compost – mainly from vegetable matter in domestic refuse – are looked at. The use of human waste in nutrient cycling is not treated directly except in brief reference to the practice by some urban farmers in Nairobi. The health aspects of urban crop–livestock interactions and of urban agriculture in general, including health risks from wastes, have been examined in parallel studies supported by Urban Harvest, as described in Chapters 9 and 12.

Nairobi Situation and How it Guided the Enquiry

Nairobi covers an area of 700 km^2 in southern Kenya, 500 km from the coast at 1670 m above sea level (Hide et al. 2001). The mean annual temperature is 17°C, with a mean daily maximum of 23°C and minimum of 12°C (Situma 1992). Nairobi receives an average of 900 mm of rainfall which comes in two distinct seasons, from mid-March to the end of May ('Long Rains') and from mid-October to mid-December ('Short Rains') (Kenya Meteorological Department, 2009). The city's population was estimated at three million in 2003, with an annual growth rate of 4.5 percent (Ministry of Planning and National Development 2003). Sixty percent of this population lives in low-income informal settlements, with poverty levels varying from 60 to 78 percent and the numbers of urban poor projected to increase 65 percent by 2015. Unemployment stands at 18, 14 percent for men and 24 percent for women (Ministry of Planning and National Development 2003).

UA in Kenya's capital is practiced in backyard farms, on open spaces under power lines, along roadsides, railway lines and river banks as well as on institutional land. In the mid-1980s, when it had a population of around one million, 20 percent of Nairobi households were growing crops and 17 percent kept livestock within the city limits (Lee-Smith et al. 1987). In the 1990s it was estimated 30 percent (150 000 households) were involved in urban farming (Foeken & Mwangi 2000). In 1998, there were about 24 000 dairy cattle in Nairobi, which produced about 42 million litres of milk, while an estimated 50 000 bags of maize and 15 000 bags of beans were also produced in Nairobi annually (Mukisira 2005).

The city generates approximately 2000 metric tonnes of solid waste daily, only 40 percent of it collected and disposed of at dumpsites (ITDG-EA 2003). Given that 60 percent of people live without services in informal settlements, heaps of garbage are a common sight. About 70 percent of the city's solid waste is biodegradable material which, if recovered, could be used either as livestock feed or for compost making. If this were recovered, it is estimated that about 2223 tonnes of nitrogen (N), 2223 tonnes of phosphorus (P) and 3700 tonnes of potassium (K), worth about US$2 million, would be generated annually from the estimated 635 000 tonnes of waste produced in Nairobi (JICA 1997). These nutrients could potentially be available to both urban and peri-urban agriculture or even sold to rural farmers currently constrained by lack of soil amendments (Njenga et al. 2004). In the wider Sub-Saharan African urban context, millions of tonnes of waste are produced annually, for example 646 780 tonnes in Dar es Salaam (Kiongo & Amend 1999), 313 900 tonnes (domestic) in Kumasi (Cofie 2003), and 765 040 tonnes in Accra (Etuah-Jackson et al. 2001), suggesting high-nutrient cycling potential.

Not all waste collection in Nairobi is done by the City Council. Three large companies and 30–40 community-based organizations (CBOs), the majority comprising unemployed youth, also collect waste, most such CBOs coming from medium- and low-income areas of the city (ITDG-EA 2002). In 2002, an umbrella body for waste-handling groups, Collectors and Recyclers (CORE), was formed with four sections, for plastics, paper, metal, glass and bones and the last one for organic materials (ITDG-EA 2002). The groups engage collectively in advocacy, marketing of recycled products and fund-raising.

Large-scale composting projects have not been well adapted to developing countries. Such composting plants in Ibadan, Nigeria and Accra, Ghana were expensive to maintain, subject to government instability, unable to meet demand for compost, and even caused environmental problems due to large heaps of accumulated waste resulting from numerous breakdowns including power shortages (Agbola 2001; Etuah-Jackson et al. 2001). Despite such failures, nutrients could hypothetically be better exploited if strategies were based on needs identified through local research and development. In particular, involving community groups and small-scale businesses in handling and processing of wastes seems more appropriate, given their current activities and the need for employment, rather than focusing exclusively on capital-intensive public or private enterprise investments. The study therefore focuses on current functioning of markets for manure and compost as well as the opportunities and constraints for integrating community level waste management. Our study benefited from the Kenya Greentowns Partnership Association (KGTPA) networks that link to such groups in the Nairobi slums.

Compost quality is an issue for community groups; scientists from Kenya Agricultural Research Institute (KARI) assessed composting methods and the quality of the product. To produce suitable compost for agriculture, the process needs to be mostly aerobic so that the organic matter is partially mineralized and humidified. Control is needed particularly of the choices of substrate, moisture content and aeration (Lekasi et al. 2003). Waste composition affects the amount and quality of compost that can be produced and the higher the organic fraction the better (Drechsel & Kunze 2001). In terms of health risks, the four main concerns that need to be considered in compost production when reusing urban organic solid waste materials are pathogens, attraction of disease vectors by heaped compost, injuries from non-biodegradable fragments and heavy metal contamination (Lock & De Zeeuw 2001).

The quality of compost as a soil amendment is measured as the ability of the compost to supply plant nutrients N, P and K. Definitions of quality have usually focused on N, the limiting nutrient in most soils in Kenya, with N content sometimes being used as the measure of compost quality (Lekasi et al. 2003). However, nutrient content may not be the most valuable benefit since compost and manure not only recycle nutrients but also improve soil structure by helping water to infiltrate, which in turn reduces soil erosion (Hollings 1995). Maintenance of soil structure provides an aerated yet moisture-retentive environment for optimum root growth while composts and their leachates also inhibit soil-borne pathogens (Hoitink et al. 1997), either by enabling the plant to overcome sub-clinical root damage or more probably by allowing increased microbial activity to take place in the rhizosphere and soil zone around the plants, thereby encouraging microfloral antagonism to the growth of pathogens. Similarly, nematode populations may also be decreased (Gallardo-Lara & Nogales 1987).

In theory there is an unlimited market for good-quality compost if the organic materials were simply recycled back to the rural areas where the bulk of urban food originates. However, the cost of production, transportation and application of composts could exceed the benefits and hence may not be competitive with manures and chemical fertilizers. Farmers' awareness of the benefits of compost and good marketing are needed for a commercial compost initiative to work well (Hoornweg et al. 1999). Marketing strategies require assessment of all existing and potential markets, information about the product, its potential uses and limitations on these, as well as a price estimate. Marketability of the finished compost is affected by local soil fertility, government policies restricting imports of chemical fertilizers or subsidizing them, availability and cost of other soil conditioners including manure as well as crop residues, transport costs, local agricultural practices, and finally the reliability, quality and quantity of compost production itself (Hoornweg et al. 1999).

The need to understand and document all these led us to structure the study in three parts:

- The dynamics of CBOs working on urban organic waste recovery and UA;
- A characterization of the quality of compost produced by these groups;
- A characterization of compost and manure marketing chains in Nairobi.

Tracking the flows of compost and manure also led us outside the city in order to understand some of the constraints to either market (compost or manure) functioning properly. A major constraint of these activities of low-income groups is undoubtedly their informality, meaning illegality, which produces many of the other effects such as poor pricing information and low-quality production.

Land access is in turn a major issue affecting performance of informal sector enterprises. With farming in urban areas generally considered illegal, there are no rules regulating farmers' access to public land, so as land values increase farmers are displaced to the urban periphery. To protect the interest of the low-income urban farmers who rely on informal access to land for subsistence farming, local governments would need to undertake localized land-use planning and guarantee adequate compensation for the loss of access to land (Poulimenos et al. 1996). Similarly, activities that interact with urban farming systems, including nutrient cycling involving compost making and manure handling, are equally affected by the absence of an enabling policy framework. Thanks to the stakeholder-based process, members of the study team were able to bring these matters to the attention of a land-policy review process in Kenya.

Methods Used

The study was designed and managed by an interdisciplinary team from the following institutions:

- Urban Harvest, a system-wide initiative of the Consultative Group on International Agricultural Research (CGIAR);
- International Livestock Research Institute (ILRI);
- World Agroforestry Centre (ICRAF);
- Kenya Agricultural Research Institute (KARI);
- Kenya Green Towns Partnership Association (KGTPA), a non-governmental organization (NGO).

The project was technically backstopped and managed by ILRI and Urban Harvest, while KGTPA led the study of group dynamics in urban organic waste recovery and urban agriculture, KARI led the study of compost quality characterization and ICRAF led the study to characterize compost and manure marketing chains.

CBOs involved in organic waste recovery and urban agriculture were identified through KGTPA contacts and secondary sources (Ishani et al. 2002; ITDG-EA 2003) as well as an iterative search based on high population density or known presence of crop and livestock farming and composting activities. Most high population density areas are informal settlements without city council services, meaning CBOs doing waste management were more likely to be found there. A further study site outside the city was included, based on the resource flow mapping that indicated manure was mostly sourced from a dry-land area in Kajiado, where Maasai cattle herders actively market manure. Fieldwork from March to December 2003 was carried out at the resulting study sites shown on the map in Fig. 10.1.

Primary data were collected using qualitative and participatory approaches. Focus group discussions (FGDs) were held with an initial set of CBOs to investigate group dynamics, compost production and marketing. This led to the identification

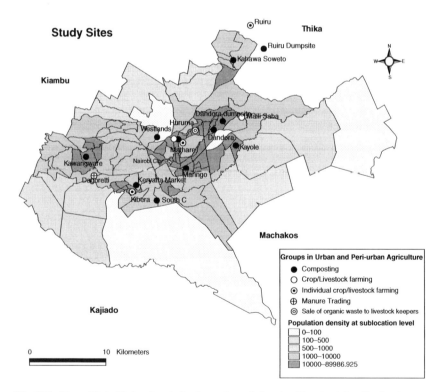

Fig. 10.1 Map of Nairobi showing study sites and population densities at sub-location level

of further CBOs and a total of 14 were included in the study. Rural, peri-urban and intra-urban resource flow patterns for organic waste and livestock manure were mapped through site visits and interviews with six manure traders and two landscapers. Two informal settlements, *Kahawa Soweto* and *Maili Saba*, were selected for in-depth study on the basis of their involvement in household waste management, crop production and livestock keeping. This began with FGDs with farmers and non-farmers as well as the CBOs identified. Participatory socio-economic and resource management mapping was done in both places, followed by questionnaire design to collect household information from 22 livestock manure producers found operating close to compost-making sites.

Although the aim was to find all CBOs involved in compost production in Nairobi, only 11 were found. While we cannot be sure this covered all such groups, we consider the findings give a good characterization and estimate of the total activity in the sector that is managed by CBOs. However, compost production by individual farmers, schools and other institutions has not been captured. A random sample survey of urban farmers in Nairobi in the 1980s found 35 percent of crop growers used compost and 29 percent used manure while 91 and 44 percent respectively were producing these inputs on their own farms. There was virtually no market in these products, only 4 and 2 percent respectively obtaining compost and manure from shops or markets, although 10 percent of farmers obtained manure from friends (Lee-Smith et al. 1987).

If the same proportions held for 2003, it can be roughly estimated that 54 500 households in the city were using compost and that 41 400 of these produced it themselves. Similarly, 37 700 households would have been using livestock manure to fertilize crops in the city, and 16 600 of these were producing it on their own mixed crop–livestock farms. Of these, about 2000 households would have been purchasing compost, and about 750 purchasing manure. Likewise, it is known that schools and other institutions in the city (prisons, orphanages and so on) engage in farming, including compost production. Since the quantities of compost and manure handled by these urban farming entities are not known, it is not possible to make a reasonable estimate of the total quantities involved. Nonetheless it is clear that the scale of this activity is much larger than that undertaken by CBOs, although the amount marketed is very little.

Thirteen compost samples and nine zero-grazing unit cattle manure samples were collected from CBOs for quality characterization through estimation of nutrients and heavy metal content. The compost and manure samples were analysed using Kjeldahl digestion procedure followed by Tecator steam distillation (Walkley & Black 1934) while organic carbon, total P, exchangeable calcium (Ca), magnesium (Mg) and K were analysed according to prescribed procedures (Anderson & Ingram 1993). Analysis for heavy metals such as lead (Pb) and zinc (Zn) was also done according to the same procedures (Anderson & Ingram 1993), all laboratory analysis being carried out at KARI.

Finally, a stakeholder workshop was held at the end of the project to share results and decide how to act on them. The workshop brought together government departments, Nairobi City Council, the National Environmental Management Authority

(NEMA), national and international research institutions, NGOs and representatives of the CBOs interviewed. Their recommendations provide the conclusions at the end of this chapter.

Discussion of the Results

Groups Involved in Organic Waste Recovery and Urban Agriculture

The 14 groups studied had diverse objectives. Eleven produced compost using organic waste, one practiced mixed farming, one kept livestock and another collected household waste (which was sorted and the organics sold to livestock farmers). All groups aimed at generating income and 11 also tried to clean up their neighbourhoods (Table 10.1). Others were involved in raising public health awareness and rehabilitating street children (Table 10.1). Seven groups were located in

Table 10.1 Objectives and gender composition of organic waste recovery and recycling groups in Nairobi

Name of group	Income generation	Environmental management	Public health	Street children rehabilitation	Gender ratio M:F	Group type
Tuff Gong Garbage Recycling Group	✓	✓	✓	✓	8:3	2[a]
Kuku Women Group	✓				2:8	4[a]
Kayole Environmental Management Association	✓	✓			5:5	4[a]
Mukuru Recycling Centre	✓			✓	0:12	3[a]
Mathare Borea Composting Group	✓	✓			4:4	4[a]
Garbage Recycling Programme (Save The Children)	✓	✓		✓	4:0	1[a]
City Garbage Recyclers S.H.G	✓	✓			9:6	2[a]
Youth United Against Environmental Pollution	✓	✓	✓		20:8	2[a]
Soweto Youth In Action	✓	✓	✓		16:5	2[a]
City Park Environmental Group	✓	✓			4:8	4[a]
Kawangware Afya Bora S.H.G	✓	✓	✓		5:15	4[a]
Ngei 1 Youth Development Group	✓	✓		✓	55:9	2[b]
Youth Foundation	✓	✓	✓	✓	12:0	1[b]
Siranga Ya Ngombe S.H.G.	✓				11:4	4[b]

1 Male youth group 2 Youth group (mixed) 3 Women's group 4 Mixed group (age and gender)
[a]composting groups
[b]non-composting groups

informal settlements, two in low-income residential estates, three in crop produce markets and two at dumpsites (Fig. 10.1). The informal settlement groups in particular engaged in organic waste recovery and recycling of other items as well as composting, thus providing services to households and the community in general as well as generating income from products other than compost.

Ten CBOs provided financial or material assistance to their rural families but four, consisting of former street people, had no rural ties. The CBOs with rural–urban linkages follow in the tradition of urban adaptation of rural self-help organization and of maintenance of ethnic and family ties through flows of goods and resources including money and food, as documented in various parts of Africa (Bryceson 2000; Maxwell et al. 2000). However, whereas women predominate in most rural associations, young men predominated in our study, as shown in Table 10.1.

Formed between 1978 and 2001 as legally registered self-help groups and providing employment to 151 poor urban people, most of whom have families with small children, most of the composting groups had declining membership, overall numbers being 55 percent less than on formation. Although one of the composting groups was growing, instability in most came as a result of frustration at the low-income generated, poor leadership, poor participation in composting activities and lack of space for doing it. Internal conflicts noted in six groups were attributed to disputes over role allocation and financial management.

These disputes were rooted in gender differences. While most of the CBOs had a mix of men and women, the general pattern was of high numbers of male youths and older women, with a gender role differentiation in waste collection, sorting and turning and financial management. Women engaged in manual work while men dominated decision making and financial management, making the women quite unhappy and in some cases resulting in conflicts between gender and age categories. These gender inequalities resulted from attitudes and cultural beliefs, as well as low levels of education, especially among the women, leading some more educated male youth to treat them dismissively. Role sharing was however observed to be less of a problem in youth-only groups.

Group Compost Operations

The groups sourced and used different organic materials and applied different composting methods. The four techniques found were pit, windrow, co-composting and dumpsite mining. In pit composting, typical of rural practices (Lee-Smith 1993), heaps of materials were contained below the soil surface. In the windrow method, materials were piled up in elongated layers on the ground, turned weekly and watered when dry (Karanja et al. 2005). Co-composting was like windrow but with the addition of livestock manure. Both windrow and co-composting heaps were covered with banana leaves, maize stalks or perforated plastic sheets, either

Plate 10.1 Women members of the City Park Environmental Group preparing compost

in the open air or under shade, except for one group using the windrow method (Plate 10.1).

These methods took 7–9 weeks to produce mature compost whereas dumpsite mining simply involved digging out already decomposed materials. Windrow and co-composting were most frequent (five and four groups respectively) while pit composting and dumpsite mining were only applied by one group each.

The CBOs sourced organic materials from household, market and food waste from canteens and hotels as well as agro-industrial waste (coffee husks) and animal manure. They generally used haphazard mixtures of these to produce their compost, with two groups using only one type of raw material, one from a market and the other from a dumpsite (which however also contained a mix). Although CBOs responded willingly to questions about composting, they did not keep good records of their compost activities such as watering and turning, or of the temperatures and time taken to mature.

Six groups transported waste 0.2–9.0 km to their composting sites using wheelbarrows, handcarts and vehicles. Only two groups used vehicles, one hiring at about US$14 per round trip while the other had a pickup acquired as a capital asset from a United Nations (UN) project aimed at promoting recycling waste from the headquarters of the two agencies located in Nairobi. The three groups located in markets moved the materials up to 1 km using wheelbarrows or handcarts. Only two groups did not transport waste as one mined decayed materials at a dumpsite while the other worked at a dumpsite where sorting and composting took place. Transport was said to be the limiting factor on the amount of materials groups were able to handle.

Only 2500 tonnes, equivalent to 0.6 percent of the total organic waste produced in the city annually, was used for compost making by the CBOs in our sample. Although very little of the city's waste is processed this way and even less is well utilized, compost making does contribute to neighbourhood cleanliness in the informal settlements where local authority services are lacking.

For example, 21 km East of Nairobi, Soweto Youth in Action collected household waste at a weekly fee of US$0.3 per bag and had about 100 customers in 2003. They used wheelbarrows to take waste to their site for sorting and after composting the organics the rest was either burned or collected for dumping through an arrangement negotiated with Nairobi City Council. The village was cleaner than other sites visited. The compost was sold on site to various buyers (Fig. 10.2), who transported it using hired vehicles for distances greater than 20 km and bicycles, wheelbarrows or carts for shorter distances. The group was planning to start collecting manure from livestock keepers and use it for co-composting for crop production at a plot they planned to hire outside the village.

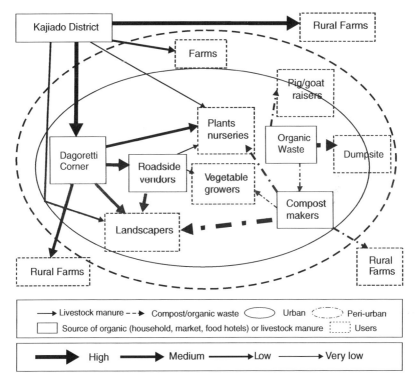

Fig. 10.2 Rural-urban manure and compost flows in and around Nairobi. Source: adapted from Njenga et al. (2004)

With less-dramatic results for the neighbourhood appearance, but no doubt of great benefit to nearby high-rise dwellers, Ngei 1 Youth Development Group collected household refuse at Huruma for Ksh 30 per household per month, then sorted and sold the organics to livestock keepers, who came to collect it on bicycles from Mathare North, 5 km away. Other items were recycled for sale.

Compost Quality

We measured the differences between urban waste compost and zero-grazed cattle manure (See Table 10.2). The nutrient content in terms of N, P and K were lower for compost than for cattle manure although there was no significant difference except in the case of carbon ($t = 1.93$; $p \leq 0.05$). The Carbon:Nitrogen (C:N) ratios for compost and manure were both well below the optimum. The N, P and K levels for compost were lower than optimum as were some for manure, while both had zinc concentrations above recommended levels (Lekasi et al. 2003; World Bank 1997).

Table 10.2 Chemical levels in compost and cattle manure samples in urban and peri-urban Nairobi

Nutrient content	Organic waste compost Mean (+sd)	Zero-grazing cattle manure Mean (+sd)	t-test	Acceptable levels
Nitrogen N (%)	1.19 + 0.31	1.70 + 0.96	−1.53	1.7[a]
Carbon C (%)	9.90 + 2.49	16.43 + 9.91	1.93[b]	
C:N ratio (%)	8.91 + 3.71	9.82 + 1.76	0.72	25[c]
Phosphorous P (%)	0.45 + 0.16	0.60 + 0.25	−1.71	1.55[a]
Potassium K (%)	1.85 + 1.08	2.05 + 0.95	−0.43	2.07[a]
Calcium Ca (%)	1.22 + 0.90	0.72 + 0.59	1.44	
Magnesium Mg (%)	0.56 + 0.18	0.68 + 0.23	−1.26	
Zinc Zn (%)	0.23 + 0.57	0.04 + 0.02	1.06	0.03[a]
Lead Pb	Trace	Trace		0.015[a]

[a]World Bank (1997)
[b]Significantly different at 0.05 probability level
[c]Lekasi et al. (2003)

The C:N ratio in compost is influenced by the proportion of green (high N) and dry (high C) materials and how they are arranged during layering (Karanja et al. 2005) and turning (Lekasi et al. 2003). Inclusion of other types of organic materials can also affect compost quality, as when household wastes are recycled to produce organic compost, for example. Further detailed studies, as well as capacity building, are needed both to document and to improve how materials are used and how practices such as watering, turning, temperature control, covering and time taken to mature are executed by CBOs.

These factors influence quality because they affect the biological decomposition process through presence of microorganisms and invertebrates, oxygen supply and aeration, pH, nitrogen conservation and moisture content. Low P values in compost could probably be addressed through fortification with suitable sources of P, such as phosphate rocks or other waste materials high in phosphorus. Co-composting organic waste with livestock manure would enhance quality by reducing N losses, a win-win opportunity that should be exploited as it would also reduce the problem of manure disposal especially in informal settlements where space is limited. Use of earthworms in making vermi-compost has been identified as a promising method for improving K value in organic waste compost production (Savala 2003). These and other options could usefully be investigated through consultation with the groups.

Zinc (Zn) contamination from items such as wires, screws, nuts, bolts and some paints in unsorted garbage needs attention as it may be inhaled or consumed through crops and poisoning can result in anaemia, reproduction and foetal growth problems and gastrointestinal upsets if daily intake exceeds 100–250 mg, which is 10–15 times higher than recommended (Opresko 1992). We found a case of compost contaminated with zinc (0.11 mg/kg) during a related study in Nakuru (see Chapter 11), while the compost samples from this study had higher than recommended Zn levels (Table 10.2). Such heavy metal contamination of compost could be mitigated through source sorting of waste involving communities, and this would reduce not only risks but also workload and enhance recovery of other types of waste such as plastic, metal, glass and bones.

Compost and Livestock Manure Marketing Chains and Flows

Most compost production took place close to a raw materials source where there was also a demand for cleaner neighbourhoods, but with transport problems limiting movement of organic waste. And while raw materials were free, labour costs pushed up compost price.

The main urban customers for the CBOs' compost were plant nursery operators, ornamental gardens, landscapers or real estate developers and urban farmers. Rural customers were small-scale farmers and large horticulture farms, mainly in the Rift Valley Province (Fig. 10.2). Amounts sold varied greatly, with groups such as the City Park Environmental Group that sells along main roads having higher demand than supply and selling up to 4 tonnes per month. Others, like the City Garbage Recycling Programme in Maringo, 6 km from the city centre (Fig. 10.1) had established strong networks with organic farmers and were able to sell 2.5–5.0 tonnes per month. Similarly, Kayole Environmental Management Association, located about 10 km from the city centre, was also able to sell 4 tonnes per month for landscaping to its regular customer, Jomo Kenyatta University of Agriculture and Technology. By contrast, groups located far from the city centre, such as Soweto Youth in Action, 21 km away, and those inside inaccessible informal settlement sites where security is an issue, such as Mathare II, only managed to sell meagre amounts.

Compost was packed in recycled bags and sold at the production sites in varied amounts based on customers' need. Only 40 percent of the total compost produced (253 tonnes per year) was traded by the groups at US$67–133 per tonne. Groups said sales were low due to customers' limited information about where to buy, low compost quality diminishing its value, customers' fears about possible health risks, inaccessibility of composting sites in the informal settlements, and insecurity of land tenure.

To this might be added the high price compared to manure, which sells at different outlets in the city at US$14–24 per tonne, two to ten times lower than that of compost. Increased market information would negatively affect compost sales since its price compared to that of manure and chemical fertilizers determines whether farmers buy the compost or other products (Nugent 1997). Another factor affecting demand for both compost and manure was wastewater farming. For example, very little compost or manure produced in *Maili Saba* was used by farmers there, as over 80 percent of them used wastewater from a nearby sewage treatment facility as a fertilizer. They, like farmers elsewhere in Nairobi, have discovered wastewater is a rich source of plant nutrients (Hide et al. 2001).

Large clients like landscapers, plant nurseries and rural horticulture enterprises, used their bargaining position to buy compost in bulk at a reduced price. High demand for compost from the large-scale horticulture farms is attributed to their need to meet international organic-farming requirements such as those in European markets as well as their awareness of compost's role in soil health.

Most animal manure used and traded in Nairobi came from the pastoral areas of Kajiado district, 60–100 km from the city. Manure bought at US$ 5–6 per tonne truck load was transported to Dagoretti Corner's manure trading centre 15 km from the city centre where it was sold at US$ 14–24 per tonne. Although customers included plant nursery operators in the city, ornamental gardens, landscapers or estate developers and farmers from urban and peri-urban areas of Nairobi, the bulk of this manure was purchased by rural farmers in the high potential areas up to 150 km away from Nairobi, mainly horticulture, coffee and tea enterprises producing largely for export. Thus Nairobi is a point of exchange in the movement of nutrients from the drier ecosystem to the moister one.

Vehicles carried the manure to rural areas while in the city transport was mainly by bicycle, wheelbarrow or handcart. Middlemen located in various parts of the city along the manure market chain either sourced directly from Kajiado or bought at Dagoretti Corner and sold to customers they had identified. Some manure users, mainly owners of plant nurseries, circumvented them and bought directly from Kajiado (Fig. 10.2).

Kajiado manure comes from cattle kept in traditional Maasai *bomas* (compounds) at night and herded in the surroundings during the day. Being completely decomposed and dry, it is considered to be of good quality. Urban manure, coming mainly from cattle kept in zero-grazing systems, was found to be not linked to the main marketing chain. At Kahawa Soweto, about 60 percent of the manure is dumped in rivers and open spaces or burned. One group, Siranga Ya Ng'ombe at Maili Saba, Dandora, produced *boma* manure from their freely grazed cattle but were unable to dispose of it through established markets like Dagoretti. Retailers there said they were unaware of the existence of this source. When so informed by the research team they said they were afraid to collect the manure due to insecurity at Dandora.

Thus, while rural manure production supported some urban livelihoods, including farming, urban manure production was de-linked from the market chain. This gap is partly due to lack of information, but product quality and urban insecurity also play a part. Livestock in urban centres are mostly kept in confinement meaning the manure produced is usually wet and bulky. The failure to collect urban livestock waste needs to be addressed in order to protect public health and maintain urban cleanliness.

Conclusions

What we found in this study is that nutrient flows in and around the city are sporadically managed by a few large and small-scale actors in an uncoordinated way, with small-scale actors operating mostly outside the market. CBOs are not currently the main actors involved in managing nutrient materials flow in Nairobi due to their small numbers compared to the volume of urban households engaged in mixed crop–livestock farming systems that recycle nutrients on farm or through non-market exchanges. Further, the materials and market flows of compost and manure are entirely disconnected from each other and both are characterized by lack of market information and ad hoc arrangements between producers and consumers. This applied to both large consumers wanting to fertilize commercial farms and landscapes and small farmers needing items like livestock fodder or trying to get rid of livestock wastes. Also a factor is the lack of security and safety in informal settlements, as highlighted by farmers at the final stakeholder workshop where the research results were discussed.

While urban farming is extensive, the poor are proportionally under-represented mainly due to their lack of space (Foeken 2006; see also Chapter 11). Since the bulk of nutrient cycling currently seems to take place on backyard urban crop–livestock farms which belong to the less-poor segment of the population, some improvements could be addressed through promotion and support of on-farm urban waste management, as is done in Kampala, Uganda, as described in Chapter 6, above.

However, over 60 percent of Nairobi's population live in informal settlements which cover only about 5 percent of the total land area (Ministry of Planning and National Development 2003). This means that wastes produced on a large scale have little chance to be recycled through local crop–livestock farms and the problem of waste management remains enormous in such places. Under these circumstances CBOs represent an important opportunity to address waste management and generate employment in the informal sector, through processing and sale of composts to customers outside the settlements. Of course, the human rights injustice of poor households living in very restricted space with little or no chance for producing food remains a major concern in Nairobi.

For community groups involved in waste management and producing commercial composts, lack of recognition of their activities and lack of space are major issues. Unless local authorities are prepared to work with these groups, their informal status, low level of knowledge and small resources will limit their contribution to city waste management, nutrient recycling, or improvement in their own incomes and well-being. Their main contribution at present is cleaning up their neighbourhoods where they sort garbage. Recognition and support from Nairobi City Council (NCC) might encourage existing groups to continue and other groups to emerge. The challenge of inaccessibility could be resolved if NCC allocated spaces to the groups next to markets and main roads.

For composts, the issue of product quality also has to be addressed. This study shows that Nairobi's urban and peri-urban composts are of lower quality than cattle manure and way below optimum levels, especially for N and P (which are important for Kenyan soils). The compost samples also had a low C:N ratio. Quality must be higher to generate demand for urban compost in competition with chemical fertilizers and manure. Further studies are needed to understand the effects on compost quality of using different types of waste such as vegetables, fruit, banana leaves, eggshells and ash. These and other possible measures to address compost nutrient content and heavy metal contamination have been proposed here. However, the main issue is whether there is political will to support informal sector enterprises. The fact that small and large rural farms are among the main customers for urban compost suggests good market potential provided other issues are resolved.

The CBOs themselves need capacity building in organic resource recovery and on how to establish sustainable governance structures, including gendered project management committees and constitutions. In this direction, several of the partners in this project have already jointly developed and run training courses on sustainable and healthy compost production for community groups and produced a manual on low-cost composting addressing compost quality and marketing challenges they face (Karanja et al. 2005).

In the case of manure, the organized market links rural herders with urban and rural crop production enterprises via a market in Nairobi. However, local production of manure by urban livestock keepers is disconnected from the organized marketing system and the product is mostly dumped or burnt. This seems to be due to lack of market information, the difficulties in accessing the manure market because of informal settlement conditions and traders' ideas about the quality of local manure. There seems to be a good potential for using urban livestock manure for co-composting with organic vegetable waste because livestock-keeping enterprises were identified less than 2 km from all the composting sites. This method would enhance compost quality without necessarily increasing production costs since manure currently dumped could be acquired with very little transport cost. Again, the main missing elements are information and institutional support. Political will from the authorities entailing allocation of space and technical advice would make composting worthwhile for groups, enhancing their livelihoods as well as contributing to environmental clean-up.

Urban farmers themselves attributed the low use of compost and livestock manure in crop farming not only to lack of secure land tenure but also to the availability of nutrients in wastewater, another form of nutrient re-use studied elsewhere (Hide et al. 2001; Karanja et al. 2008) and observed in all informal settlements visited. While addressing the health issues surrounding this practice is beyond the scope of this article, it is touched on in Chapter 9, above, and in Cole et al. (2008). Briefly, safe practices for re-use of wastewater in urban food production do need to be developed and promoted (IWMI 2002).

Associations, networks and forums such as CORE and the Nairobi and Environs Food Security, Agriculture and Livestock Forum (NEFSALF), which include public sector participation, are already doing advocacy, support and training for CBOs involved in nutrient cycling in Nairobi. Political will from NCC and government to work with the informal sector and low-income groups in general is essential however. Absence of an enabling legal framework was found to limit compost demand in Nairobi, composting groups having no formally acquired space to operate and farmers showing reluctance to invest in fertilizing soils on land that did not belong to them, meaning they focus on short-term profitability since the risk of investment loss is high.

A task force to develop an urban agriculture policy in Kenya was formed in 2007, led by the Ministry of Agriculture's Provincial Agriculture Board and KARI. The Ministry of Agriculture convened the group to begin drafting in early 2009. Nutrient recycling is expected to be part of the policy initiative. Public awareness campaigns, media articles and official statements could then be developed, promoting compost production as a strategy for cleaning up the city, which would improve the demand for compost and its business potential. An urban agriculture policy also needs to be localized through bylaws for enhanced regulation and control at town level, as is already underway in Nakuru as described in Chapter 11. Some of the considerations for legalizing urban agriculture and organic waste management have also been incorporated in a draft Land Bill for Kenya (Ministry of Lands & Settlement 2006).

In addition to these policy steps, further research is required, especially through pilot community projects, to explore, evaluate and demonstrate strategies such as source sorting of organic materials and co-composting as options for quality enhancement. These could be combined with communities learning about heath risk management. An economic study on cost–benefit analysis, willingness to pay, perceptions and attitudes, market threats and opportunities of community-produced compost would help groups develop business plans and arrive at unit prices to compete effectively in the market. Finally, there is need for more detailed quantification of the amounts and quality of organic resources involved in rural–urban flows and their role in natural resource management in both urban and rural ecosystems, based on this preliminary mapping.

These measures would enhance urban organic waste recovery hence reducing nutrient mining in food production areas while also reducing environmental pollution in consumption areas, helping to close the nutrient cycle.

References

Agbola, T 2001, 'Turning municipal waste into compost: the case of Ibadan', in Drechsel, P & Kunze, D (eds) *Waste composting for urban and peri-urban agriculture: closing the rural-urban nutrient cycle in Sub-Saharan Africa,* CABI Publishing, IWMI & FAO, New York, NY, pp. 69–83.

Anderson, JM & Ingram, JSI 1993, *Tropical soil biology and fertility: a handbook of methods,* CAB International, Wallingford.

Bryceson, D 2000, *Rural Africa at the crossroads: livelihood practices and policies,* ODI, London.

Cofie, O 2003, *Co-composting of faecal sludge and solid waste for urban and peri-urban agriculture in Kumasi, Ghana,* IWMI, SANDEC, KNUST & KMA, Kumasi.

Cole, DC, Lee-Smith, D & Nasinyama, GW (eds) 2008, *Healthy city harvests: Generating evidence to guide policy on urban agriculture,* CIP/Urban Harvest and Makerere University Press, Lima, Peru.

Drechsel, P & Kunze, D (eds) 2001, *Waste composting for urban and peri-urban agriculture: closing the rural-urban nutrient cycle in Sub-Saharan Africa,* CABI Publishing, IWMI & FAO, New York, NY.

Etuah-Jackson, I, Klaassen, WP & Awuye, JA 2001, 'Turning municipal waste into compost: the case of Accra, Ibadan', in Drechsel, P & Kunze, D (eds) *Waste composting for urban and peri-urban agriculture: closing the rural-urban nutrient cycle in Sub-Saharan Africa,* CABI Publishing, IWMI & FAO, New York, NY, pp. 84–95.

Foeken, D 2006, *"To subsidise my income": urban farming in an East-African Town,* Brill Academic, Leiden.

Foeken D & Mwangi A 2000, 'Increasing food security through urban farming in Nairobi', in Bakker, N, Dubbeling M, Gündel S, Sabel-Koschella, U & de Zeeuw H, *Growing cities, growing food: urban agriculture on the policy agenda: a reader on urban agriculture,* DSE, Feldafing, p. 303.

Gallardo-Lara, F & Nogales, R 1987, 'Effect of the application of town refuse compost on the soil-plant system: a review', *Biological Wastes,* vol. 19, pp. 35–62.

Hide, JM, Kimani, J & Thuo, JK 2001, *Informal irrigation in the peri-urban zone of Nairobi, Kenya: an analysis of farmer activity and productivity,* Report OD/TN 104, HR Wallingford & DFID, Wallingford.

Hoitink, HAJ, Stone, AG & Han, DY 1997. Suppression of plant diseases by composts. HortScience. 32:184–187.

Hollings, CS 1995, 'Sustainability: the cross-scale dimension', in Munasinghe, M & Shearer, W (eds) *Defining and measuring sustainability: the biological foundations,* The International Bank for Reconstruction and Development/World Bank, Washington, DC.

Hoornweg, D, Thomas, L & Otten, L 1999, *Composting and its applicability in developing countries,* World Bank Working Paper Series 8, World Bank, Washington, DC.

Ishani, Z, Gathuru, PK & Davinder, L 2002, *Scoping study of urban and peri-urban poor livestock keepers in Nairobi,* Mazingira Institute, Nairobi.

ITDG-EA (Intermediate Technology Development Group-East Africa) 2002, *Sustainable waste management strategy for Nairobi,* Workshop Proceedings Report UN Complex, Gigiri, 24 September 2002.

ITDG-EA (Intermediate Technology Development Group-East Africa) 2003 *Nairobi solid waste management network* http://www.wastenet.or.ke/ [Accessed 2 Dec 2004].

IWMI (International Water Management Institute) 2002, *The Hyderabad declaration on wastewater use in agriculture,* IWMI, Colombo.

JICA (Japan International Cooperation Agency) 1997, Economic Infrastructure. *Master plan study of Nairobi.*

Kahindi, JHP, Karanja, NK, Alabaster, G & Nandwa, S (eds) 2001, *Proceedings of the workshop on enhancement of productivity and sustainability of urban/peri-urban agriculture (UA/PUA) through efficient management of urban waste,* UN-Habitat, Nairobi.

Karanja, NK, Kwach, O & Njenga, M 2005, *Low cost composting manual: techniques based on the UN-Habitat/Urban Harvest-CIP community based waste management initiatives*, UN-HABITAT, Nairobi.

Karanja, NK, Njenga, M, Prain, G, Munyao, P, Kang'ethe, E, Kironchi, G, Kabiru, C, Gathuru, K, Muneri, C, Kinyari, P, Githuku, C, Kaluli, W, Gathenya, J, Home, P, Githigia, S & Bebora, L 2008, *Assessment of benefits and risks in wastewater reuse for agriculture in urban and peri-urban areas of Nairobi*, Project Report, supported by IDRC.

Kenya Meteorological Department (KMD), 2009. Climatological Data. (http://www.meteo.go.ke/customer/climat/rain.html).

Kiongo, S & Amend, J 1999, 'Linking (peri) urban agriculture and Organic waste management in Dar es Salaam', paper presented at *International IBSRAM-FAO workshop on urban and peri-urban agriculture*, Accra, 2–6 August 1999.

Lee-Smith, D 1993, *Peasants, plantation dwellers and the urban poor: a study of women and shelter in Kenya*, Thesis 4, Lund University, Sweden.

Lee-Smith, D & Lamba, D 1991, 'The potential of urban farming in Africa', *Ecodecision*, vol. 3, pp. 37–40, Montreal, November.

Lee-Smith, D, Manundu, M, Lamba, D & Gathuru, PK 1987, *Urban food production and the cooking fuel situation in urban Kenya – national report: results of a 1985 national survey*, Mazingira Institute, Nairobi.

Lekasi, JK, Ndungu, KW & Kifuku, MN 2003, 'A scientific perspective on composting', in Savala, CEN, Omare, MN & Woomer PL (eds) *Organic resource management in Kenya: perspectives and guidelines*, FORMAT, Nairobi, pp. 65.

Lock, K & De Zeeuw, H 2001, 'Health and environmental risks associated with urban agriculture' in *Annotated bibliography on urban agriculture*, Sida & ETC-RUAF, CD-ROM, RUAF, Leusden.

Maxwell, D, Levin, C, Armar-Klemesu, M, Ruel, M, Morris, S & Ahiadeke, C 2000, *Urban livelihoods and food and nutrition security in Greater Accra, Ghana*, International Food Policy Research Institute (IFPRI) Research Report 112, IFPRI, Washington, DC.

McIntire, J, Bourzat, D & Pingali, P 1992, 'Crop-livestock interactions in Sub-Saharan Africa', in Baltenweck, I, Staal, S, Ibrahim, MNM, Herrero, M, Holmann, F, Jabbar, M, Manyong, V, Patil, BR, Thornton, P, Williams, T, Waithaka, M & de Wolff, T (eds) *SLP project on transregional analysis of crop-livestock systems level 1 report: broad dimensions of crop-livestock intensification across three continents*, ILRI, CIAT, IITA, University of Peradeniya & BAIF, Nairobi.

Ministry of Lands & Settlement 2006, *Draft national land policy*, Ministry of Lands, Nairobi.

Ministry of Planning and National Development 2003, *Economic survey*, Ministry of Planning and National Development, Nairobi.

Mougeot, LJA 1993: Overview-urban food self-reliance: significance and prospects. Ottawa: International Development Research Centre. IDRC Report Volume 21(3).

Mougeot, L 2000, 'Urban agriculture: definition, presence, potential and risks, main policy challenges', paper presented at an international workshop on *Growing cities growing food: urban agriculture on the policy agenda*, La Habana, 11–15 October 1999.

Mukisira, E 2005, 'Opening speech', in Ayaga, G, Kibata, G, Lee-Smith, D, Njenga, M & Rege, R (eds) *Prospects for urban and peri-urban agriculture in Kenya*, Urban Harvest, International Potato Center, Lima.

Njenga, M, Gathuru, K, Kimani, K, Frost, W, Carsan, C, Lee-Smith, D, Romney D & Karanja, N 2004, *Management of organic waste and livestock manure for enhancing agricultural productivity in urban and peri–urban Nairobi*, Project Report, Mimeo.

Nugent, RA 1997, 'The Significance of Urban Agriculture', in Baumgartner, B & Belevi, H (eds) *A systematic overview of urban agriculture in developing countries*, 2001, Swiss Federal Institute for Environmental Science (EAWAG) & Technology and Department of Water and Sanitation in Developing Countries (SANDEC), Zurich.

Opresko, DM 1992, *Toxicity summary for zinc and zinc compounds*, Chemical Hazard Evaluation and Communication Group, Biomedical and Environmental Information Analysis Section, Health and Safety Research Division, Oak Ridge National Laboratory, Tennessee, http://www.eprf.ca/ebi/contaminants/zinc.html [Accessed 12 August 2008].

Poulimenos, G, Doutsos, T, & Maxwell, DG 1996, 'Highest and best use? Access to urban land for semi-subsistence food production', *Land Use Policy,* vol.13, no. 3, pp. 181–195.

Republic of Kenya, Central Bureau of Statistics, Ministry of Finance and Planning 2001, *The 1999 population and housing census,* vol. 1.

Savala, CEN 2003, 'Using earthworms to make vermicompost', in Savala, CEN, Omare, MN & Woomer, PL (eds) *Organic resource management in Kenya: perspectives and guidelines,* FORMAT, Nairobi.

Savala, CEN, Omare, MN & Woomer, PL (eds) 2003, *Organic resource management in Kenya: perspectives and guidelines,* FORMAT, Nairobi.

Situma, FDP 1992, 'The environmental problems in the City of Nairobi, Kenya', *African Urban Quarterly* vol. 7. nos. 1 and 2, February.

Staal, S 2002, 'Livestock and environment in peri-urban ruminant systems', presentation at the *LEAD/ILRI Seminar*, International Livestock Research Institute, Nairobi, May 2002.

Urban Harvest 2002, 'Stakeholder meeting on food security and urban agriculture in Nairobi', *The Strategic Initiative on Urban and Peri-Urban Agriculture (SIUPA)*, CIP, Nairobi, 6 August 2002, Mimeo.

Walkley, A & Black, IA 1934, 'An examination of Degtjareff method for determining soil organic matter and proposed chromic acid titration method', *Soil Science*, vol. 37, pp. 29–38.

World Bank 1997, *The use of compost in Indonesia: proposed compost quality standards*, Infrastructure Operations, Country Department III, East Asia and Pacific Region, Washington, DC.

Chapter 11
Crop–Livestock–Waste Interactions in Nakuru's Urban Agriculture

Nancy Karanja, Mary Njenga, Kuria Gathuru, Anthony Karanja, and Patrick Muendo Munyao

Introduction

As a complement to the research in Nairobi presented in the previous chapter, which mapped materials and market flows of nutrients in Kenya's capital city, this chapter presents a more in-depth picture of sources and types of waste generated by farmers in an urban area and the management practices involved. Both studies are aimed at informing policy. Whereas the Nairobi study focused on the handling of nutrients by community-based organizations, this study of Nakuru focuses on how urban farming households handle waste, including that generated by livestock. Some of the health risks involved are examined in Chapter 12 of this book.

Nakuru is close to the Equator, about 60 km northwest of Nairobi in the Rift Valley, a major geological feature of the African continent. It lies on the north shore of Lake Nakuru, a protected World Heritage site adjoining a National Park. At 1700–1850 m above sea level, the town has a sub-humid equatorial climate with bi-modal rainfall of about 950 mm per annum and had a population of 239 000 in 1999 growing at the rate of 4.3 percent annually (Republic of Kenya 2000). Its main economic activities are commerce, industry (including a Union Carbide factory), agriculture and related tertiary services. Commerce is mainly concentrated in the town centre, with informal commercial activities on the increase. Vendors and small-scale businesses crowd transport termini and the reserves of major roads (MCN 1999).

Nakuru has both large- and small-scale farming within its boundaries. Large farms are located in the west of the town and include the giant farm owned by the Rift Valley Institute of Science and Technology (RVIST). Small farms are steadily increasing in numbers, especially in the peri-urban areas. Many farms have been sub-divided into urban residential plots where smallholder farming is practiced. Together, these urban farms supply 22 percent of the basic food intake of farming households, and 8 percent of the overall food and nutritional needs of the town, with most of the rest coming from the rich agricultural hinterland (Foeken 2006).

N. Karanja (✉)
Urban Harvest, c/o CIP, P.O. Box 25171, Nairobi 00603, Kenya
e-mail: nancy.karanja@cgiar.org

An estimated 35 percent of Nakuru's population engaged in urban farming in 1998 with 27 percent of all households growing crops and 20 percent keeping livestock (Foeken & Owour 2000; Foeken 2006), figures very similar to those from a study of six Kenyan towns in the 1980s by Mazingira Institute (Lee-Smith et al. 1987). Common crops in Nakuru are maize, kale (*sukuma wiki*), beans, onions, spinach, tomatoes and Irish potatoes while chicken, cattle, goats, ducks and sheep are common livestock (Foeken 2006).

We explored crop–livestock–waste interactions in the town so as to generate useful data for urban agriculture environmental management and policy development in Nakuru and similar urban areas in the region. We were able to situate our data, which emphasized the situation of livestock keepers, in relation to that from a random sample survey of urban farmers carried out 6 years previously, in 1998 (Foeken 2006).

The 1998 Nakuru study explored the relationship between UA and poverty, finding that the poor are proportionally less represented among urban agricultural producers than the better off, who derived greater benefit from agriculture mainly due to their more secure access to land (Foeken 2006). Livestock keeping was associated with commercial orientation and with being better off. Whether rich or poor, farming households were larger and healthier, but it was harder for Nakuru's poor to get some of the benefits from farming, with poor women-headed households benefiting least. The study showed that most of the town's poor – those who needed food and would have farmed if they could – did not in fact do so (Foeken 2006). Our study builds on these insights by taking a gendered approach to how crop–livestock–waste interactions are managed. Its findings are linked to ongoing policy and legal review in the Municipal Council of Nakuru.

Methods Used

Study Sites

Nakuru is divided into 15 administrative-electoral wards that constitute the residential areas and business district. For both this study and the one described in the following chapter, four wards were purposively selected for sampling, based on their having significant livestock populations along with crop farming, and representing a range of income and human population densities. Our aim was to sample the range of crop–livestock–waste systems (with an emphasis on cattle keepers) rather than to compare them, although a few comparisons are drawn in our discussion of the data.

Kaptembwo and Kivumbini are low-income urban areas to the West and South West respectively, with high human population densities and low livestock populations and small land sizes. Peri-urban Nakuru East has middle-income farm households with medium-sized land holdings, human and livestock population densities, most farmers having backyard farms. Peri-urban Menengai to the north has higher income farmers with larger pieces of land and lower human population densities with a higher ratio of livestock to people.

Participatory Urban Appraisals and Household Interviews

The baseline survey began with a participatory urban appraisal (PUA) in each ward, where the objectives, benefits and sampling frame of the study were first discussed with the farmers. Agricultural extension officers had a list of livestock and crop farmers, to which missing names were added during the PUAs. The focus group discussions (FGDs), held with men and women separately, covered crop–livestock–waste interaction and health risk assessment. The latter subject is dealt with in the next chapter.

To suit the purposes of the study of health risks described in Chapter 12, random samples of 40 cattle farmers at each site were selected from the lists generated. For this chapter's study of crop–livestock–waste interactions, 10 more farmers were added to this sample at each site. These farmers, who grew crops and kept other types of livestock (including poultry, sheep and goats), were randomly selected from the lists to make up a sample size of approximately 50 crop and livestock farmers in each ward, for a total of about 200. This means that our findings are biased toward cattle keepers.

For this chapter, a semi-structured questionnaire was administered to the selected households to gather information on household characteristics, food consumption, characteristics of crop and animal agriculture, waste generation and re-use, income sources and levels and gender issues. Both male and female interviewees responded to the questions on gender in the questionnaire. Data on gender issues were also generated through the FGDs held with men and women separately.

Estimating Organic Waste Production and Utilization

The types and quantities of waste generated were estimated based on figures obtained from the household questionnaires, validated by participatory investigation with selected households. As a by-product of the investigation, these households were also sensitized on source sorting and waste reuse and recycling. Each household received two garbage collection bags of different colours in which to place their organic and inorganic waste for a period of 24 h. The organic component was then hand-sorted into different crop types such as peelings of potato, banana, sweetpotato and so on, permitting estimation of the dry weights produced for each and determination of nutrient content. Production and utilization of organic waste from other sources such as institutions, markets and hotels were also established through the use of separate questionnaires, as well as by taking actual measurements where possible.

Data Analysis

The household survey data was coded, entered, screened and cleaned in a relational database designed using the survey questionnaire (Microsoft Office Access

2003[R] – Microsoft Corporation, USA). Links between hypothetically related variables and outcome measures of interest from different data tables were tested using MS Access and results were exported to statistical software (Instat[+] for windows V 3.029- 2005[R]) for descriptive analysis, cross-tabulations and relationship hypothesis testing using either Chi-square test, Z-test and /or student's t-test as appropriate. Outcomes were compared across the sex of household heads, gender division of labour and the four study sites, and inferences drawn accordingly.

Results

Characterizing the Sample

The total number of respondents was 213, of whom 56 percent were women and 44 percent men. Most of the respondents were either household heads (44 percent) or their spouses (40 percent), while 11 percent were children over 20 years old and 5 percent were other adults, mainly workers and in-laws living with the household (Table 11.1). Out of the 213 households, 169 were mixed farmers, 11 grew crops only and 33 kept livestock only.

Many characteristics of our sample were similar to those from a random sample of Nakuru urban farmers drawn in 1998 (Foeken 2006), such as the proportion of women-headed households (16 percent) and the large size of farming households (six persons) compared to the norm for Nakuru (four). However, the much higher incidence of house ownership (86 percent compared to 22 percent) supports the correspondence between property ownership and livestock (especially cattle) farming, indicating that such farmers are relatively better off. Our household heads were also slightly older on average, at 49 years compared to 41 (Foeken 2006).

We were also able to confirm Foeken's finding that livestock keepers were more likely than crop growers to be farming for income purposes. Almost half (47 percent) of household heads had farming as their primary economic activity, and there was a significant gender difference ($p<0.05$), the large majority (63 percent) of women-headed households (defined as a household with no male head) engaging in farming as their main economic activity compared to 44 percent of the male-headed households. Only 23 percent in our sample had formal employment as their main occupation, about half the norm for Nakuru farmers and non-farmers in 1999 (Foeken 2006, p. 182), indicating a high level of dependence on farming as a source of livelihood (Table 11.1).

The fact that we found 32 percent of household heads had only primary school or no formal education, suggests that such households may select urban farming as a livelihood strategy. More men than women households heads were formally employed (25 percent compared to 9 percent, $p<0.01$) possibly associated with such women's lack of formal education (24 percent) compared to 4 percent of men heading households ($p<0.01$) and consistent with findings from a similar study carried out in Dagoretti, Nairobi (Kimani et al. 2007). Furthermore, there was a significant difference ($p<0.01$) in household heads having college education (19 percent of men heading households compared to only 6 percent of the women). The average annual contribution of urban agriculture to the income of an urban farmer in our sample was KShs 86 850 (US$ 1240) from both livestock and vegetable production, representing 43 percent of their annual income.

Table 11.1 Population characteristics

Characteristic	Category	Statistics (% of N)
Sex of HH head	Male	84.0
(N = 213)	Female	16.0
	Total	100.00
Mean age of HH head	All households	49.4
(N = 210)	Male	49.8
	Female	47.2
Level of education	No formal	7.1
of HH	Primary	24.5
(N = 212)	Secondary	48.1
	College	17.0
	University	3.3
	Total	100.00
Residential status	Own house	86.0
(N = 212)	Tenant	9.4
	Govt./Co-op.	1.4
	Squatter	2.8
	Others	0.4
	Total	100.00
Main occupation of HH	Business	15.5
Head (N = 206)	Farmer	46.6
	Formal employment	22.8
	Informal employment and others	15
	Total	100.00
Mean HH size	All households	6.1
(N = 213)	Male-headed HH	6.3
	Female-headed HH	5.4

HH = household

Access to and Location of Urban Farming Plots

As with the finding on high house ownership, the large majority (81 percent) of the urban farm plots were owned by the farmers (Fig. 11.1). This compares with only 33 percent of Nakuru urban farmers in general being plot owners (Foeken 2006, p. 186) and 46 percent in Kampala (see Chapter 6, above) again suggesting our sample of mostly cattle keepers was biased toward the better off. Land ownership was significantly different ($p<0.05$) for male and female household heads. Traditionally, women in Kenya cannot inherit land, increasing their vulnerability to poverty.

Over 60 percent of the farmers studied used their backyards or compounds, while others farmed on roadsides (30 percent) or other open space such as under power lines or on institutional land (Fig. 11.2), a common pattern observed in the region

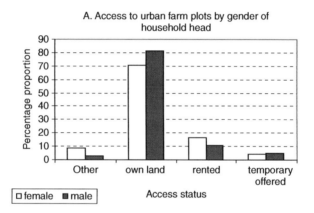

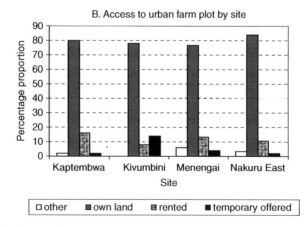

Fig. 11.1 Urban farming plots by gender of household head (A) and site (B)

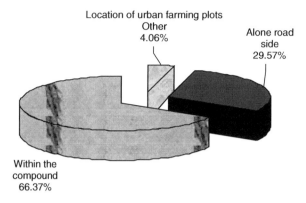

Fig. 11.2 Proportion of urban farming plots by location

(see the Kampala study, Chapter 6, above). The ten rural plots identified were all owned by men heading households, again attributable to restrictions on women's ownership of property.

Comparing neighbourhoods, it is worth noting that significant proportions of farmers used roadsides in higher-income Menengai with its larger plots, as well as in low-income high-density Kaptembwa, (24 and 36 percent respectively), indicating the opportunity even well-off farmers take to graze their livestock where publicly owned pasture is to be found.

Household Food Consumption

Households purchased 70 percent of food items they consumed, disaggregation showing that more than 50 percent of kale and spinach consumed was sourced from households' farms while tomatoes, oranges and cabbages were purchased. The commonest foods consumed in 1 week were Irish potatoes and maize (38 and 25 percent of all items noted respectively), indicating a potential source of household organic waste. We estimated a weekly per capita consumption of 17 kg of Irish potatoes and 13 kg of maize in households we studied and a total of KShs 1033 (US$ 14) per household being spent on food. Most money was spent on kale (21 percent), followed by dry maize (14 percent), tomatoes (17 percent) and onions (8 percent). These results illustrate the contribution of urban agriculture as a source of food as well as saving income through home production, taking kale as an example.

Urban Livestock Production

Due to the bias in our sample, 83 percent of households studied kept cattle, compared with 12 percent for Nakuru urban farmers in general (calculated from Foeken

2006, tables 3.1 and 5.1). The gender difference among household heads was not significant ($p>0.05$) despite cattle keeping being traditionally associated with men. Most cattle keepers (85 percent) were mixed farmers and kept a range of livestock, similar to what is observed in other urban centres in the region, as reported by Tegegne et al. (2002) for Addis Ababa, Ethiopia; Ishani et al. (2002a, b) for Kisumu and Nairobi, Kenya; and Njuki and Nindi (2000) for Dar es Salaam in Tanzania.

Nakuru farmers have found zero-grazing more costly and time consuming, while keeping animals free-range runs more risk of theft or contamination of feed (Foeken 2006, pp. 70–71). Perhaps due to farmers juggling these constraints, we found that the proportions of cattle and chickens kept in these ways were more or less reversed from Foeken's earlier study, with fewer cattle keepers (48 percent) zero-grazing their animals and more chickens (56 percent) being kept in chicken houses. Sheep and goats were mainly left to range freely at least part of the time. Ducks and pigs were kept by only a few households, both being confined or semi-confined.

Types of Livestock Feeds and Their Sources

Measured by weight, more than half (53 percent) of the fodder fed to all livestock was grass, 30 percent was concentrates and 17 percent was organic refuse. Gathering grass for cattle – mainly from roadsides or other open public land but also from their own plantings of Napier grass – provides opportunities for small business operators on bicycles (Foeken 2006, pp. 72–73). However, we found that nearly half of the fodder for all livestock (41 percent) was obtained from the households' own urban farms or from their neighbours, 30 percent was purchased from these vendors, 21 percent came from households' own nearby rural farms, while the rest was purchased in rural areas. According to Tegegne et al. (2002) 87 percent of urban livestock keepers in Addis Ababa depend on purchased hay from residues of crops like teff, wheat and lentils.

Concerning cattle feed specifically, 42 percent consisted of concentrates (mainly dairy meal, all of which was purchased), 42 percent was grass, especially Napier grass, and 16 percent was organic waste. Almost two-thirds of the organic waste fed to cattle in our sample was from the farmers' own sources – mainly farm and kitchen waste – the rest being purchased. In Addis Ababa it was observed that 93 percent of the farmers gave supplements to the livestock while 47 percent fed them household organic waste (Tegegne et al. 2002).

By contrast, concentrates were said to constitute 99 percent of the total feed given to chickens (33 percent layers mash, 32 percent growers mash and 24 percent chick mash), and virtually all of this was purchased in urban areas. It must also be borne in mind, however, that the numerous free-range chickens also foraged on grass and refuse dumps in the neighbourhoods. It should also be remembered that farmers generally consume these free-range chickens themselves, preferring the taste, while they sell the chickens fed on concentrates as a source of income.

Purchased fodder and grass were mainly sourced by men while women were more often involved in sourcing the organic refuse, a division of labour observed in many studies where men prefer being involved in organized, less tedious and time consuming but tidy activities (Njenga et al. 2004). In the urban farms, the purchased fodder was mainly for zero-grazed cattle. Farmers transported fodder using bicycles (40 percent), humans or animals walking (38 percent) or vehicles (22 percent). Concentrates bought from feed stores, agro-vets, maize mills, shops or kiosks were similarly transported.

Use of Raw Organic Household Waste as Animal Feed

On average, 91 percent (by weight) of the total amount of raw household waste produced by these farming households was re-used, the largest proportion of this (96 percent) being fed to livestock. This indicates the useful role played by farming households, and in particular livestock keepers, in managing urban waste and recycling it for productive purposes. The practice is widespread (Njenga et al. 2004; Tegegne et al. 2002).

Gender analysis revealed another important dimension here, in that women played a greater role than men in such waste management. Sixty percent of adult women were involved in waste re-use compared to only 20 percent of adult men, while 62 percent of adult women made the decisions on how waste was to be re-used compared to 20 percent of adult men. Only 4 percent of the raw household waste was not fed to livestock, and was re-used by being given to neighbours, thrown in a pit to decompose into compost or scattered on the urban plots. Again this seemed to be up to the women, who managed the disposal of 83 percent of this residual waste, men taking care of only 16 percent. These findings are in line with others from the African region, indicating women are more likely than men to be the waste handlers (Lee-Smith 1999, 2006).

Farmers' Attitudes to and Use of Crop Inputs

Previous studies in Kenya have suggested urban farmers use more organic than chemical inputs in their food production (Lee-Smith & Lamba 1991). In 1985, 35 percent of Nairobi farmers used vegetable matter compost and 29 percent farmyard manure, while only 18 percent used chemical fertilizer (Lee-Smith et al. 1987, pp. 125–126, p. 129). Nakuru showed higher use of chemicals in 1999, 36 percent of farmers using them, while 38 percent used compost and 53 percent manure (Foeken 2006). Our study found that women-headed households were more likely to irrigate their crops, and that 84 percent of male-headed households used chemical fertilizer compared to 78 percent in the female-headed households, although the difference was not significant ($p>0.5$). While cost and crop yields were the major factors in farmers' choice of inputs, the amount of work involved in processing organic inputs may hinder their use. Irrigation is associated with higher yields as is the use of more than one input, whether organic or chemical (Lee-Smith et al. 1987; Foeken 2006, pp. 55–61). We noted a slight increase in use of chemical fertilizers (39 percent of urban crop farmers) compared to 36 percent found by Foeken and Owour (2000). Of the 110 crop farmers who did not use chemical fertilizer, 78 percent said they had enough manure while 13 percent thought it was too expensive.

Our study examined farmers' attitudes as well as their practices and found almost all (96 percent) perceived low soil fertility as a constraint to crop production with no significant difference between men and women respondents ($p>0.05$). When asked how soil fertility could be improved, 78 percent favoured using manure while a further 10 percent favoured mixing manure and fertilizer (Fig. 11.3). Other methods of improving soil fertility included use of crop residues, fallowing or crop rotation.

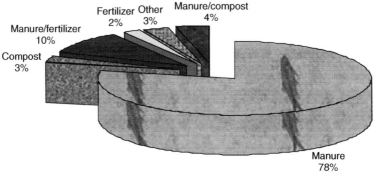

Fig. 11.3 Soil fertility management options ($N = 176$)

Decisions about the use of different crop inputs were generally made by adult males in the households when it came to chemical fertilizer and compost (64 and 57 percent male decisions respectively, compared to 30 and 36 percent decisions by adult women). Decision making over the use of crop residues was equally by men (48 percent) and women (50 percent).

Nakuru crop farmers using chemical fertilizer used 29 kg on average, spending KShs 853, with no real difference between men and women farmers. The proportion of crops grown using chemical fertilizer were maize (35 percent), kale (15 percent), maize intercropped with beans (13 percent) and potatoes (7 percent). Of the 17 households using vegetable compost, 15 used material from their own farms while two obtained it from community groups who produced compost.

All the crop residues used for soil fertility improvement were obtained from individual plots with no significant differences by gender of household head ($p>0.05$). Most of the crop residues, like organic domestic waste, were fed to livestock (Table 11.2).

Table 11.2 Uses of crop residues by Nakuru farmers

Gender	Fed to livestock	Fuel	Left on farm	Other	Sold	Total
Female- headed HH	14 (48%)	1 (3.4%)	9 (31%)	2 (6.9%)	3 (10%)	29
Male- headed HH	100 (55.6%)	14 (7.8%)	42 (23.3%)	15 (8.3%)	9 (5.0%)	180
Total	114 (54.5%)	15 (7.2%)	51 (24.4%)	17 (8.1%)	12 (5.7%)	209

N = 209 responses

In assessing knowledge of compost making we found that 58 percent of the farmers knew how to make compost and the difference between men and women was not significant ($p>0.05$). The knowledge was gained from schools, seminars, agricultural shows and farmer field schools. However, only 10 percent of respondents made compost on their farms while a few belonged to a group that was involved in compost making. The reasons given by those who did not make compost were that they had enough animal manure, or that it was too labour intensive, or they lacked space or did not have enough knowledge on how to make it.

Manure Production and Use

Earlier work in Nakuru found that almost half (48 percent) of livestock farmers used all or some of the manure produced by their animals for crop cultivation on their own mixed farms, thus recycling the nutrients effectively for food production purposes. However, while another 15 percent gave away or sold some of it for the same useful purpose, almost half (45 percent) also simply dumped some of the waste in the streets or on refuse dumps (Foeken 2006, p. 77, table 5.6).

This study attempted to go further into the utilization of farm and domestic wastes than the earlier study where emphasis was more on the pollution aspects of wastes, especially manure. Foeken and his colleagues noted that there was higher utilization of manure from large livestock, including cattle, than of small livestock, mainly chickens (98 percent compared to 56 percent using it on their own farms or giving it to their neighbours). That study also found a strong connection between waste re-use and mixed farming incorporating cattle and better waste re-use by farmers who had more space in their compounds. Getting rid of manure seemed to be more of a problem for the farmers in high-density areas with little space (Foeken 2006, p. 77). Our findings support these conclusions and examine in greater detail the quantities of manure produced in each area and its re-use.

Table 11.3 shows that growing their own crops in town was the main use to which livestock farmers in all areas put manure. Usually, growing their own rural crops was next most important, followed by sales of manure and a mix of other uses. Our data confirm the lower re-use of manure in areas with high population density and small farming plots, namely Kivumbini and Kaptembwa, with very high efficiency of re-use in the middle-income backyard mixed farms of Nakuru East.

Table 11.3 Total annual production of manure (tones) by sampled households in each area, and their ways of re-use and disposal

Ways of re-use and disposal of manure	[b]Kivumbini ($n=50$)	[b]Kaptembwa ($n=43$)	[c]Menengai ($n=59$)	[d]Nakuru East ($n=50$)	All areas ($n=202$)
Own crop production in urban areas	116.9	149.0	375.1	487.0	1128.0
Own crop production in rural areas	38.4	19.9	4.5	48.6	111.4
Sold	10.5	12.9	18.3	9.7	51.4
Other uses[a]	17.0	53.7	202.2	292.7	565.6
Subtotal Re-used	**182.8**	**235.5**	**600.1**	**838.0**	**1856.4**
Dumped or disposed of	914.1	538.2	651.2	119.9	2223.4
Total manure produced	**1096.9**	**773.7**	**1251.3**	**957.9**	**4079.8**
% Re-use	(16.7)	(30.4)	(48.0)	(87.5)	(45.5)
Average produced/ HH	**21.9**	**18.0**	**21.2**	**19.2**	**20.2**

[a] Biogas, planting flowers or trees, growing pasture, giving for free, making poultry feed
[b] low income
[c] high income
[d] medium income

Since average household production of manure in the four areas was similar, the overall average was used to project a total manure production for Nakuru of 282 800 tonnes of manure annually, using Foeken's estimate of 14 000 households keeping livestock in the town in 1998 (Foeken 2006, p. 39). The figure of 20.2 tonnes of manure per household refers to wet weight.

The breakdown of data by men and women farmers in each area, shown in Table 11.4, allows for gender analysis of the findings.

It is immediately apparent from these figures that, while manure production is consistently lower for women-headed farming households, efficiency of re-use is generally higher. The exception is the high-income area of Menengai, where efficiency of re-use is slightly lower for women-headed households, in contrast to low-income Kaptwemba, where such households are considerably more efficient in manure re-use than those headed by men.

The higher production of manure by men-headed households is probably due to the stronger association of men with cattle, which produce larger amounts of manure, whereas even women who keep cattle may more often also keep poultry which produce smaller amounts. The manure that was not re-used was thrown away – heaps were a common site in the streets, especially in the high-density residential areas.

Kale, maize and bananas were the crops most often treated with manure as a fertilizer by farming households, with chicken manure being preferred for kale and cattle manure for growing maize. The detailed figures are shown in Table 11.5.

Table 11.4 Gender breakdown of annual manure production and use in tones, by site

Ways of re-use and disposal of manure	Kivumbini (n=50)		Kaptembwa (n=43)		Menengai (n=59)		Nakuru East (n=50)		Grand Total
	Female (13)	Male (37)	Female (4)	Male (39)	Female (6)	Male (53)	Female (8)	Male (42)	(202)
Total manure produced	228.7	868.2	60.7	713.0	58.8	1192.5	82.0	875.9	4079.8
Own crop production in urban areas	22.6	94.3	6.1	142.9	10.6	364.5	38.7	448.3	1128.0
Own crop production in rural areas	23.4	15.0	0.0	19.9	0.0	4.5	0.0	48.6	111.4
Sold	10.5	0.0	6.0	6.9	0.0	18.3	0.6	9.1	51.4
Other uses[a]	0.3	16.7	26.0	27.7	15.9	186.3	37.3	255.4	565.6
Subtotal re-used	56.8	126.0	38.1	197.4	26.5	573.6	76.6	761.4	1856.4
Dumped	171.9	742.2	22.6	515.6	32.3	618.9	5.4	114.5	2223.4
% Re-use (re-use/total×100)	24.8	14.5	62.8	27.7	45.1	48.1	93.4	86.9	45.5
Average produced/ hh	17.6	23.5	15.2	18.3	9.6	22.5	10.3	20.9	20.2

[a]Other uses include biogas, planting flowers or trees, growing pasture, giving for free, making livestock feed.
Female = female headed HH, male = male headed HH

Table 11.5 Comparison of cattle and chicken manure use for various crops

Crop	Manure type (% of household)		Chi sq.	P value
	Chicken	Cattle		
Kale	34	18	9.1	0.002
Maize	19	27	1.8	0.17
Banana	21	6	9.6	0.002
Beans	12	16	0.7	0.42
Others	11	33	14.1	0.0002
	97	100		

Sources of Manure for Crop-Only Farmers

Of the eleven farmers in our sample who produced crops only, three were female-headed households. These farmers used 28 urban plots, all in the low-income high-density neighbourhoods of Kaptembwa and Kivumbini, where they mostly grew vegetables, beans and maize intercropped with other traditional vegetables, which clearly contributed to their livelihoods through sales in addition to food. The three women-headed households owned a quarter of these plots, with male-headed households predominantly renting the rest. Manure collected from urban residential areas provided crucial inputs to these production systems.

Policy Influencing and Technology Transfer in Nakuru Municipality

The findings of this study, along with others carried out earlier, were used as inputs to a formal process of the Municipal Council of Nakuru (MCN) aimed at developing urban agriculture bylaws. Following resolutions of a workshop for Nakuru councillors in May 2005, MCN's Department of Environment led a consultative law-making process to enable and regulate farming within the municipality. In fact, the review process, consultative meetings and drafting of the bylaws were supported by the project because they were activities in line with its stated objectives. The research project inception and feedback workshops, held in Nakuru in December 2004 and March 2006 respectively, were seen as a way of creating awareness on urban agriculture as a productive sector and strengthening dialogue for the bylaw development process. An earlier, parallel process in Kampala, Uganda, also supported by Urban Harvest, was likewise used as a model in Nakuru (Lee-Smith et al. 2008).

In 2009, the draft urban agriculture bylaws (MCN 2006) were awaiting in-depth farmer consultations from all the 15 wards. Meanwhile, collaborative training courses on crop–livestock production and waste management for urban agriculture and mushroom production using agricultural waste, including marketing, were held in December 2005. Partners involved were the Ministries of Agriculture and Livestock and Fisheries Development, the Nairobi and Environs Food Security, Agriculture and Livestock Forum (NEFSALF) and the German Society for Technical Cooperation (GTZ). Thirty-four women and 24 men involved in urban agriculture and organic waste re-use were trained, along with 11 agriculture and livestock extension staff (Karanja et al. 2006).

Conclusions

Our study confirms earlier findings from the end of the 1990s in Nakuru and shows that mixed crop–livestock farmers in the town tend to be among the better off, owning their own housing and land, and to be more oriented than urban farmers in general towards income generation from farming, though they also use it to feed themselves. However, we also established that this group of farmers is slightly older and less educated, and less often engaged in wage employment than farmers in general as measured in the late 1990s, suggesting that persons with these characteristics can more easily find a means of livelihood in farming, and that this applies especially to women heading households.

The farmers run small-scale enterprise farm systems that are efficient in nutrient recycling, using domestic organic waste as fodder and manure as fertilizer, especially where they have backyard space. Thus we confirm the earlier findings that the lower income groups find it harder to farm effectively. With an average of 20 tonnes of manure produced by an urban farmer in Nakuru, those in a middle-income area with backyard mixed farms achieved a very high re-use rate of 88 percent, mostly applied to their own crops, while those in a low-income area with higher density and less space only achieved 17 percent re-use, resulting in dumping and environmental contamination. Some intensive vegetables producers in these low-income areas were making good use of this manure and the practice could be expanded. Low re-use of manure could also have been as a result of inadequate technical skills and knowledge on the benefits of closing the nutrient loop in crop–livestock farming systems.

Using a gender analysis, we also established that women tended to be more involved than men in managing the nutrient cycling of domestic organic waste as livestock fodder, and, further, that they had higher rates of efficiency of re-use of manure from livestock in all but the high-income areas of Nakuru studied. Further, while use of chemical fertilisers appears to be higher and increasing in Nakuru (39 percent) compared to some other towns, we found women farmers tend to use these chemicals less often than men, perhaps due to their high cost. Men were also noted to sell manure more than women.

Overall, the farmers in our study recycled nearly all their domestic organic waste, mostly as livestock fodder, and this must be seen as a benefit to the town in terms of waste management and efficient food production. However, using our data and that of earlier studies, we were able to project that about 283 000 tonnes of wet manure is produced annually in the livestock and mixed farms of Nakuru, and that just over a half is not re-used in farming.

Thus our study suggests that urban agriculture in the town would work much better if the lower income farmers were encouraged to farm more efficiently using crop–livestock systems on land set aside for the purpose. Alternatively, or in addition, the systematic collection and re-use of livestock wastes from low-income farms in high-density areas would greatly increase efficiency of food production as well as waste management. This could be done through organized collection and distribution points and effective information to crop farmers through the official channels of agricultural extension services as well as the Municipal Council's Environment Department. The dumped manure could also be co-composted with other types of organic waste and packaged as a bio-fertilizer.

Apart from better urban waste management, this fairly simple institutional innovation would enhance agriculture productivity as well as incomes for an important group of urban residents, those with fewer jobs and less education, and especially women heading households. In turn, such a measure would contribute to the achievement of the Millennium Development Goals of alleviating hunger and poverty.

References

Foeken, D 2006, *"To subsidise my income" urban farming in an East-African Town*, Brill, Leiden, Boston, MA.
Foeken, D & Owour, S 2000, *Urban farmers in Nakuru, Kenya*, ASC Working Paper no. 45, Africa Studies Centre, Leiden and Centre for Urban Research, University of Nairobi.
Instat + for windows V 3.029- 2005®; Statistical Services Centre, The University of Reading, UK and Genstat Release 4.2 Discovery Edition, 2005® –VSL International Limited, Waterhouse Street, Hemel Hempstead, UK.
Ishani, Z, Gathuru, PK & Lamba, D 2002a, *Scoping study of interactions between gender relations and livestock keeping in Kisumu*, Mazingira Institute. A report prepared for Natural Resources International.
Ishani, Z, Gathuru, PK & Lamba, D 2002b, *Scoping study of urban and peri-urban poor livestock keepers in Nairobi*, Mazingira Institute, Nairobi.
Karanja, N, Njenga, M, Gathuru, K, Kangethe, E, Karanja, A, Njehu, A, Odumbe, M, Kiarie, S, Kimani, I, Wokabi, S, Getanda, C, Kingori, P, Mugo, P, Njenga, A & Kirima, F 2006, *Local Participatory Research and Development on Urban Agriculture and Livestock Keeping in Nakuru*. Supported by DFID-LPP. Project Report.
Kimani, VN, Ngonde, AM, Kang'ethe, EK & Kiragu, MW 2007, 'Gender perceptions and behaviour towards health risks associated with urban dairy farming in Dagoretti Division, Nairobi, Kenya', *East African Medical Journal*, Nov, vol. 84, no. 11 (Supplement), S57–S64.
Lee-Smith, D (ed) 1999, *Women managing resources: African research on gender, urbanisation and environment*, Mazingira Institute, Nairobi.
Lee-Smith, D 2006, 'Risk perception, communication and mitigation in urban agriculture: community participation and gender perspectives', in Boischio, A, Clegg, A & Mwagore D (eds) *Health risks and benefits of urban and peri-urban agriculture and livestock (UA) in Sub-Saharan Africa*, Urban Poverty Environment Series Report #1, IDCR, Ottawa, pp. 75–84.
Lee-Smith, D, Azuba, SM, Musisi, JM, Kaweeza, M & Nasinyama, GW 2008, 'The story of the Health Coordinating Committee, KUFSALCC and the urban agriculture ordinances', in Cole, D, Lee-Smith, D & Nasinyama, GW (eds) *Healthy city harvests: Generating evidence to guide policy on urban agriculture*, CIP/Urban Harvest and Makerere University Press, Lima, Peru.
Lee-Smith, D & Lamba, D 1991, 'The potential of urban farming in Africa', *Ecodecision*, December, pp. 37–45.
Lee-Smith, D, Manundu, M, Lamba, D & Gathuru, PK 1987, *Urban food production and the cooking fuel situation in urban Kenya*, Mazingira Institute, Nairobi.
MCN (Municipal Council of Nakuru) 1999, *Strategic structure plan. Action for sustainable urban development of Nakuru town and its environs*, Municipal Council of Nakuru (Final Draft), Nakuru.
MCN (Municipal Council of Nakuru) 2006, *Urban agriculture by-laws*, draft three. Laws of Kenya (Cap. 265).
Microsoft Office Access 2003, Microsoft Corporation, USU.
Njenga, M, Gathuru, K, Kimani, K, Frost, W, Carsan, C, Lee-Smith, D, Romney D & Karanja, N 2004, *Management of organic waste and livestock manure for enhancing agricultural productivity in urban and peri–urban Nairobi*, Project Report.
Njuki, JM & Nindi, SJ 2000, *Gender and urban agriculture: cattle raising and its implication for family welfare and the environment in Dar es Salaam, Tanzania*, Mazingira Institute, Nairobi.
Republic of Kenya 2000, *Economic survey 2000*, Nairobi, Government Printer.
Tegegne, A, Tadesse, M, Alemayehu, M, Woltedji, D & Sileshi, Z 2002, *Scoping study on urban and peri-urban livestock production in Addis Ababa, Ethiopia*. International Livestock Research Institute and Ethiopia Agricultural Research Organization, Addis Ababa, Ethiopia.

Chapter 12
Benefits and Selected Health Risks of Urban Dairy Production in Nakuru, Kenya

Erastus K. Kang'ethe, Alice Njehu, Nancy Karanja, Mary Njenga, Kuria Gathuru, and Anthony Karanja

Introduction

Health risks from food production in urban areas are attracting increased international attention, especially in poor countries with rapid urbanization where urban farming is widely practiced to mitigate hunger and poor nutrition as well as reduce food expenditures. This study examines a selected range of health risks as compared to the benefits for an urban population for which a considerable quantity of background data are available, namely Nakuru municipality in Kenya. The research was carried out in conjunction with a related survey of crop-livestock-waste interactions in the same town, described in the previous chapter.

Smit (1996) predicted that urban agriculture (UA) would be contributing one-quarter to one-third of world food production, including half of vegetables, meat, fish and dairy products consumed in cities by 2005. In the cities of eight East and Southern African countries (Eritrea, Ethiopia, Kenya, Mozambique, Tanzania, Uganda, Zambia and Zimbabwe), Denninger et al. (1998) estimated that nearly 25 million out of 65 million people in urban areas got some of their food from UA and that by 2020 at least 35–40 million urban dwellers in the same countries would depend on UA to feed themselves. Lee-Smith et al. (1987) showed that two-thirds of urban households in Kenya grew part of their food, while 29 percent did so on urban land. Similar figures have recently been confirmed for the town of Nakuru, where three-quarters of households were farming in 1998, 35 percent on urban land. A quarter and a fifth of Nakuru households grew crops or kept livestock respectively, with the livestock population estimated to be 160 000 poultry, 25 000 head of cattle, 3000 goats, 3500 sheep and 1500 pigs (Foeken 2006, pp. 38, 67).

The Nakuru data confirmed findings of earlier studies that producing food is the main reason people engage in urban farming, as stated by 93 percent of crop farmers and 74 percent of livestock keepers in the town (Foeken 2006, p. 80). However, these

---

E.K. Kang'ethe (✉)
Department of Public Health, Pharmacology and Toxicology, Faculty of Veterinary Medicine, University of Nairobi, Nairobi, Kenya
e-mail: ekiambi@yahoo.com

data also revealed that the poor are not well represented among urban farmers – they constitute only 33 percent of farmers but 64 percent of non-farmers. Furthermore, although nearly all urban farmers kept small livestock, only the better-off kept large livestock such as cattle (Foeken 2006, pp. 41, 69, 137–151).

A typical "high-income" farming household in Nakuru owns a stone or block house with a large backyard where crops are grown and small and large livestock kept. Cattle are generally kept in stalls and fodder is brought to them (zero grazing). A middle-income urban household would usually be renting their house (built mainly of block, stone, wood, wattle or mud) on a smaller plot of land, and they would follow similar practices, although cattle may also be grazed along roadsides or on vacant land. A low-income household would be renting rooms in a densely populated area, and have no backyard as such, but may grow crops and keep small livestock, including chickens and goats, in their immediate surroundings. Small livestock are kept indoors at night in structures built next to the house or in the house itself. Cattle are kept in specially built sheds, either all the time or at night, and taken out to graze during the day. Water supply is generally available on-plot in high- and middle-income areas, and at water points in low-income areas. Sanitation is usually by sewer or septic tank in the high- and middle-income areas and by pit latrine in the low-income areas.

UA production feeds more than just the farm households, providing 22 percent of food needs of the farming households and 8 percent of all food consumed in the town (Foeken 2006, p. 155). In fact, dairy production systems in urban and peri-urban regions are becoming a dynamic and a fast-growing sector in response to ever-increasing urban demand for milk and milk products in particular. A number of surveys in African cities have shown that livestock keeping is a significant activity (Mosha 1991; Lee-Smith & Lamba 1991; Egziabher et al. 1994). The Maseru dairy plant in Lesotho processed 3000 litres of milk produced by urban farmers daily, accounting for 40 percent of the town's milk demand (Greenhow 1994). The value of livestock kept in urban areas in Tanzania was estimated at US$ 2.2 million, generating an annual income of US$ 3.4 million from sale of milk alone (Sawio 1993), while 75 percent of the 23 000 cattle in Nairobi, Kenya were dairy cows producing milk worth an estimated US$13 million annually (Lee-Smith & Lamba 1991). By 1998 there were about 24 000 dairy cattle in Nairobi producing about 42 million litres of milk, while an estimated 50 000 bags of maize and 15 000 bags of beans were also produced in Nairobi annually (Ministry of Agriculture 2002 in Ayaga et al. 2005). In a recent study in Dagoretti, a Nairobi neighbourhood with a high number of urban farmers, the average daily production of a household dairy farm during a 7-month lactation period was 3000 l of milk (Kang'ethe et al. 2005).

The amount of income generated from sale of milk, animals and manure helps urban farming households release income for other household needs like medical expenses and education. Livestock enterprises, no matter how small, offer new opportunities to families living in material poverty. With the milk, manure and income they provide, people are able to pursue new goals and, by achieving them, enhance their quality of life. Building on the data available for Nakuru, our study confirms the picture of urban dairy production as an emerging commercial activity for small-scale producers who also use it as a food security strategy.

Notwithstanding the benefits accrued from urban livestock keeping, intensification of dairy production and poor animal husbandry practices allow the emergence of zoonotic infectious diseases, which include brucellosis, bovine TB, *E. coli* O157:H7, and non-typhoidal salmonellosis. This is especially risky since approximately 90 percent of the total volume of milk produced in Sub-Saharan Africa is consumed fresh or fermented and only a small proportion is marketed through official channels (Cosivi et al. 1995). In Kenya, 84 percent of the total milk produced is sold at farm gate and only 16 percent is pasteurized (Omore et al. 2002). In addition to zoonoses, allergies, noise, and odour from manure, environmental pollution from manure uncollected by local authorities and accidents caused by wandering animals also pose risks in urban livestock keeping.

The aim of our study was not to do a risk analysis but rather an identification of hazards urban dairy farmers could be exposed to considering their closeness to livestock. We also aimed at identifying urban farmers' perceptions of the benefits and risks of dairy keeping and possible mitigation strategies as well as assessing the prevalence of brucellosis, bovine tuberculosis and aflatoxin M1 in milk and the potential risk of cryptosporidiosis through handling of animal manure. The choice of the hazards characterized was deliberate and was made in order to provide data for comparison with that available from Nairobi. It was hoped that the results of this study would generate data on some potential risks that exist and for which there is need to institute specific mitigation measures. One outcome is that a process of awareness creation was begun through sharing the output of the study among many stakeholders, both in Nakuru and other towns in Kenya.

Approaches and Methods Used

Health Risks Examined

Brucellosis is a bacterial zoonosis (i.e. a disease that can be transmitted from animals to humans) caused by *Brucella abortus* that is found mostly in cattle and *Brucella mellitensis,* an organism mostly associated with goats though it has recently been shown to infect cattle. The bacteria colonize the reproductive system of the animals and are excreted through milk. Humans are infected through contact with uterine secretions of aborted foetuses, or through drinking raw or unpasteurized milk.

Bovine tuberculosis is a bacterial zoonosis caused by *Mycobacteria bovis.* It causes wasting and coughing in animals and the bacteria are excreted through milk. In humans, most cases do not involve the lungs (extra pulmonary). Humans get infected through consumption of raw or unpasteurized milk, infected meat or through aerosols from infected animals. Human cases of tuberculosis have increased and it not known how many may be due to *Mycobacteria bovis* in this era of

HIV/AIDS. Determination of the reactor rate in cattle would help to assess the potential risk posed to humans by infection with tuberculosis due to *Mycobacteria bovis*.

Cryptosporidiosis is a protozoan infection of humans with Cryptosporidia species, many of which exist in many different animal hosts, and some of which have not been shown to infect humans except in immune-compromised individuals with HIV/AIDS. Cattle are known to harbour four species of cryptosporidia (*C. bovis, C.rynae* (deer-like genotypes), *C. andersoni and C. parvum*). It is only *C. parvum* that has been shown to be zoonotic. The parasite inhabits the gastrointestinal tract and infective oocysts are excreted in the feces. These oocysts are resistant to environmental conditions and common disinfectants used to treat drinking water. The major route of human infection is fecal-oral or through drinking water contaminated with infected cattle feces. While cryptosporidiosis is a self-limiting diarrhoea in immuno-competent individuals, it can be life threatening in those who are immuno-compromised. Urban dairy management practices can expose humans to constant contact through raw cattle dung and manure.

Aflatoxin B1 is one of many mycotoxins produced by certain fungal species under the right environmental conditions of moisture and temperature when they grow on cereals, nuts and pulses. When grain contaminated with aflatoxin B1 or feed made from contaminated grain is fed to animals, they break down the aflatoxin B1 and release aflatoxin M1 that is excreted through milk. Aflatoxins in general have been shown to be carcinogenic, mutagenic, teratogenic, immunosuppressant and cause depressed growth and loss of productivity. Aflatoxin B1 is ten times more toxic than aflatoxin M1. Aflatoxin B1 is associated with acute poisoning and M1 with chronic disease. At the time of study, Kenya reported acute human aflatoxicosis attributed to contaminated maize. Attempts to rid the market of grossly mouldy maize were instituted, but accompanied by fears that these grains could find their way into animal feeds. Urban dairy production uses more commercial feeds than its rural counterpart, making it more prone to using contaminated animal feed. Kenyans consume about 80 kg/capita of milk on average. If this milk is contaminated with aflatoxin M1 then there is a high exposure to aflatoxin M1 among the most vulnerable age groups in urban areas where the milk is consumed.

The Study Population and Data Collected

The same four wards were used for this study as for the one described in the previous chapter, giving a bias toward livestock keepers, but also a useful basis for comparison with the earlier data. The overall profile of urban farming households in Nakuru established by earlier studies showed that, while urban farmers are to be found more in the low density (high income) than the high density (low income) areas, this tendency was more pronounced for crop growing than livestock keeping. This is because growing crops is space-intensive whereas livestock require little space if the animals are kept in stalls or allowed to roam freely. All low-income urban livestock keepers kept small livestock, mostly chickens, but only 23 percent of them kept large livestock compared to 70 percent of the better-off livestock farmers (Foeken 2006, pp. 40, 143).

A preliminary participatory urban appraisal (PUA) began our studies in each of the four wards as described in Chapter 11 above. This was followed by focus group discussions (FGDs) on benefits and health risks of urban livestock keeping, taking 2 days in each ward and with about seven men and seven women farmers meeting separately. Proportional piling was used to depict the group's consensus based on individual households' perceptions. The discussions elicited the knowledge, attitudes and husbandry practices affecting participants' exposure to the risks posed by the selected health hazards and the mitigation strategies they used to reduce the risks. Ranking was done by piling stones. For benefits for example, each participant was given 50 stones and asked to distribute them on each benefit, depending on their perception of how cattle contribute to these in their household. The stones were counted and the mean for each benefit from each gender group obtained.

From the total list of farmers generated during the FGDs, a random sample of around 40 cattle farmers per site was drawn as described in the previous chapter. The actual numbers were 38, 33, 50 and 45 cattle-keeping households at Kivumbini, Kaptembwa, Menengai and Nakuru East, respectively. The first two are low-income areas, while Menengai is high income and Nakuru East middle income. The household head or spouse, a child over 18 or a worker, was interviewed in each of the 166 households to deduce their knowledge of the disease or condition in their animals and the risk pathways involved in each, as well as mitigation measures taken to reduce the selected hazards. All questionnaires were cross-checked for consistency, accuracy and omissions. Where any of these were detected the enumerator returned the next day to cross-check the information. Further, bio-sampling and laboratory analysis were conducted to assess the actual presence of the selected hazards in these livestock-keeping households.

The *prevalence of brucellosis in cattle* was assessed by detecting the presence of antibodies to brucella organisms in milk. Ten ml of unboiled milk pooled from lactating animals in each sampled household was frozen until analyzed using indirect ELISA (BrucELISA 160 M, VLA, UK). Buffers and reagents were reconstituted according to the manufacturer's recommendations and a microtitre plate pre-coated with *Br. melitensis* antigen. Fifty µl of the diluting buffer was added and immediately 50 µl of milk sample added. The plate was incubated at room temperature for 30 min on a rotary shaker or at 37°C for one hour without shaking. The plate was then washed using the wash solution and dried using absorbent paper. The conjugate 100 µl per well was added and the plate further incubated at 37°C for one hour. The plate was washed five times with the wash solution. The substrate was added 100 µl per well, and the plate incubated at room temperature for 15 min and then the reaction stopped by adding the stop solution. The optical densities were read at 405 nm using a multiskan manual reader (Labsystems). Positive, medium and negative controls were included in each assay. The positive/negative cut-off value was taken as 10 percent of the mean of the optical density of the eight positive control wells. Any test sample having an optical density equal to or above this value was considered positive. In order to validate the assay, the binding ratio was calculated as follows:

$$\text{Binding ratio} = \frac{\text{Mean of eight positive controls}}{\text{Mean of three negative controls}}$$

If the binding ratio was greater or equal to 10, the results were accepted and if less than 10 the results were rejected.

Milk samples collected for the detection of brucellosis were also used for the *detection of Aflatoxin M1* in milk. Aflatoxin M1 in milk was determined using Charm SL Aflatoxin test kit for raw and commingled bovine milk. The Charm SL Aflatoxin test is a rapid receptor lateral flow assay. The test detects Aflatoxin M1 at the USA action level in milk (500 ppt). On addition of milk, visible binding agents react with any Aflatoxin in the flowing sample. The T line stops flow of unbound reacted binder and the C line stops reacted binder (Table 12.1).

Table 12.1 Concentration response of Aflatoxin M1

Concentration (ppb)	% positive Rosa reader	% positive visually	Average ± SD from optional ROSA ppt calibration
0	0	0	−242 ± 63
0.125	7	7	−110 ± 162
0.2	23	23	264 ± 180
0.25	27	27	299 ± 163
0.3	50	50	383 ± 191
0.4	87	90	586 ± 62
0.5[a]	98	98	636 ± 74

Source: Charm Sciences Inc. USA.
[a] USA action level equivalent to 500 ppt Aflatoxin M1

The detection of M1 was done on fresh cooled milk samples. Three hundred microlitres of the milk sample was added to the test strip already fixed in the Rosa incubator. The tape was resealed and incubated for eight minutes. Thereafter, the strips were removed and the results read by comparing the intensity of the T and C lines. Positive samples had the test line or T line lesser than C or totally absent while in negative samples the T line was the same as or darker than the Control C line.

For the *detection of Bovine tuberculosis (BTB)*, the sampling unit was an individual animal selected for testing in each household. The question was whether the household was at risk. The oldest animal was selected and a single comparative intra-dermal tuberculin test administered at the cervical region and 0.1 ml of bovine purified protein derivative was injected intra-dermally. Fifteen centimetres below the site of bovine tuberculin, 0.1 ml of avian purified protein derivative was also administered. The sites of injection were marked with indelible ink and the undulation measured 72 h later. A difference of 4 mm or greater between the swelling of the bovine and avian tuberculin sites indicated a positive reaction.

Detection of Cryptosporidia oocysts. Farmers in Nakuru used cow dung in their gardens. But first they would heap it to dry before applying it to the fields as a soil improver. Cattle are known to harbour zoonotic cryptosporidia. If the cattle are infected, they pass these oocysts in the dung. If such contaminated dung is not properly handled by composting or allowing it to dry, manure could be a source of infection for people handling either raw cow dung or incompletely dried or composted cow dung. About 100 g of cattle manure was collected from cow dung heaped to dry before being applied to the gardens in each household. A portion of each of these manure samples was stored in 70 percent alcohol until staining to demonstrate the oocysts. A wet smear of the sample was spread on a microscope slide and air-dried. The slides were fixed in absolute alcohol for 5–10 min, stained using basic carbol fuchsin for seven minutes (Zeihl – Nelsen method), washed under tap water and decolourised using 3 percent acid alcohol for five minutes. The slides were counterstained using malachite green for 7–10 min, washed, air dried and observed microscopically under oil immersion for oocysts that stained red against a blue–green background.

Data were analyzed using Ms Excel for the participatory data and Ms Access for the household survey and a part of the participatory data. This was then exported to Instart® for both descriptive and analytical analysis. Mean scores for benefits, health risks and practices were obtained by dividing the sum of scores by the number of participants in each group in the four sites and results compared across gender. Inferences were made accordingly. The apparent prevalence of bovine tuberculosis, brucellosis, AFM1 and cryptosporidiosis were calculated as the number of samples testing positive divided by total samples at 95 percent level of confidence.

Results

Farming Household Characteristics

As stated in Chapter 11 above, our survey of 213 urban farmers revealed 16 percent women-headed households, similar to the earlier surveys by Foeken (2006). We found the highest proportion of women heading household (26 percent) in low-income Kivumbini while the lowest (10 percent) was in high-income Menengai (Fig. 12.1). As the previous chapter also showed, urban farmers in our sample were somewhat older, slightly less educated and more focused on farming as a livelihood strategy and source of income than was the case with the random sample survey of Nakuru farmers. We also found gender differences, women farmers having less education and being less often in formal wage employment as an alternative livelihood strategy than were the men.

The large majority of urban farmers surveyed (169 of 213) were mixed farmers, 11 grew crops only, while 33 kept livestock only. Of the 202 farmers keeping livestock, chicken were the most common, kept by 56 percent, which may be compared with Foeken's (2006, p. 67) finding that 80 percent of all urban livestock farmers

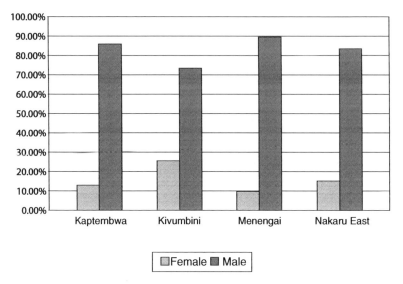

Fig. 12.1 Distribution of household head by gender in all sites

kept chickens. In terms of livestock numbers, the most common type of livestock was also chickens, followed by cattle and sheep (Fig. 12.2). Households kept an average of three cattle, with no significant difference across gender [$p=0.94$]. Cattle contributed a much higher proportion of livestock market value however, at 66 percent, with poultry, the most kept species, contributing only 15 percent followed by

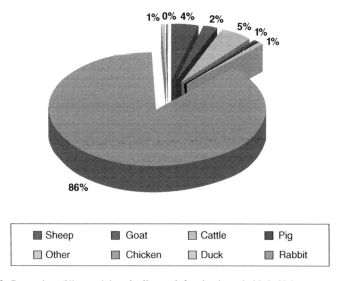

Fig. 12.2 Proportion of livestock kept by livestock farming households in Nakuru

sheep (8 percent), goats (4 percent), pigs (2 percent), ducks (1 percent) and rabbits only 0.2 percent. Kang'ethe et al. (2007e) reported similar findings in Kisumu where cattle contributed 70 percent of total livestock value and chicken 9 percent though again they were the most kept species. In contrast, in Kibera where cattle were not recorded, chickens contributed 25 percent of the total value of livestock.

Gendered Perception of Risks and Benefits

FGD participants listed both the benefits and health risks associated with urban dairying. The most important benefits were food and nutrition, self-employment leading to improvement of living standards, education for children, increased income, security, poverty alleviation, organic waste recycling, uplifting social status and cultural reasons. Men and women scored these differently, men ranking income as most important overall, while employment benefits were ranked first by women (Fig. 12.3). Across the four sites the ranking of the benefits by gender followed a similar trend except for the low-income area Kivumbini, where men ranked food and nutrition first, as did women in the high-income area of Menengai.

The results are interesting to compare with Foeken's (2006) survey showing food provision being the chief reason for urban farming (93 percent for crop growers and 74 percent for livestock keepers). Commercial reasons predominated for only 8 percent of crop farmers and 26 percent of livestock farmers, if both categories of income generation and income diversification are combined (Foeken 2006, p. 80). By showing that both men and women tended to rank income and

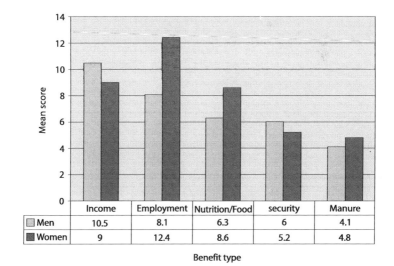

Fig. 12.3 Ranking of benefits of livestock keeping by gender in Nakuru

employment first, our survey thus suggests that cattle keepers are on the leading edge of commercialising urban farming in the town.

Though we classified food and nutrition as one benefit, during discussion women viewed milk as a source of food while men saw livestock keeping as having both a direct nutritive value as well as enabling purchase of other foods from income generated from the sale of animal products. This is suggestive of the gender role separation, with women more responsible for household provisioning and men for earning income (Lee-Smith et al. 1987). Financial security and manure were the other benefits listed and ranked by the participants.

Health risks listed included diseases, odour, environmental degradation and pollution, injuries or accidents, multiplication of disease vectors, allergies, noise, drug residues, stress and skin problems from handling animal waste. Again there were gender differences in the overall scoring (Fig. 12.4).

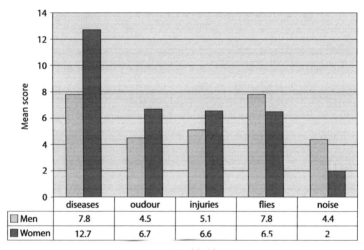

Fig. 12.4 Overall gender ranking of listed health risks in Nakuru-municipality

Only in high-income Menengai did men rank disease as the major health risk. In low-income Kaptembwo and Kivumbini and middle-income Nakuru East, men ranked odour, flies and injuries as the major health risks. Women on the other hand ranked disease as the major health risk in all areas. The diseases referred to were brucellosis by both men and women in low-income Kaptembwo and by women in middle-income Nakuru East, while men in Nakuru East referred to rabies and tetanus and women in Kaptembwo referred to tuberculosis.

Results of the household survey of 166 cattle farmers indicated that 70 percent of the respondents associated livestock keeping with risks other than infectious disease (odour 25 percent, injuries 23 percent, pollution 19 percent, and breeding of insects such as flies 10 percent, noise 5 percent, allergies 4 percent and others 6 percent). However, about three-quarters (76 percent) also said they were aware of the

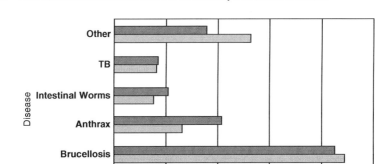

Fig. 12.5 Disease health risks associated with livestock keeping by gender

various zoonotic diseases, with no significant difference between men and women as shown in Fig. 12.5. "Other" diseases mentioned were allergies, cancer, amoeba, anaemia, poisoning and gout, mastitis, milk fever, new castle diseases, and fowl typhoid, cysticercosis, diarrhoea, ECF, rabies, salmonellosis and stomach ache.

Practices in Livestock Husbandry

Animal husbandry practices which farmers associated with health hazards through transmission of disease included milking, feeding, treating, dipping and watering the animals, cleaning of cattle sheds and disposal of animal waste. Men and women had different perceptions of who carried out most of these activities, as shown in Figs. 12.6 and 12.7.

Women perceived themselves as the main ones at risk of exposure to health hazards associated with dairy farming, followed by hired male workers and boys in the household. Men household heads on the other hand scored themselves as at greater risk than their womenfolk, along with hired male workers, with boys in the family exposed to a lesser extent. Hired female workers were rarely involved in attending to livestock. These results reveal the gender convention of livestock keeping being thought of as a male task. However there is a clear disagreement among men and women in livestock-keeping households, with men claiming they work on various tasks and the women saying they do little such work. This can be attributed to men being influenced by conventional interpretations of gender roles, while those more involved in the actual tasks (women) give a more accurate interpretation. In other studies (Ishani et al. 2002; Kang'ethe et al. 2005), similar disagreements were observed. In Dagoretti for instance, men argued that although they are not directly

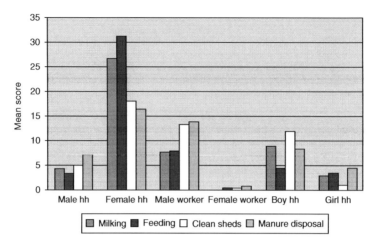

Fig. 12.6 Women's perceptions of persons involved in animal husbandry practices

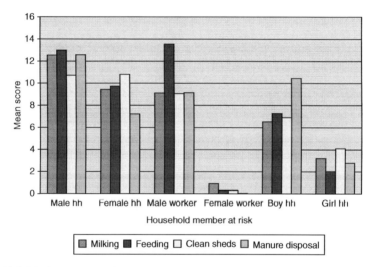

Fig. 12.7 Men's perceptions of persons involved in animal husbandry practices

involved in daily farming activities involving cattle, they are responsible for giving out money for feeds and treatment of sick animals as well as supervising. They argued that this made them as much involved as were the women.

Selected Health Hazards Investigated

A total of 117 milk samples were collected and analyzed for brucella antibodies. Three samples (3 percent) one each from Kaptembwo, Nakuru East and Menengai, were found to be positive. Sixty-one percent of the participants in the FGDs had heard of brucellosis. Of these 62 percent were women. The gender difference on knowledge of brucellosis was highly significant [$p = 0.0004$]. Of the women who had heard of brucellosis, 40 percent were from Menengai ward, where most residents have higher incomes and only 20 percent of households have women with low levels of education. Out of the eight FGDs, 14 percent (4/28) of the women participants knew how brucellosis was transmitted between cattle and humans compared to only 7 percent (2/28) men. A significantly higher proportion of women respondents boiled the milk used to make traditionally fermented milk [$p = 0.01$]. This may have been due to the fact that 62 percent of the women respondents in the household survey had heard of brucellosis and 14 percent of the women in the FGDs knew how brucellosis was transmitted. During the FGDs (conducted before the household survey), where the participants did not know about the hazards or risks, the transmission pathways were explained and participants' knowledge improved, likely contributing to more people being aware of both risks and mitigations during the household survey. Farmers are known to share information gained from educational meetings and this information may have reached even those who did not participate in the FGDs.

Five percent of the household survey respondents had a family member who had suffered from brucellosis and this was not statistically significant across gender. Twenty-seven percent of respondents reported having seen at least one person in the community suffering from brucellosis. Of those who had seen a case of brucellosis, more men (94 percent) knew the major clinical symptoms of brucellosis in humans with a significant proportion of men associating fatigue with brucellosis [$p < 0.05$] (Table 12.2). Fifty percent of respondents in the FGDs considered themselves at risk of contracting brucellosis. Ninety-nine percent of the respondents in the household survey listed boiling of milk as the most important mitigation strategy for preventing brucella infection. Other measures listed included use of artificial insemination, screening and vaccination of animals and testing of milk.

Table 12.2 Association of clinical symptoms of brucellosis in humans by gender

Symptoms	Men		Women		p-value
	No.	%	No.	%	
Headache	8	50	13	40.6	0.54
Joint ache	6	37.5	5	15.6	0.09
Fever	6	37.5	9	28.1	0.5
Fatigue	7	43.8	0	0	<0.05
Body weakness	1	6.3	17	53.1	0.002
Inappetance/Nausea	4	25	9	28.1	0.82
Don't know any symptom	1	6.3	7	21.8	0.17
Total knowing a case of brucellosis	16		32		

Note: Multiple symptoms identified by some respondants

Concerning aflatoxin in milk, AFM1 was detected in six samples out of 117 (5.1 percent). During the FGDs, 67 percent of the participants from the four sites had heard of aflatoxicosis, most of them from Menengai and Kaptembwo and 65 percent women. All those who had not heard of it were men (all the men from Kivumbini, Menengai and Nakuru East). However only 21 percent of farmers (12 out of 56) associated the intoxication with consumption of mouldy grains, while none knew how aflatoxins could be transmitted from animals to humans or how aflatoxicosis manifests in affected animals. It was therefore not surprising that they did not consider themselves at risk of intoxication and were not aware of any preventive measures against it.

By contrast, while 41 percent of respondents to the household survey had heard of aflatoxin poisoning and a small proportion thought the cause was the chemicals used to preserve grains, 58 percent knew how cows could pass on the intoxications to humans, and a high proportion of these were men. About 99 percent said the route of transmission was by consumption of mouldy grains, mainly maize. As was the case with brucellosis, the apparent disagreement in results between the participatory and household surveys may be attributed to the knowledge gained during the FGDs, with administration of the household survey shortly afterward.

About 42 percent of the feed given to cattle in the study households was in the form of concentrates. The rest was composed of grass and organic waste (42 and 16 percent respectively). Concentrates therefore appear to be the most likely source of aflatoxin in cattle. Forty-eight percent of the animals were reared in zero-grazing units and it was this proportion that had a high risk of being fed contaminated feeds with Aflatoxin B1 (Fig. 12.8).

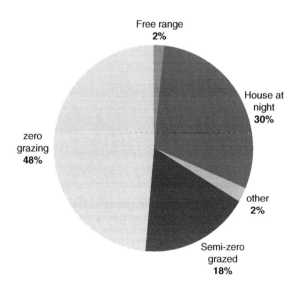

Fig. 12.8 Cattle rearing systems in the sampled households (other types of rearing included tethering)

Eighteen percent of the households compounded their own feed rations and the milk from the cows belonging to one of these households tested positive for aflatoxins. The cocktail often contained dairy meal, maize germ, poultry waste and salt. Sixty-eight percent of the respondents did not take any precautionary measures against aflatoxins or did not know of any. Of the six households whose milk tested positive for aflatoxins, three kept the animal feeds in stores while the others kept them in the house. One of the stores was assessed to be humid with conditions favouring mould growth that could lead to aflatoxin production.

Out of 97 animals tested for bovine tuberculosis, 17 animals (18 percent) had positive tuberculin test results. Sixty-one percent (34/56) of FGD participants had heard of human tuberculosis, 71 percent of them women (24/34), but nobody knew the cause or was aware bovine tuberculosis could be transmitted to humans. Some symptoms of tuberculosis in humans mentioned by three out of the eight groups included emaciation or wasting, coughing, weakness, light hair colour, sweating and chest pains. Participants in four of the groups knew of at least one person who had coughed blood, but none of the participants knew whether they were at risk of bovine tuberculosis, nor did they know any risk factor associated with the disease or how it could be prevented.

As with the other diseases, the household survey showed that later on people knew more because of knowledge gained during the focus group discussions. Thirty-four percent knew that cattle could be infected with tuberculosis, with no significant gender difference. Its routes of transmission from animal to humans were given as consumption of raw milk from infected animals (41 percent), consumption of meat and milk from infected animals (26 percent), air borne transmission (13 percent) and close contact (9 percent). Three percent of the respondents had a family member suffered from tuberculosis in the past year. The symptoms of tuberculosis in humans mentioned included chest pains (8 percent), coughing (13 percent), general body weakness (9 percent), difficulty breathing (17 percent) and loss of appetite (4 percent). However, a big proportion, 30 percent, did not mention any sign, possibly due to not having any close relatives suffering from the disease.

On measures taken to prevent risk of bovine tuberculosis, varied responses were given (Table 12.3). Among those who perceived that human health could be affected by livestock, a higher proportion of female (85 percent; 23/27) respondents knew of the mitigation measures compared to their male counterparts (76 percent; 22/29), though this difference was not significant ($p=0.4$). Differences between the proportion of men and women mentioning the various mitigating measures were also not significant.

Only one out of the six respondents whose animals tested positive replied to the question on tuberculosis transmission and indicated that transmission was through consumption of raw milk. On mitigation measures, one reported boiling of milk while two did not know of any measure. Two did not respond to the question and one reported taking no measure at all.

Out of 128 manure samples, 33 were positive for cryptosporidia oocysts. This represents a prevalence of 26 percent, higher than the 17 percent reported in cattle waste in Dagoretti (Kang'ethe et al. 2007a).

Table 12.3 Mitigation measures aimed at preventing exposure to bovine tuberculosis by respondents who perceived the disease a risk to human health

Mitigation measure	Men		Women		p-value
	No.	%	No.	%	
Boil milk	18	62.1	17	63	0.95
Meat inspection	1	3.4	2	7.4	0.5
Cook meat well	3	10.3	4	14.8	0.6
No precaution	5	17.2	3	11.1	0.5
Don't know	2	6.9	4	14.8	0.3
Total perceiving human health affected by livestock	29		27		

Note: Multiple measures identified by some respondants

Conclusions and Recommendations

The study confirms the emergence of cattle rearing and dairy production as commercial activities for urban farmers, who also engage in it in order to feed their families. This was common in all the four wards having people of different income levels. Knowledge of disease transmission pathways and mitigation strategies were found to be important factors in heightening risk from the types of livestock-rearing practices engaged in by all income levels of farmers, with some variations between men and women suggesting direction for intervention approaches. Information and training for urban farmers, women and men, are needed, through various channels, including agricultural extension services.

The brucellosis prevalence of 3 percent found in Nakuru is slightly higher than the 1 percent reported for Dagoretti (Kang'ethe et al. 2007b), although lower than that of 0–12 percent reported in a study of formally and informally marketed milk in Kenya generally (Arimi et al. 2005). However, the milk samples in that study were not from urban households but from areas with high and low market access and different cattle production systems. Nasinyama & Randolph (2005) reported a higher prevalence of brucellosis in Kampala (44 percent) using the milk ring test. The lower prevalence in Nakuru may be attributed to very small herd sizes with little to no inter-household mixing of cattle and to better husbandry practices as a result of high levels of education of the household heads.

The major routes of transmission of brucellosis include consumption of inadequately boiled milk, and contact with aborted foetuses or uterine secretions from infected animals. From this study, only 1 percent reported drinking unpasteurized or raw milk, less than the 4 percent found in Dagoretti. The 43 percent who made sour milk (*lala*) without first boiling it is comparable to 31 percent found in the Dagoretti study. It is these groups that are at higher risk of contracting brucellosis.

Of the measures to mitigate the risks, boiling of the milk by 99 percent of the respondents compares well with the 96 percent found in Dagoretti and would take care of the risk posed by brucellosis and bovine tuberculosis. Wearing of protective clothing while cleaning the cattle shed was however not considered a mitigation measure by respondents in Nakuru and the fact *brucella* organisms can penetrate intact skin means that those involved in cleaning the shed (mainly women, hired men and boys) are at risk. The study showed that although women perceived disease as a risk more than did men, men were generally more knowledgeable than women about the specifics, with the one exception of the women's knowledge on the mitigation strategies for brucellosis. Considering that women are more involved with the daily livestock-keeping activities than men, providing knowledge is vital and would assist in mitigating the risks to this vulnerable group.

Aflatoxin M1 was detected at the level of 5 percent, lower than the 44 percent reported in the Dagoretti study (Kang'ethe et al. 2007c). A possible explanation is that only 48 percent of the farmers were practicing zero grazing in Nakuru (52 percent used other methods) with minimal use of commercial feeds; only 42 percent of households gave them to cattle. In view of the fact that only 21 percent of FGD participants associated the intoxication with consumption of mouldy grains and none of the farmers knew how aflatoxins could be transmitted from animals to humans, farmer training on the routes of transmission is an important mitigation strategy. Another risk factor that may predispose them to using contaminated feed is the compounding of feed rations at home.

The prevalence of bovine tuberculosis at 18 percent was higher than the 10 percent reported for Dagoretti (Kang'ethe et al. 2007d). Although there was no knowledge of tuberculosis in cattle and how it could be transmitted to humans in the FGDs, this proportion rose to 34 percent during the household survey, indicating that such discussions can create some awareness. However, knowledge of bovine tuberculosis is too low, with farmers unaware of the risk they may expose themselves to when carrying out routine dairy-related activities. Boiling of milk was given as a mitigation strategy by 24 percent of respondents while 29 percent did not have any mitigation strategies to prevent exposure to bovine tuberculosis.

Farmers' traditional practice of heaping cow dung or slurry produces anaerobic decomposition due to the high moisture content (Gichangi et al. 2006), and does not generate heat or ammonia that would kill pathogenic Cryptosporidia oocysts. The presence of these in manure spread on crops may present a risk to consumers of the food produced if it is eaten raw as salad or not properly cooked. Farmers applying manure without protective clothing and who fail to observe hand washing after such activities may be at risk if the manure contains viable infective oocysts. Both participatory discussions and the household survey revealed poor knowledge of the risks associated with dairy production and utilization of manure among both men and women in Nakuru. This portends badly for the farmers who may be at risk of exposure to a number of health risks in the course of their daily duties.

It is imperative that training on causation and transmission pathways of health hazards associated with dairy and manure be carried out in order to raise awareness among vulnerable persons. In addition, ways of mitigating these health risks must be taught to dairy farmers. Government agencies responsible for controlling zoonotic diseases should redouble their efforts to provide these services for the public good in order to protect consumers and ensure food safety. Public information as well as extension information for farmers would be appropriate.

The private sector is contributing to farmer training in animal feeding, housing and milking hygiene and delivery of other forms of services (artificial insemination and treatment of animal diseases). The private sector should widen its scope of service delivery with the aim of increasing the knowledge base of the farmers on health risks associated with dairy production. This would complement and support the role of the public sector in mitigating these risks. Finally, more research is needed, especially risk analysis of all the hazards, in order to improve understanding of the health burden posed by these risks in the health sector. This would allow proper budgetary allocations to tackle these risks.

References

Arimi, SM, Koroti, E, Kang'ethe, EK, Omore, A & McDermott, J 2005, 'Risk of infection with *Brucella and Escherichia coli* 0157:H7 associated with marketing of unpasteurized milk', *Acta Tropica*, vol. 96, pp. 1–8.

Ayaga, G, Kibata, G, Lee-Smith, D, Njenga, M & Rege, R 2005, *Policy prospects for urban and peri-urban agriculture in Kenya: results of a workshop organized by the Kenya Agricultural Research Institute (KARI), Urban Harvest – CIP & International Livestock Research Institute (ILRI)*, Policy Series Dialogue #2, KARI Headquarters, Nairobi, 15 July 2004.

Cosivi, O, Meslin, FX, Daborn, CJ & Grange, JM 1995, 'Epidemiology of *Mycobacterium bovis* infection in animals and humans, with particular reference to Africa', *Revue Scientifique et Technique, Office International des Epizooties*, vol. 14, pp. 733–746.

Denninger, M, Egero, B & Lee-Smith, D 1998, *Urban food production, a survival strategy of urban households, report of a workshop on East and Southern Africa*, Workshop series 1, Regional Land Management Unit (RELMA) & Mazingira Institute, Nairobi.

Egziabher, AG, Lee-Smith, D, Maxwell, DG, Memon, PA, Mougeot, LJA & Sawio, C 1994, *Cities feeding people: an examination of urban agriculture in East Africa*, IDRC, Ottawa.

Foeken, D 2006, *"To subsidise my income": urban farming in an East-African Town*, Brill Academic, Leiden.

Gichangi, EM, Karanja, NK & Wood, CW 2006, 'Composting cattle manure from zero-grazing systems with agro-organic waste to minimize nitrogen losses in smallholder farms in Kenya', *Tropical and Subtropical Agro-ecosystems*, vol. 6, pp. 57–64.

Greenhow, T 1994, *Urban agriculture: can planners make a difference?*, Cities Feeding People Report 12, IFHP-CIP, Edmonton.

Ishani, Z, Gathuru, PK & Lamba, D 2002, *Scoping study of interactions between gender relations and livestock keeping in Kisumu*, Mazingira Institute, Nairobi.

Kang'ethe, EK, Randolph, TF, McDermott B, Lang'at AK, Kimani VN, Kiragu, MW, Ekuttan CE, Ojigo D, Onono J, Ngonde A & M'Ibui, GM 2005, *Characterization of benefits and health risks associated with urban smallholder dairy production in Dagoretti Division, Nairobi, Kenya*, a project report submitted to IDRC.

Kang'ethe, EK, McDermott, B, M'Ibui, GM, Randolph, TF & Lang'at, AK 2007a, 'Investigation into the prevalence of bovine cryprosporidiosis among smallholder dairy households in Dagoretti Division, Nairobi, Kenya', *East African Medical Journal* 84, S76–S82.

Kang'ethe , EK, Ekuttan, CE, Kimani, VN & Kiragu, MW 2007b, 'Investigation into the prevalence of bovine brucellosis and the risk factors that predispose humans to infection among urban dairy and non dairy farming households in Dagoretti Division, Nairobi, Kenya', *East African Medical Journal* 84, S96–S100.

Kang'ethe, EK, M'Ibui, GM, Randolph, TF & Lang'at, AK 2007c, 'Prevalence of Aflatoxin M1 and B1 in milk and animal feeds from urban smallholder dairy production in Dagoretti Division, Nairobi, Kenya', *East Africa Medical Journal* 84, S83–S86.

Kang'ethe, EK, Ekuttan, CE & Kimani, VN 2007d, 'Investigation of the prevalence of bovine tuberculosis and the risk factors for human infection with bovine tuberculosis among dairy and non dairy farming households in Dagoretti Division, Nairobi, Kenya', *East African Medical Journal*, vol. 84, S92–S95.

Kang'ethe, EK, Kimani, TM & Gathuru, KP 2007e *Determining users, research priorities to translation of research outcomes into tangible benefits: a scoping study of urban livestock keepers in Kibera, Nairobi and Kisumu*, report submitted to Natural Resources International LTD (UK).

Lee-Smith, D & Lamba, D 1991, 'The potential of urban farming in Africa', *Ecodecision*, 3 Montreal.

Lee-Smith, D, Manundu, M, Lamba, D & Gathuru PK 1987, *Urban food production and the cooking fuel situation in urban Kenya*, Mazingira Institute, Nairobi.

Mosha, AC 1991, 'Urban farming practices in Tanzania', *Review of Rural and Urban Planning in South and East Africa (RUPSEA)*, vol. 1, pp. 83–92.

Nasinyama, GW & Randolph, TF 2005, *Provisional technical report: characterizing and assessing the benefits and risks of urban and peri-urban (UPA) livestock production in Kampala City, Uganda*, Mimeo.

Omore, A, Arimi, S, Kang'ethe, EK, McDermott, J, Staal, S, Ouma, E, Odhiambo, J, Mwangi, A, Aboge, G, Koroti, E & Koech, R 2002, 'Assessing and managing milk-borne health risks for the benefit of consumers in Kenya', in *MoA/KARI/ILRI/UoN/KEMRI Collaborative Research Report*, Smallholder Dairy (R&D) Project, International Livestock Research Institution (ILRI), Nairobi.

Sawio, C 1993, *Feeding urban masses. Towards an understanding of the dynamics of urban agriculture and land use change in Dar es Salaam*, PhD Thesis, Clark University, United States of America.

Smit, J 1996, *Urban agriculture: progress and prospects 1975–2005*, Cities Feeding People Series Report 18, IDRC, Ottawa.

Chapter 13
Urban Agroforestry Products in Kisumu, Kenya: A Rapid Market Assessment

Sammy Carsan, Dennis Osino, Paul Opanga, and Anthony J. Simons

Introduction

As researchers from the World Agroforestry Centre (ICRAF) and the National Museums of Kenya, we undertook the work described here as a contribution to Kenya's national development agenda, articulated in policies such as the Poverty Reduction Strategy Paper (PRSP) and Economic Recovery Strategy for Wealth and Employment Creation in Kenya after the change of government in December 2002. Those policies' priorities of ensuring food security and livelihood support systems require the identification of activities that can provide a basis for localized action. Urban hunger and poverty are major issues, and urban agriculture (UA) is one such activity that can play an important role in reducing poverty and enhancing opportunities for wealth creation. In a region experiencing declines in agricultural productivity and competitiveness resulting in national-level food insecurity (Cunningham et al. 2008), producing food and non-food agroforestry products nevertheless provides many with a meaningful sense of food and income security.

Agriculture in the larger Kisumu district is carried out at a subsistence level – maize, beans and sorghum commonly being grown for household consumption. However, vegetables such as kale, tomatoes and local vegetables are increasingly being cultivated for the urban market although the district remains a net food importer (UN-Habitat 2006) with food supplies coming mainly from neighbouring districts. Urban agriculture, including livestock keeping, though widely practiced, is restricted and penalized through repressive implementation of outdated urban bylaws, being seen as a nuisance and a threat to public health (Ishani et al. 2002). Still, a recent livelihood survey of Kisumu showed that 60 percent of households living on the peri-urban fringes of Kisumu were involved in some form of urban agriculture and livestock keeping, with a total of 737 households found to be holding an average of KShs 150 000 (US$2150) worth of livestock per household (Onim 2002). However, urban crops often fail due to poor farming practices, unreliable

---

S. Carsan (✉)
World Agroforestry Center (ICRAF), Nairobi, Kenya; University of the Free State, Bloemfontein, South Africa
e-mail: s.carsan@cgiar.org

rains, drought and frequent floods in low-lying areas, posing a direct challenge to poverty-reduction opportunities for the city.

UA, as defined in previous chapters, includes integrated agriculture and forestry systems. Growing trees can be combined with food crops or mixed cultivation and animal husbandry, which together diversify income and produce. The potential of urban and peri-urban agriculture to alleviate food poverty in Kisumu is recognized in the City Development Strategy 2004–2009 (UN-Habitat 2003). In fact, the city has a strong agricultural tradition and intensification of UA and livestock keeping is taking place. As well as being a place of production, a city like Kisumu acts as a trading hub for produce from surrounding areas and there is considerable variety in agroforestry food and non-food products traded there, with great potential to benefit many low-income urban dwellers, either from production or marketing or both. However, systematic studies to understand the value chain for key agroforestry products have been lacking, hindering the development of better marketing strategies. There is a lack of even a basic understanding of where things are coming from and going to, despite it being established that such key products as fruit, timber, medicinal plants and local vegetables have potential to contribute significantly to farmers' and small urban traders' income and food objectives. An enhanced role for these products is depressed by lack of data on local markets, whose size or even availability is unknown.

It is against this backdrop, and based on the need to better inform national policies on food security and other livelihood support systems, that we carried out a rapid market assessment on agroforestry food and non-food products in urban and peri-urban Kisumu. The study aimed to situate urban agroforestry within the wider system of production of agroforestry products by identifying local demand and supply patterns for existing products and their embedded value chain characteristics. Key food and non-food products surveyed were fruit, local vegetables, medicinal plants, tree nurseries and timber.

This chapter specifically discusses marketing prospects for key identified agroforestry products, whether from urban, peri-urban or rural production, as well as current challenges and opportunities to enhance their enterprise value. This is intended to be of benefit to urban food security and livelihoods through production and marketing and to be of assistance to Kisumu City, which faces the task of integrating urban agriculture and livestock keeping into its broad urban development plans, albeit within a carefully regulated framework to minimize conflicts and health problems such as addressed elsewhere in this book.

The Study Area

Situated on Lake Victoria, Kisumu City is Kenya's third largest urban area and headquarters for both Kisumu District and Nyanza Province. Founded as a railway terminus and internal port in 1901, it later became the leading commercial, trading, industrial, communication and administrative centre in the Lake Victoria basin (UN-Habitat 2003). It covers an area of approximately 417 km^2 and had an estimated population of 345 312, growing at a rate of 2.8 percent per annum, according to the 1999 Kenya Government census. At 1100 m above sea level, the city has a humid climate with bimodal rainfall averaging 1245 mm per annum.

Kisumu has suffered from neglect under successive political regimes in the country, lacking both national policy support and investment. Remarkably, despite its great natural resource potential, it has one of the highest food poverty levels in Kenya. Fifty-three percent of its population lived below the food poverty line in the early 2000s compared to 8 percent in Nairobi, 39 percent in Mombasa and 30 percent in Nakuru (CBS & MPND 2003). Its status as an opposition stronghold during national economic decline in the 1980s and 1990s eroded the city's economic potential, resulting in high rates of poverty. About 60 percent of Kisumu's population lives in un-serviced settlements and over 15 percent has HIV/AIDS (UN-Habitat 2006). Over half the city's population is engaged in the informal sector, with increasing street trading in the Kisumu Central Business District accompanied by an insurgence of street children. Along with rapid urbanization this means many low-income households have extremely insecure livelihoods (Guendel & Richards 2003). Further, residents suffer from lack of decent and affordable housing and the city's transport and communication potential is unexploited. There are great disparities in the distribution of urban infrastructure and basic services between slums and higher income areas. The road network is deficient, excluding many slums from easy access to the city centre and other strategic areas. The poor are forced to rely on the informal sector for survival.

Methods

Several survey techniques were combined to carry out the assessment. Lacking any secondary data on traded agroforestry products to inform the study design, we made a cross-section of urban and peri-urban Kisumu using area maps, key individual interviews and chain referrals, to discover what products exist and their market activities. Most such products were found in the urban informal settlements where they are commonly traded in open-air markets. Using a sketch map of Kisumu, a total of sixteen such urban and peri-urban markets were identified for assessment. These markets were classified as urban or peri-urban based on their distance from the central business district and the observable trade volume. The following four markets were classified as urban: Jubilee, Kibuye, Kondele and Mamboleo, while nine markets, namely Kiboswa, Otonglo, Kibos, Nyamasaria, Kisian, Obambo, Ojola, Chiga and Rabour, were classified as peri-urban. Figure 13.1 shows the four urban and the innermost of the peri-urban markets visited. Three World Bank-funded open-air markets in the informal settlements of Migosi, Manyatta and Nyalenda were found to be not in use, reportedly due to their being poorly sited in terms of strategic attraction of good trade volumes and lacking proper road links to facilitate transactions. Permission was sought from the relevant local authorities for the market visits and interviews, and in the case of urban markets the visits were also carried out with the agreement of the market superintendents.

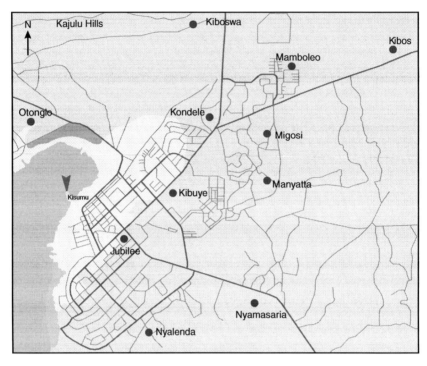

Fig. 13.1 Map of Kisumu city showing key open air markets

Two or three individuals found selling the products being investigated were interviewed in each open-air market. A total of 25 respondents were interviewed, either from the open-air markets or at these individuals' workshops or timber yards. Four focus group discussions (FGDs) were carried out, two in urban markets (Kibuye and Kondele) and two in peri-urban markets (Kiboswa and Kibos) using the same interview questionnaire tool, with open-ended questions leading to discussion. FGD's were easy to set up as the traders for each product were found stationed together at the market place. Finally, five nursery operators and three timber yard operators were also interviewed to complete the product survey. Although minimal, the numbers were keyed in and analyzed using Microsoft Excel for simple descriptive statistics. Simple value chain assessment was performed for each product market (Thompson et al. 2005).

Agroforestry Food Products

Fruit

The fruits most traded in Kisumu at the time of survey included tamarind (*Tamarindus indica*), guava (*Psidium guajava*) and mango (*Mangifera indica*). Fruit such as pawpaw (*Carica papaya*) are important, although these were not surveyed as they were not in season. A more extensive study would be needed to get a comprehensive picture, although the three types of fruit found are indicative of market conditions (Thompson et al. 2005).

Tamarind is a common fruit in many urban and peri-urban market outlets in Kisumu. Two cultivars, readily identified by their taste and size, are traded locally. Fruit pods come from the neighbouring regions of Alego, Gem, Ugenya and Asembo, where the wholesale traders gather them from homesteads in 50 and 100 kg size sacks. Traders have to meet the costs of collection (plucking and packing) and transport to urban wholesale markets like Kibuye. Here, bulk loads of tamarind pods are broken down and traded in smaller units called *gorogoro* or tin-load – a standard local measuring unit. It was estimated that about two to three 100 kg size sack loads are sold per week per trader in the urban market outlets such as Kibuye. Table 13.1 details the levels of transaction.

Retail traders, mainly women coming from the peri-urban markets around Kisumu, such as Kondele, Nyamasaria, Kibos, Jubilee and Otonglo, source their tamarind supplies from the urban markets. At these peri-urban outlet markets, which are mainly in the informal settlements, the pods are sold in small bundles or by the single pod. The retail volume per trader at the small peri-urban outlets like Kibos and Nyamasaria is normally about one to two tin-loads per week. There is no processing or special packaging being used, and the consumers of tamarind are mainly urban dwellers. Interestingly, the tamarind chain was found not to extend beyond Kisumu, with nothing being transported to Nairobi or other cities, although the

Table 13.1 key fruit type transactions in urban and peri-urban Kisumu

Produce	Produce source	Collection cost/unit (in US$)	Farmer SP/unit (in US$)	Trader SP (US$)/unit (urban outlet)	Selling price/unit (peri-urban outlet)
Tamarind	Alego, Gem, Ugenya Asembo	0.5 – 0.8 per 50 – 100 kg size sack	2 – 2.6 for 50 kg sack 5 – 7 for 100 kg sack	0.3 – 0.4 per 2 kg tin-load	KSh 5 – 10 Bundle Or KSh 1 per piece
Mango	Nyahera, Kiboswa Imports from Nairobi	Farmers collect	0.25 – 0.40 per basket	6.6 per 100 kg size sack	KSh 2 – 5 per piece
Guava	Widely available in peri urban Kisumu	Farmer/trader collection	5 – 6 per basket	6 – 7 per basket load	KSh 1 per piece
Local vegetables	Kisumu–Kakamega route, Kiboswa, Gambogi, Serem, Sui & parts of Nandi Escarpment	Farmer/ trader	2.6 – 4 for 100 kg size sack	n/e	High trader mobility between markets can make up to US$13 per day

Key: n/e: not established; SP: selling price; Exchange rate US$1 equivalent to KSh.75

survey could not conclusively identify all possible buyers at the farm level. The interviewed buyers did not indicate that there was any form of competition for farm produce. Table 13.1 provides a summary of key costs in the tamarind trade.

A particular local variety of mango is sourced from peri-urban Nyahera and Kiboswa along the Kisumu–Kakamega road and is marketed in small basket-loads during the hot season. Farmers seem more engaged in the mango value chain than the tamarind one, either by selling their produce at strategic points along the busy Kisumu–Kakamega route or by trading it at peri-urban outlets such as Kibos and Kondele. Table 13.1 summarizes mango transactions in Kisumu. During the hot season (January to February), mangoes of different varieties are imported from outside Kisumu and often traded in sacks at Kibuye urban market, interestingly at a lower bulk price of KShs 500 ($6.6) for a 100 kg sack.

Guavas are not widely traded, even though they grow abundantly, especially in peri-urban Kisumu. The depressed market demand is indicative of a widely available product. However, traders in certain peri-urban places like Kiboswa take advantage of their proximity to the main road to sell to travellers.

Local Vegetables

Women dominate all levels of the vegetable supply chain. Peri-urban traders often source vegetables from their own home gardens situated in close proximity to the market outlet whereas the more urban traders, particularly those in Kibuye, may not necessarily be farmers. However, the majority of the women traders hail from Kakamega District, and at the retail outlets they operate in groups of five to ten. The main source of the vegetable is the Kisumu–Kakamega peri-urban route, from farms scattered in Kiboswa, Gambogi, Serem, Sui and parts of Nandi Escarpment.

Numerous species of vegetables were found to be traded in Kisumu, many of them local, traditional foods. The most common were wild spider flower, creeping foxglove, cowpeas, mustard collard and black nightshade, but there were many others. Table 13.2 provides a list of some of the vegetable species found in the city.

In outlets such as Kondele, Kibos and Jubilee, vegetables are sold in bundles, often by the producers themselves, the size of bundle varying with outlets and consumers, while at the retail level the different traditional vegetable species are sold by the traders. This wide distribution responds to consumers' needs to prepare the variety of local dishes that are largely accompanied by these vegetables. The vegetable trade seems to offer more returns to traders than fruit and medicinal plants. Traders rotating among the different open-air markets can make an average of $13 per day. A sack-load of leafy vegetables costs $2.6–4.0 at the point of origin; transport for a 100 kg size sack to the urban market outlets costs $1.3–2.6 (Table 13.1). Traders source the vegetables from their own and from neighbours' farms.

In addition, a complementary vegetable-seed trade has emerged as a result of the booming vegetable trade, suggesting that urban and peri-urban production may be expanding through home gardens. Kiboswa market is now one of the largest points for local vegetable seeds. Seeds for most vegetable species were found to

Table 13.2 Local vegetable species traded in Kisumu markets

Common name	Botanical name	Seed trade?
Akeyo/dek/chisaga (wild spider flower)	*Cleome gynandra*	Yes
Apoth/murenda(e) (Nalta jute)	*Corchorus olitorius/trilocularis*	Yes
Atipa (creeping foxglove)	*Asystasia gangetica*	Yes
[a]Awayo	*Oxygonum sinuatum*	n/f
Cowpea	*Vigna unguiculata*	Yes
Dodo/chiboga/muchicha	*Amaranthus* sp.	Yes
Kanthira (mustard collard)	*Brassica carinata*	Yes
Managuu, Osuga (black nightshade)	*Solanum nigrum*	Yes
Mitoo (sunhemp)	*Crotolaria brevidens*	Yes
[a]Moringa, ben tree	*Moringa oleifera*	n/f
[a]Rosemary	*Rosemarinus officinalis*	n/f

[a] n/f- not found: trade in seed for this vegetable species not found

be traded except for *Oxygonum sinuatum*, *Moringa oleifera* and *Rosemarinus officinalis* (Table 13.2). Seeds are sold in cups (200 ml size), with prices varying by the species. A cupful of *Crotolaria brevidens* and *Cleome gynandra* were the most costly seed traded at $2 and $1.5, respectively (Table 13.3).

Table 13.3 Key plant species used as remedies for different health problems

Local name	Botanical	Main and/or combination uses
Ajua	*Ceasalpinia volkensii*	Malaria
Aloe	*Aloe* spp.	Ulcers, arthritis, as a laxative, wounds, rashes, burns
Arupiny	*Commiphora africana*	Typhoid fever, stomach problems, malaria
Murembe/Orembe	*Erythrina. abyssinica*	Gynaecological conditions, diarrhoea, malaria
Mbao/Ndege	*Eucalyptus* sp.	Headaches resulting from cold
Manyasi	*Cotula anthemoids*	Cold, colic, rheumatism
Mukombero	*Mondia whytei*	Aphrodisiac, appetizer
Mwarubaini	*Azandiracta indica*	Malaria
Nyaluetkwach	*Toddalia asiatica*	Gastrointestinal, fever, malaria, cholera, rheumatism, lung diseases
Nyatigotigo	*Chenopodium opulifolium*	Leaves (cooked as spinach) are used as a nutritional supplement – no medicinal value reported
Obolo	*Annona senegalensis*	Respiratory infections, de-worming, toothaches, snake bites, cuts, wounds,
Ochol	*Euclea divinorum/ racemosa*	Purgative
Odolo	*Chrysanthelium americanum*	Urinary tract infections
Ombasa	*Tylosema fassoglense*	Gastrointestinal, anaemia, healing uterus after birth
Ombulu	*Abrus precatorius*	Respiratory problems
Mswaki	*Salvadora persica*	Toothbrush
Onera	*Terminalia brownii*	Yellow fever in children
Tido	*Ekebergia capensis*	Emetic, dysentery, headaches, chronic coughs
Yago	*Kigelia africana*	Respiratory and gynaecological conditions, malaria, ulcers, sores, syphilis

Moringa pods were found only in Jubilee market, where demand – mostly from the Asian urban community – was reported to be dwindling. The pods, mostly harvested in August and September, are sourced from Kajulu Division and sections of Kano plains or from small-scale individual farmers in Kisumu town. One trader sourced them from his own farm in Kano plains while others were reported to be supplied from Nairobi.

Medicinal Plants

The World Health Organization (WHO) estimates up to 80 percent of the world's population, mainly in poor countries, rely on traditional healthcare, and recent studies have aimed at documenting and promoting better incorporation of ethnomedicine in Kenya and Uganda into government regulation (Aduma 1999; GAN 1999). An estimated 2500 species of medicinal and aromatic plants are traded worldwide. Most of these medicinal plants are still collected from wild sources (Cunningham et al. 2008). During this survey over 20 plant species were recorded as useful for treatment of different ailments (Table 13.3), either singly or in combination therapies.

The survey confirmed the existence of an important medicinal plants trade in Kisumu, enhanced by a strong cultural attachment as well as belief in their efficacy. Kibuye and Kondele urban markets are major outlets for medicinal plant products in the city. Much of the trade takes place in stalls, especially in Kondele. The products are sold either as raw extracts (roots, leaves, bark) or in powdered form. A significant number of consumers were said to prefer raw extracts, as they are sure of the ingredients rather than powdered formulations where it is difficult to distinguish the active ingredients included. The prices for products are flexible and often based on negotiations with customers. However, customer bargaining for lower prices was reported to increase when it comes to locally sourced and well known species. Products are prepared at home as an accompaniment to meals for many home-based therapies, including malaria and HIV/AIDS management. Many herbal medicine hawkers made claims for the efficacy of their formulations to cure HIV/AIDS. We made the assumption that the many ailments treated were actually opportunistic diseases from HIV. Most plant species are harvested from the surrounding peri-urban hilltops of Kajulu. However, plant material for product formulations may also be sourced from diverse regions, including neighbouring districts.

Interestingly, we found that cultivation of medicinal plants on farms was rare and impeded by certain cultural and traditional religious beliefs. Many believe these plants can only be obtained from common property such as the surrounding community hills. Both men and women, old and young, are engaged in the trade in medicinal plant products, but it is limited to particular family circles due to traditional beliefs. To overcome some of the challenges in traditional medicine practice, training of ethno-healers on record keeping and plant analysis, and promotion of cultivation of medicinal plants by both women and men, are among measures being proposed for a Regional Center for Medicinal Plants and Biodiversity Research, with an emphasis on the Lake Victoria Basin in which Kisumu lies (Aduma 1999).

Products Transport

Transportation for agroforestry products to key urban and peri-urban markets was identified as a major factor determining volumes of trade and even the nature of trade. Strategically located open-air markets like Kibuye and Kondele receive higher volumes as they are easily accessible by main paved road from different parts of the city and are also linked to the high-potential areas of Kakamega District. Road transport by public means (taxi) or hired truck is a common form of transport for the more bulky products. Charges are per distance covered and determined on negotiation. Table 13.4 indicates transport costs for key agroforestry products traded in Kisumu.

Table 13.4 Transport cost for various products around urban and peri-urban Kisumu

Product	Quantity	Locations	Distance (km)	Maximum cost (KSh)
Avocado	100 kg sack	Kiboswa-Kibuye	7	50 ($0.70)
Mango	100 kg sack	Kibuye-Kondele	2	50 ($0.70)
	100 kg sack	Kiboswa-Kibuye	7	50 ($0.70)
Tamarind	50 kg sack	Kibuye-Kondele	2	20 ($0.3)
	100 kg sack	Alego-Kibuye	50	2450 ($35)
Local vegetables	100 kg sack	Kiboswa-Kibuye	7	200 ($2.7)
		Serem-Kibuye	15	100 ($1.4)
	100 kg sack	around Kiboswa Gambogi-Kondele	10	50 ($0.70)
Poles	7 tonne lorry	Gambogi-Kibuye	12	3500 ($50)
		Hamisi-Kibuye	30	4500 ($64)

Non-food Agroforestry Products

Increased agroforestry programs could provide fuelwood, building poles, and timber for the Lake Victoria region (Kairu 2001). There are many open spaces in the urban and peri-urban areas of Kisumu, indicating the potential to produce many tree-based products, even though much of this land appears to be privately owned. This survey assessed the marketing activities for eucalyptus poles, timber, firewood and tree nurseries.

Kibuye and Kondele urban markets were found to be the main outlets for wood products such as poles and timber. Commonly traded timber species in Kisumu include cypress (*Cupressus lusitanica*), pine (*Pinus patula*), eucalyptus (*Eucalyptus saligna*) and podo (*Podocarpus* sp.). A respondent in Kondele revealed that cypress, pine, eucalyptus and podo are commonly sourced from the distant North Rift areas of Nandi, Kapsabet, Marakwet and Nakuru, including Molo and Elburgon. The respondent sources 8000 feet of various sizes of cypress that is supplied by seven-tonne lorries. The gate price of the wood (inclusive of transport) is about KShs 130 000 ($1800). On average, he places orders for timber every 3–4 weeks, depending on demand. The common sizes of timber sold include 6 × 2, 8 × 1, 4 × 2, 3 × 2, 2 × 2 and 6 × 1. The premises are rented at a fee of KShs 5000 ($70) per month, while municipal licensing fees are paid annually and range between KShs 6000 and 8000 ($83–111). Currently, cypress is the most popular timber in the city.

The large timber traders in Kisumu also serve neighbouring districts. However, there is shortage of wood supply owing to a logging ban on gazetted forests. Environmental degradation due to destruction of forests is widespread in Kenya while erosion due to deforestation is affecting Lake Victoria in particular. Thus the persistent timber shortage is feared to worsen in the near future, while agroforestry and reforestation initiatives have not been intensified.

The supply of cypress and other wood types was, for instance, reported to have decreased in the last 5 years, while the consumer behaviour of the Luos – the majority of customers within and around the city – is resistant to other wood types. One trader cites the example of *Podocarpus*, which he believes is a superior wood, although locals still prefer cypress. The price of timber has been increasing tremendously, with a 100 percent price increment in the last 5 years, cypress selling at KShs 28 for 6 × 2 size, KShs 25 for 8 × 1 size, KShs 17 for 4 × 2 size, KShs 15 for 3 × 2 size, KShs 11 for 2 × 2 size and KShs 17 for 6 × 1 per foot.

It was reported to be difficult to operate in the timber value chain due to government regulations and huge transport costs. Transporters sometimes twin as suppliers of timber products, gaining competitive advantage by combining both functions. Timber enterprises involve huge financial investments. There was an indication that all timber traded from one neighbouring district was sourced from farms.

Most of the poles traded in Kisumu, unlike for sawn wood, come from farm plots. The poles are mainly harvested from *Eucalyptus* spp. whose coppicing ability provides poles on at least a yearly basis per stump. Preliminary investigations indicate that eucalyptus poles, used mainly in construction as scaffolding, are sourced around the border of Kisumu and Kakamega or in peri-urban locations such as Kiboswa. The normal practice is that a dealer based in an urban market (mainly Kondele), operates both as retailer and wholesaler, buying eucalyptus poles from farmers with woodlots on their farms. The price is based on negotiation of the value of standing trees in a farm woodlot. Roughly, it costs about KShs 15 000–18 000 ($208–250) to fill a seven-tonne truck with eucalyptus poles. The buyer pays for the cost of felling trees, with loading estimated at KShs 2250 ($31). Such a task requires about 15 workers, who are paid at a rate of KShs 150 ($2) per person per day for the work done. Transport costs KShs 3500–4500 ($50–64) from the peri-urban areas of Gambogi and Hamisi to Kibuye, respectively (Table 13.1). On average, the traders require three seven-tonne truckloads per month in the Kondele outlets surveyed.

In Kondele, the poles fetch a good price when they are green, but when dry they are sold cheaply for firewood. When still green, the poles are mainly used in construction work. Prices vary with diameter. The smallest size is sold at KShs 15 ($0.2), while the biggest is sold at KShs 60 ($0.8) per pole. Consumers and some retailers come all the way from Busia, Bondo and Siaya towns. Further, investigation indicates the dealer requires the area chief's permit to enable him to transport forest product (poles) from one district to another. Some of the dealers have more than one pole outlet.

Tree Nurseries

Tree nurseries were identified as an important emerging business in urban Kisumu. The local authorities also seem comfortable with it as it contributes to the city beautification and environmental programs. There are a number of nurseries scattered along the roads leading into and out of Kisumu city. Young men aged 25–40 run tree nursery businesses along the main access roads, selling tree seedlings to passers-by, who are residents of Kisumu and neighbouring districts. From the five nurseries surveyed, it was evident that fruit-tree seedlings form a prominent part of the whole nursery enterprise, the most common being mango (*Mangifera indica*), orange (*Citrus* spp.), pawpaw (*Carica papaya*), avocado (*Persea americana*), jackfruit (*Artocarpus heterophyllus*), Indian blackberry (*Syzingium cuminii*) and passion fruit (*Passiflora edulis*). Seeds for establishing tree nurseries are reportedly sourced from Nakuru region, especially for timber species such as *Grevillea robusta* and *Eucalyptus*.

Few indigenous species and quantities were found in the nurseries as compared to the exotics. Some of the indigenous species found included *Spathodea campanulata* and *Markhamia lutea* whereas common exotic species identified in the nurseries were *Dovyalis caffra*, *Eucalyptus* spp., *Casuarina equisetifolia*, *Cupressus lusitanca*, *Grevillea robusta* and *Cassia siamea*.

Another feature of urban tree nurseries in Kisumu is the production of species with ornamental value, as earlier studies have demonstrated. Some key ornamental species traded include *Dovyalis caffra, Ficus benjamina, Palm washingtonia, Spathodea campanulata, Callistemon citrinus, Hibicus* spp. and *Bougainvillea* spp. In the early nineties it was reported that the sale of seedlings is a common informal sector business with the majority of plants and seedlings on sale being trees and decorative shrubs (Smith 1998). More recently it has been shown that gross incomes for urban and peri-urban community nurseries are largely supported by the large-sized ornamental seedlings produced (Muriuki & Carsan 2004).

While there are huge price differences between seedling species, the prices do not differ much from one nursery to another. The ornamentals are usually sold to new homeowners and for landscaping functions. They fetch marginally higher prices compared to fruit trees or any other type of seedling. Surveyed tree nurseries can stock up to 20 000 seedlings per year, earning an income range of KShs 5000–10 000 ($69–140) per month.

Market Chain Analysis

In our study we found three types of market channel for agroforestry products in Kisumu (Fig. 13.2). In the first type, farm products are taken to urban wholesale outlets, mainly Kibuye and occasionally Kondele, and then further to retailing outlets in peri-urban market centres such as Kibos, Kiboswa, Nyamasaria and Otonglo. An example of this kind of arrangement is the tamarind chain. The second channel applies to peri-urban farm products taken directly to retailing outlets, mostly skipping the wholesaling step. Products supplied in this manner include local vegetables, medicinal plant products, construction poles and fruit selling at bus stops.

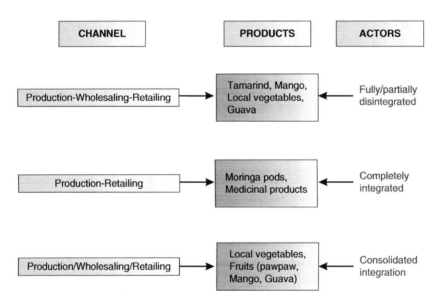

Fig. 13.2 Diagram of market chains for agroforestry products in Kisumu

In most instances, those dealing in products along the first channel are fully or partially disintegrated, which means that there are no direct connections or links between actors in the production and retail functions of the market chain. Transport providers or agents often facilitate these links. In case of the second channel, there is a possibility of integration among actors. This is illustrated when dealers make a personal effort to go to the source area and find the product, for example for the supply of local vegetables. The third channel represents consolidated integration, where the functions of production and retailing are consolidated, as in the case of peri-urban vegetable vendors. The most completely integrated case is that of the nursery enterprises, where production and retailing roles are performed by the same trader.

These preliminary findings indicate that fruit such as tamarind have a complete market chain involving wholesaling and retailing even though they are seasonal, available only during the period August–October. Overall, our study shows that profit margins for agroforestry food products are generally low, although, with targeted improvements in the value chains of local vegetables and fruit varieties, there is potential for bigger margins.

These commercial agroforestry activities provide income and employment for numerous people, especially women and youth. Young men aged 20–35 were identified as key actors in the tree nursery enterprises, while the traditional vegetable supply chain is dominated by women, all the way from the farms to the Kisumu urban markets.

We identified the following marketing challenges during this assessment:

- *Product value chain* – there is an enormous disparity in incomes between actors in the value chain, with agents and "middlemen" who have better market networks reaping the most benefits. Small-scale producers often failed to achieve better market conditions and prices owing to their limited power;
- *Product differentiation* – farmers often failed to differentiate their products so as to attain a competitive edge and maximize income. They were unable to cushion themselves against frequent price fluctuations for certain products;
- *External factors* – there is a failure of policy and institutional support. Poor urban planning that doesn't incorporate urban agriculture and agroforestry as a means of livelihood constrains product development. In addition, sub-optimal infrastructure means there is an increased cost to trade. The inaccessible road networks compound high transport costs and limit trade volume.

Our study revealed that urban markets were more likely to source products from outside Kisumu while peri-urban market outlets were more likely to source some of their products from the surrounding farmlands. We found that Kibuye is an important centre for trade in agroforestry food and non-food products in Kisumu City and constitutes a major pathway for the wholesale and retail trade in food. Although there was this distinction between urban and peri-urban markets – with the latter sourcing products from peri-urban production – key urban markets such as Kibuye also sold products from outside Kisumu to the peri-urban markets.

The Kisumu–Kakamega route is a key link for the supply of agroforestry food products entering Kisumu City, including traditional vegetables, construction poles, and fruits such as mangoes, pawpaw and tamarind. At present, almost all the agroforestry food and non-food products used in Kisumu come from the bordering districts and very little from the urban area, even though there is potential for production of local vegetables, fruits, medicinal and timber trees in peri-urban or even urban sites. The community-owned hills within Kisumu that surround the city – places like Kajulu and Nyabondo – are an important source area for medicinal plants. However, this trade is taking its toll in the form of land degradation and even of species' survival. Thus an important opportunity is presented to rescue this area ecologically and for the benefit of the community through carefully managed agroforestry, in coordination with efforts to conserve indigenous knowledge and promote ethno-medicine (Aduma 1999).

Conclusions

It is evident that there is potential for market development for many agricultural and agroforestry products in Kisumu, provided there is more attention paid to the products' value chains, right from the producer level, to enhance their marketability. However, more research is required to identify opportunities and pathways through which smallholders could wield more power in market channels to boost their profits. An improvement in market outlets for tamarind and guava seems like a useful opportunity to boost the products' trade. Further agroforestry extension information is needed to help farmers increase and diversify their choices of tree species for trade products. However, both cultural and ecological issues need attention in designing these interventions.

There was very little value addition or processing in the product chains studied, the exception being some herbal products that were powdered and packaged in bottles. Fruit pods such as tamarind have complete wholesale and retail market structures even though they remain seasonal, available only during the August–October period. Subject to sustainability assessment, the trade in construction poles offers opportunities for income generation due to the high demand from building and construction work in Kisumu and the neighbouring districts of Nyando, Bondo and Siaya. There is need to scale up action research and foster policy dialogue on urban agroforestry products along the lines of the Kampala study of urban agriculture described in this volume (see Chapter 8 in particular), and to carry out a comprehensive market survey on products that would offer a competitive advantage to farmers and traders in urban and peri-urban Kisumu. Opportunities for value addition will need to be explored and intervention points identified to enhance production of agroforestry products that give sustainable incomes to urban and peri-urban residents. Such a study would include products such as firewood, baskets, charcoal, tool handles, woodcarvings and grass, that were not included here. All these products need to be assessed also in relation to sustainability and the requirements of the National Environmental Management Act (NEMA) and policies that aim to conserve Kenya's valuable biological diversity and forest cover. Urban agroforestry offers considerable opportunities for improving the environmental management and sustaining the biological diversity of a city like Kisumu, and of simultaneously improving the livelihoods of the urban poor. As a first step, efforts should be made to enhance awareness of the opportunities and then to linking research to the needs of farmers in urban and peri-urban Kisumu.

References

Aduma, PJ 1999, *Medicinal plants and biodiversity in the Luo and Suba communities of Lake Victoria*, Report of a Workshop at Maseno University College, Archived Project Report 633.88 (676.2) A3, IDRC, Ottawa.

CBS & MPND (Central Bureau of Statistics & Ministry of Planning and National Development) 2003, *Economic survey 2003–2007*, Government of Kenya, Nairobi.

Cunningham, AB, German, L, Paumgarten, F, Chikakula, M, Barr, C, Obidzinski, K, van Noordwijk, M, Koning, Rd, Purnomo, H, Yatich, T, Svensson, L, Gaafar, A & Puntodewo, A 2008. Sustainable trade and management of forest products and services in the COMESA region: an issue paper. Bogor, Indonesia. Centre for International Forestry Research (CIFOR).

GAN (Green Africa Network) 1999, *Ethnomedicine and medicinal plant conservation in Rachuonyo District*, Pilot Study Archived Project Report 633.88:615.89 (675.2) G7, IDRC, Ottawa.

Guendel, S & Richards, W 2003, 'Peri-urban and urban livestock keeping in East Africa — a coping strategy for the poor?', paper presented at *Deutscher Tropentag 2003*, Georg-August-University, Göttingen, 8–10 October 2003.

Ishani, Z, Gathuru, PK & Lamba, D 2002, *Scoping study of urban and peri-urban poor livestock keepers in Nairobi*, Mazingira Institute, Nairobi.

Kairu, JK 2001, 'Wetland use and impact on Lake Victoria, Kenya region', *Lakes & Reservoirs: Research & Management*, vol. 6, no. 2, pp. 117–125.

Muriuki, J & Carsan, S 2004, 'Assessing the merits of community level seedlings production and distribution', in Ngamau, D, Kanyi, B, Epila- Otara, J, Mwangingo, P & Wakhusama, S (eds) *Towards optimizing the benefits of clonal forestry to small scale farmers in East Africa, ISAAA Briefs No. 33*, ISAAA, Ithaca.

Onim, JF 2002, *Scoping study for urban and peri-urban livestock keepers in Kisumu City, Kenya*, Lowland Agricultural and Technical Services Limited (Lagrotech), Kisumu.

Puntodewo, A 2008, *Sustainable trade and management of forest products and services in the COMESA region: an issue paper*, Center for International Forestry Research (CIFOR), Bogor, Indonesia.

Smith, DW 1998, 'Urban food systems and the poor in developing countries', *Transactions of the Institute of British Geographers*, New Series, vol. 23, no. 2, pp. 207–219.

Thompson, AA, Strickland, AJ & Gamble, JE 2005, *Crafting and executing strategy, the quest for competitive advantage: concepts and cases, 14th edition*, McGraw-Hill Companies, New York, NY.

UN-Habitat 2003, *Kisumu City development strategies*, http://hq.unhabitat.org/programmes/ump/kisumuCDS.asp [Accessed 29 February 2008].

UN-Habitat 2006, *Kisumu urban sector profile*, United Nations Human Settlements Programme.

Part IV
Urban Agriculture and Institutional Change

Chapter 14
IDRC and Its Partners in Sub-Saharan Africa 2000–2008

Luc J.A. Mougeot, Francois Gasengayire, Diana Lee-Smith, Gordon Prain, and Henk de Zeeuw

Introduction

The peer-reviewed studies in this book are part of a larger set of ongoing urban agriculture initiatives in Sub-Saharan Africa. In this chapter we highlight the role played by international collaboration in supporting these, focusing on a group of institutions and how their interactions have evolved over time. Even though these institutions only formalised their programs on urban agriculture over the last decade or so, several were already active in the field earlier. The ideas and concepts associated with urban agriculture have been around for even longer, while the reality of urban agriculture itself dates as far back as humanity's first urban settlements. For one, the "garden city" concept is rooted in modern urban planning (Smit et al. 1996; Lee-Smith & Cole 2008).

Urban agriculture became a topic of research and policy interest in the 1990s not only because by then it had become increasingly visible in towns and cities, it had also moved onto the radar screen of an environmental movement that was paying growing attention to urban areas from the 1970s. Still, what was missing was for urban agriculture to graduate from being a topic of interest for the mass media and activism to becoming a field of professional and institutional endeavour. A decade ago, urban agriculture lacked multi-disciplinary scrutiny of its claimed benefits and alleged adversities. It lacked the dedication of qualified technical expertise to overcome its constraints and realize its potential. It also lacked public policy attention aimed at helping tackle a range of urban challenges from malnutrition to unemployment, environmental degradation and public insecurity. This, therefore, was the process that Canada's International Development Research Centre (IDRC) and its partners endeavoured to promote in Sub-Saharan Africa over the last decade, following some initial research supported in Kenya, Uganda and Tanzania in the 1980s.

L.J.A. Mougeot (✉)
International Development Research Centre (IDRC), P.O. Box 8500, Ottawa, ON, Canada
e-mail: lmougeot@idrc.ca

When the Rio Earth Summit of 1992 addressed urban agriculture in its "Agenda 21", IDRC had already supported several research projects on the subject over the previous decade, mainly in Sub-Saharan Africa, through its Social Sciences Division. It was the first international agency to recognize the importance of this livelihood activity and to channel resources toward investigating it. Following the Earth Summit, urban agriculture became one of four topics of IDRC's Urban Environment Management Program established in 1993. This was replaced in 1997 by a Program Initiative entirely focused on urban agriculture, called "Cities Feeding People", named after a book published by IDRC in 1994. The book contained data on East African cities revealing that one-third or more of urban populations were feeding themselves through their own urban production. In 1996, the United Nations Development Programme (UNDP) published its review of urban agriculture. Based on a global survey that included data from the early IDRC-supported studies, it estimated that as many as 800 million people were then involved in this economic activity worldwide (Smit et al. 1996).

Two things influenced IDRC's decision to move its urban environmental programming toward a greater focus on urban agriculture in the 1990s. Firstly, during a period of reduced Canadian and Organization for Economic Co-operation and Development (OECD) official development aid, including reduced public funding for international development research, IDRC saw a need for narrowing the scope of its urban environmental programming. Its stock of research in urban agriculture as a new field of scientific and policy endeavour offered IDRC and its partners the opportunity of making a greater difference than could be had with the same amount of resources applied to more mainstream areas of urban development research such as housing, employment, water and sanitation or solid waste management. Secondly, drawing on research it had supported on urban food systems in the 1980s and early 1990s, IDRC recognized that, no matter how efficient were food supply systems in developing country cities, the poor continued to be systematically sidelined. Demand was more an issue than supply but the poor's dire situation was obviously made worse when supply was disrupted. Given a persistent mix of factors, both at micro and macro levels, self-provisioning strategies had been on the rise in developing country cities for over a generation. This was not to go away anytime soon. IDRC saw the urgency of understanding and tackling the constraints associated with urban and peri-urban agriculture practices because of their potential to alleviate hunger and poverty, if managed properly, as much as because of their potential to aggravate environmental and human health hazards if not managed properly. The macro crisis in overseas development aid (ODA) funding, which ended up affecting IDRC and similar agencies, was turned into an opportunity to creatively re-focus its programming by supporting the emergence of what the Centre saw as a legitimate and necessary field of scientific expertise and institutional capacity in developing countries.

The need was particularly felt in Sub-Saharan Africa, where, based on project findings going back to the mid-1980s, IDRC had begun not only to document urban agriculture activities but was also building up a cadre of specialists on urban agriculture in different parts of the continent. Following a series of descriptive and analytical surveys of urban agricultural activities in cities such as Dakar, Nairobi, Kampala, Harare and Dar es Salaam, the Centre built on these to support more action-oriented research. Such projects would be geared toward informing multi-stakeholder consultations and policy analysis and interventions to address challenges faced by some of the more important urban agriculture systems and their practitioners. In many instances, researchers who had led the original baseline projects played key leadership roles in the larger action-oriented projects. Given the emphasis of these projects on improving capacities for the longer term sustenance of activities initiated, collaboration with new and existing institutions working on the same priorities became an important part of the IDRC strategy. The Centre teamed up with new partners for training, improving research capacity and research-to-policy linkages, as well as direct support to research projects. By comparison with conditions it had met in Latin America and the Caribbean, IDRC was conscious of the fact that networking in Africa would have to account for a less-robust regional institutional environment.

In order to carry forward its strategy in Sub-Saharan Africa, one well-established network of scientific expertise with which IDRC sought to partner was the Consultative Group on International Agricultural Research (CGIAR). Over its 40-year history, the Centre has had a strong association with the CGIAR in a wide range of areas from commodity specific to environmental concerns. The CGIAR clearly held strategic potential for the development of knowledge and capacity in urban agriculture because of its links to a wide range of institutional partners and its reputation for sound and innovative research. During a review of the CGIAR system in 1998, it was recognized that growing urbanization and the increased reliance of city dwellers on urban farming had not yet been incorporated into its agenda. In response to that review and through a competitive call, a system-wide initiative on urban and peri-urban agriculture (now known as Urban Harvest) was launched in late 1999, with the International Potato Center (CIP) as the convening Center.[1]

The international network of Resource Centres on Urban Agriculture and Food Security (RUAF Foundation) was another key organization with which IDRC partnered in this field. Building up over time to eight regional partnerships spread across the globe, RUAF today supports and promotes activities on urban agriculture, including action research and policy intervention. It also acts as an information clearinghouse, its Urban Agriculture Magazine providing an important platform for reporting new work in the area.

Both separately and together, IDRC, Urban Harvest (UH), the RUAF Foundation and other international partners such as UN-Habitat, the UN Food and Agriculture Organization (FAO) and the International Water Management Institute (IWMI – also part the CGIAR) collaborated with government and non-government bodies in Africa and other developing regions to collectively bring greater research and policy

---

[1] The initiative was originally named The Strategic Initiative on Urban and Peri-urban Agriculture or SIUPA, and was renamed Urban Harvest in 2003.

support to existing and new local and national urban agriculture initiatives during 2000–2008. The following sections describe some of the activities undertaken by these three main partners and how they interacted one with another.

Building Capacity – IDRC'S Regional Courses and Individual Research Awards (AGROPOLIS)

In the period 2000–2008, IDRC used several modalities to carry forward its programming in the field of urban natural resource management and food security. Running from 1997 until early 2005, the Program Initiative known as "Cities Feeding People" (CFP) was focused on urban agriculture. Following a positive review, urban agriculture was retained as one of the four research topics of larger Program Initiative which succeeded CFP in 2005, called "Urban Poverty and Environment" (UPE). At the time of this writing an external evaluation of Phase I (2005–2010) of the UPE had just been completed.

CFP grants supported major research studies in the African region on urban livestock and on wastewater use and treatment for urban horticulture, all with an emphasis on policy influence. Building on project results in Rosario and Kampala, a multi-regional project on producing the "Edible Landscape" was launched in 2004, coordinated by McGill University in Canada. This focused on design and planning innovations to integrate urban agriculture into the built environment. In addition to Kampala, Uganda and Rosario, Argentina, activities were supported in Colombo, Sri Lanka and Montréal, Canada. Over the 2000–2005 period CFP and partner organizations in West Africa, Latin America and the Caribbean, East Africa and the Middle East, developed a series of four regional training courses on urban agriculture. The courses were at first targeted for researchers (West Africa), then for city teams composed of representatives of research organizations, local government and civil society organizations. Each new course benefited from the experience of previous ones and each served as the incentive for systematizing methodologies tested in individual projects carried out in the region until then. The courses also served as platforms for regional networking and forums for debate and improvement of city proposals that either strengthened urban agriculture actions underway or introduced new ones in participating cities. The 2004 course is described below in detail.

In 1998 CFP introduced a ground-breaking scheme to develop capacity among younger researchers in urban agriculture and related fields and to ensure a more bottom-up input into defining relevant research questions. The AGROPOLIS awards programme offered graduate field research grants to master's and doctoral students, expanding to include post-doctoral awards in 2002. In all, more than 60 such awards were made to graduate students all over the world, the majority of them for research in Africa. Twenty-one of the studies have been published in two books produced by IDRC (Mougeot 2005; Redwood 2009) and two-thirds of these are from Africa. At least four others – all from Africa – have been published elsewhere, in refereed journals or as mostly refereed chapters in Urban Harvest's book on health and urban agriculture (Nabulo et al. 2006, 2008a, b; Sebastian et al. 2008; Serani et al. 2008; Yamamoto 2008; Yeudall et al. 2007, 2008).

One of the benefits of the AGROPOLIS programme was its identification of research questions requiring resolution in order for actors to better address issues of hunger and poverty alleviation. Questions emerged from the younger generation of researchers and led to the search for answers about why people were producing urban food as well as how they were doing it (Mougeot 2005). For instance, a major concern of many of the studies was the contamination of urban-produced food. These included the examination of pesticide use in Lomé, Togo, in the first AGROPOLIS book, as well as studies of wastewater use in Ghana, Senegal and Congo in the second (Mougeot 2005; Redwood 2009). Likewise, the influential study of food security in Namibia in the first book was followed up by those in Kenya, Zimbabwe and Malawi in the second (Frayne 2005; Redwood 2009). The strong gender focus of the programme – a gender dimension was required for all studies and several made it their main question – provided much new knowledge that informed later studies (Hovorka 2005; Gabel 2005; Hovorka et al. 2009).

Two issues – the potential risks from contaminants and the positive potential for food security of urban agriculture (UA) – have also been a major focus of related work by IDRC's partners, RUAF and Urban Harvest, in their research and publications. Both themes have also been taken up by other organizations. Urban food security in particular is likely to be an enduring concern for many, as economic and food crises disproportionately affect poor countries, especially those in Africa.

The interactive nature of the programmes of IDRC and its partners is very clear through the functioning of the AGROPOLIS awards programme. Urban Harvest's book on Healthy City Harvests (Cole et al. 2008) relied on no less than four such awards that made possible the research reported in five of its chapters. AGROPOLIS enabled a greater number of graduate students to take part in the Urban Harvest-supported research on health and UA in Kampala and to strengthen the quality of the findings.

AGROPOLIS required supervision of the research from a reputable university, as well as a formal association of the awardee with a local implementing institution that was likely to make use of the research findings (Mougeot 2005, p.23). This worked particularly well in the case of the links to Urban Harvest, where quality supervision came from several Canadian universities as well as Makerere University in Uganda; at the same time, implementation was strongly grounded through the development of local research capacity linked to Kampala City Council. Thus the outcome was that several refereed articles appeared in scientific journals that emerged from the awards, and the findings also influenced legal reforms in Kampala concerning urban agriculture (Lee Smith et al. 2008). Other examples of benefits accruing to local actors from awardees' association with local organizations during their fieldwork are summarised in (Mougeot 2005, pp. 23–24).

In 2006, the ECOPOLIS program, designed as part of the UPE Program Initiative, replaced AGROPOLIS and continued to build on its lessons and successes IDRC, 2009. Taking a broader urban poverty and environment perspective, ECOPOLIS now offers research and design awards to young researchers (master's and doctoral level) enabling them to undertake research on (and designs for) water and sanitation, solid waste management, vulnerability to natural disasters, housing and land tenure, in addition to urban agriculture. Examples of projects supported through ECOPOLIS on urban agriculture include a doctoral award on health impacts of urban agriculture in Dakar, Senegal (University of Geneva), and a master's award on urban design guidelines to facilitate integration of waste management with urban agriculture in Cagayan de Oro City, Philippines (University of Calgary).

Inserting the Urban into International Agricultural Research – Urban Harvest

Originally focused almost exclusively on rural agriculture, the CGIAR research agenda began to pay attention to peri-urban issues in the 1990s, especially livestock production, livestock feed and dairy, agro-processing enterprises and the re-use of urban wastewater in agricultural production. In order to account for existing CGIAR research programs already active on urban issues and in order to engage their scientists in planning the new Urban Harvest's global programme, regional stakeholder meetings were held during 2000. The Sub-Saharan Africa meeting was attended by 10 of the 16 CGIAR Centres, plus four other international research organizations and a wide range of regional and national stakeholders. IDRC's active involvement in the meeting and in launching the CGIAR initiative in the region led to the award of a grant to Urban Harvest in 2002 in support of its African program, providing partial funding for the position of a Regional Coordinator and enabling a strong linkage with its own regional project portfolio. Urban Harvest's programme emphases on sustainable urban livelihoods, health, nutrient cycles and research-to-policy linkages meshed well with those of IDRC.

The new Urban Harvest Africa office set up in Nairobi in 2002 made a systematic assessment of urban and peri-urban agriculture (UPA) activities in the research programs of the CGIAR centres in the region, using face-to-face or e-mail interviews and documentary sources. Of the 16 existing centres at that time, nearly all had African offices and two were headquartered in Nairobi, namely the International Livestock Research Institute (ILRI) and the World Agroforestry Centre (ICRAF). Urban Harvest's Africa office was itself housed at the International Potato Center (CIP), on the ILRI campus. It was found that the largest amount of UPA work in the region was being done by ILRI and IWMI, both of which would subsequently make UPA more explicit within their new research structures (King'ori 2004).

Several CGIAR research centres have been actively involved in new Urban Harvest-supported studies reported on in this book (Table 14.1). Other international agricultural research centres, including the World Vegetable Center (AVRDC), the International Plant Genetic Resources Institute (IPGRI) and IWMI, collaborated through other activities, such as training courses on aspects of urban agriculture. Thus, the engagement of the CGIAR with urban issues steadily gained momentum in the region over the period.

Table 14.1 CGIAR centres' participation in urban harvest-supported Africa research

City	Lead centre	Other CGIAR participation
Yaoundé	IITA	World Agroforestry (ICRAF), World Fish (ICLARM)
Kampala	CIAT	CIP, IITA, ILRI, IPGRI
Nairobi	ILRI	World Agroforestry, CIP
Nakuru	CIP	ILRI

Another sign that urban agriculture was increasingly recognized as part of the CGIAR agenda was its presence on the program of CGIAR Annual General Meetings in Nairobi in 2003 and Mexico in 2004. In Nairobi, a tour of urban and peri-urban agriculture and smallholder dairy was jointly organized by Urban Harvest and ILRI. This was linked to a special networking meeting that attracted numerous participants and featured presentations by Nairobi City Council and the Kenyan Ministry of Agriculture, as well as IDRC. Urban Harvest's work in Sub-Saharan Africa and in Latin America was also featured at the Mexico meeting, through a special presentation during the field day and a stand in the exhibition hall.

The period between 2002 and 2008 was characterized by efforts not only to integrate different CGIAR Centres into the Urban Harvest programme, but also to strengthen collaboration with IDRC-supported research in the region and with civil society and community organizations.

A good example was IDRC's attention to urban livestock-health linkages, supported by both its Cities Feeding People and its Ecohealth Program Initiatives. This research involved teams of scientists and local government personnel investigating the health risks of urban livestock keeping. Another was the project funded by IDRC and implemented by Urban Harvest in collaboration with universities and farming communities in Nairobi. This was to assess the benefits and risks of wastewater reuse for agriculture in urban and peri-urban areas. The purpose was to understand the type, source and level of wastewater-associated pollution, and to work with key stakeholders on developing possible strategies for reducing these burdens without compromising the benefits of wastewater reuse. The Nairobi project built on IWMI's experience with a similar project in Ghana, also supported by IDRC. Urban Harvest's support to health research in Kampala, which included risks and benefits of livestock keeping (Cole et al. 2008), complemented these IDRC initiatives.

Urban Harvest took part in, and at times helped to run, related workshops in 2003 and 2004. It also facilitated the award of small grants from the IDRC Regional East and Southern Africa Office (ESARO) to Urban Harvest partners. These included:

- laboratory analysis of urban zoonotic diseases in Kampala;
- research on gender and urban dairy farming in Addis Ababa;
- development of a website on urban waste management for UPA in Nairobi;
- the start-up of community-based UPA activities in Nakuru;
- feasibility of utilizing commercial food and urban and peri-urban pig production in Kampala; and
- analysis of market opportunities and threats for urban and peri-urban farmers in Nairobi.

The synergy between IDRC and Urban Harvest also became apparent through Urban Harvest's methodological advice to a 2004 IDRC-sponsored livestock health workshop in Nigeria, and to the Scientific Committee of a sub-regional project on "Access to Land for Urban Agriculture for the Urban Resource Poor in East and Southern Africa". This project was coordinated by the Municipal Development Partnership for Eastern and Southern Africa (MDP-ESA) based in Harare, Zimbabwe and supported by IDRC.

Important links with other international partners, especially United Nations entities such as UN-Habitat and FAO, were also forged during these years, enabling urban agriculture to gain considerable political momentum in the region. This collaboration mainly occurred through joint organization of, or participation in, several influential meetings and the contribution of the Urban Harvest Regional Coordinator to drafting various declarations from these meetings, including:

- the Declaration on "Feeding Cities in the Horn of Africa" (Addis Ababa, May 2002);
- the Harare Declaration on UPA Policy (August 2003);
- the report of the meeting on "Urban Policy Implications of Enhancing Food Security in African Cities" held at UN Headquarters, Nairobi in May 2002;
- the Nairobi Declaration on Treated Wastewater Use in Urban Agriculture (February 2006).

These had more than a regional impact and also gave added support to activities within countries, as is described for the city of Kampala by Lee-Smith and colleagues (2008, p. 225). The 2002 Nairobi meeting brought together representatives of IDRC, UN Habitat, RUAF, Urban Harvest and MDP-ESA, thereby also enabling them to brainstorm on developing a regional course for Anglophone Africa. Further, results from research on several types of urban producer organizations around the world (co-funded by IDRC and UN-Habitat) were discussed, offering a platform of information from which FAO, with support from and in close consultation with IDRC, later reviewed different types of urban producer organizations and published a practical guide for working with low-income urban and peri-urban producers' organizations (FAO 2007).

Collaborating with local public sector and civil society organizations was another crucial aspect of Urban Harvest's mainstreaming work, and representatives from these attended the various regional meetings. The use of stakeholder platforms to integrate research with policy and institution building is described in some detail for the three countries dealt with in this book in Chapter 15.

Knowledge Management and Capacity Development for Policy and Action – the RUAF Foundation

Collectively, RUAF activities aim at reducing urban poverty and food insecurity by stimulating participatory governance and better urban environmental management. They do so by enabling and empowering urban and peri-urban food producers, both men and women, and by helping local governments, non-governmental organizations (NGOs) and private enterprises to integrate urban agriculture in their policies and action programmes.

The initiative to set up RUAF was taken by ETC (the Netherlands) in response to needs from actors in the South and North for effective mechanisms to exchange research data and practical experience on urban agriculture. The case for such an initiative was developed at a meeting of the International Support Group on Urban Agriculture, convened by IDRC in 1996 in Ottawa. But it took up to 1999 before the first RUAF programme could take off with DGIS (Dutch cooperation agency) and IDRC core funding, plus additional contributions from the Swedish International Development Agency (SIDA), GTZ and others. RUAF is an international network of seven regional Resource Centres on Urban Agriculture and Food Security (three of them in Africa) plus one global resource centre. RUAF was established as a Foundation in March 2005 and operates as an international network, providing training, technical support and policy advice to local and national governments, producer organizations, NGOs and other local stakeholders. The global resource centre, ETC-Urban Agriculture in Leusden, the Netherlands, is charged with programme coordination and strategy development; it also manages the Foundation's website and produces the Urban Agriculture Magazine in collaboration with its regional centres.

RUAF's outreach in Sub-Saharan Africa is led by its following regional centres:

- IAGU, Dakar – regional coordination for Francophone West Africa;
- IWMI-Ghana, Accra – regional coordination for Anglophone West Africa;
- MDP, Harare – regional coordination for Southern and Eastern Africa.

The Foundation's programme for 2005–2008, titled *Cities Farming for the Future,* included capacity building, policy development, action planning, production of knowledge in the form of documents and electronic and web-based materials, learning and reflection and gender mainstreaming. Among the books produced by RUAF, one titled *Cities Farming for the Future* was co-published in 2006 with IDRC and the International Institute for Rural Reconstruction (IIRR). It captures much of the results of this work and the concepts and learning that have emerged. The book was produced in close collaboration with many partners engaged in urban agriculture, including IDRC and Urban Harvest.

RUAF activities are focused on 20 cities worldwide and are run by teams of local stakeholders from various sectors, including municipalities, NGOs, producer organisations, research and training institutions and governmental organisations. RUAF's 2005–2008 programme was carried out intensively in those selected cities. In each, there were courses for training of trainers, policy workshops and project design and implementation. A multi-stakeholder forum and interdisciplinary working group in each city carried out a joint situation diagnosis, followed by joint policy formulation and action planning, resulting in a City Strategic Agenda on Urban Agriculture, as well as the implementation and monitoring of pilot projects.

So far, RUAF's work in Sub-Saharan Africa is being carried out in the following cities:

- *West and Central Africa* – Pikine (Dakar - Senegal), Bobo Dioulasso (Burkina Faso) and Porto Novo (Benin);
- *West Africa (Anglophone)* – Accra (Ghana), Freetown (Sierra Leone) and Ibadan (Nigeria);
- *Southern and East Africa* – Bulawayo (Zimbabwe), Cape Town (South Africa) and Ndola (Zambia).

While Urban Harvest contributed to the design and monitoring of new IDRC-supported projects in Eastern and Southern Africa, RUAF in West Africa built on existing IDRC-funded projects, including a West African network project on sustainable urban agriculture. This project investigated land tenure and wastewater reuse issues in addition to establishing baseline data on urban agriculture in seven countries: Benin, Burkina Faso, Ivory Coast, Mali, Mauritania, Niger, and Senegal. The project, coordinated by IAGU (African Institute for Urban Management), was followed up by IDRC support for an investigation of micro-credit mechanisms for urban agriculture as well as building urban farmers' business skills and relationships with financial institutions in Benin, Burkina Faso, Mali and Senegal.

To mainstream gender in urban agriculture, RUAF and Urban Harvest joined forces. In 2003 RUAF ran a regional gender course in South Africa, attended by a staff member from Urban Harvest in Nairobi. She later became the Gender Focal Point for Urban Harvest's parent organization, CIP, running gender courses and activities in several countries and partner organizations. In September 2004, Urban Harvest and RUAF jointly ran an inter-regional workshop on *Women Feeding Cities – Gender Mainstreaming for Urban Agriculture* in Accra, Ghana. It was hosted by IWMI's regional office, with extra support from IDRC and the CGIAR's programme on Participatory Research and Gender Analysis (PRGA). Material from the workshop was subsequently reworked, augmented and eventually compiled into a book, *Women Feeding Cities: Mainstreaming Gender in Urban Agriculture and Food Security*, which serves as a set of guidelines and a manual for people doing urban agriculture projects worldwide (Hovorka et al. 2009).

Relevant Knowledge for Action in African Urban Agriculture – the 2004 Anglophone Africa Course

Too often, training in African and other poor countries simply recycles knowledge produced in an alien environment. But IDRC and its partners developed a different approach, devoting their collective energies toward producing and disseminating regionally relevant knowledge in urban agriculture. The urban agriculture African regional course in English, held in 2004, offered the chance for the three major partners in this region – IDRC, Urban Harvest and RUAF – to work together as a collective unit. Led by Urban Harvest, the course was the third in a series supported by IDRC through regional bodies, the first held in Francophone Africa in 1998 under the coordination of the Institut sénégalais de Recherches agricoles (ISRA), and the second in Latin America in 2001, coordinated by the regional office of UN Habitat's Urban Management Program. A fourth course, for the Middle East and North Africa, was held in 2006, coordinated by the American University in Lebanon.

A technical committee set up in May 2002 became the central element for course organization, as well as a powerful networking tool for coordinating and sharing information on each member organization's urban agriculture activities in the region. The members were Urban Harvest, IDRC, RUAF, International Water Management Institute (IWMI) Ghana Office, Municipal Development Programme for Eastern and Southern Africa (MDP-ESA Regional Office, Harare), UN-Habitat's Training and Capacity Building Branch and the Urban Management Program's Latin America and Caribbean Office (UMP-LAC). The last mentioned also provided assistance to Urban Harvest's Africa office, having organized the Latin America course. The Government of Kenya's Research Development Department later joined the committee and contributed actively to the course.

The committee identified seven areas of regional priority around which to develop learning modules:

- UA history, concepts and dynamics;
- health impacts of UA;
- UA crop production systems;
- UA livestock production and marketing systems;
- solid waste management and UA;
- wastewater re-use in UA;
- integration of UA into urban planning.

The best sources of knowledge and expertise in the region on each theme were contacted and brought together as a team to develop the course content. Though complex to operate, this approach contributed greatly to identifying and structuring regional knowledge and was an improvement over that used in Latin America where the course coordinator assisted key experts individually responsible for the various modules.

The organization of the course for Anglophone Africa was based on the common perception of IDRC, Urban Harvest and RUAF that sustaining effective urban agriculture depends on concerted actions by three sets of actors – those involved in local and national policy, those capable of undertaking urban agriculture research, and civil society organizations that support social mobilization and development. Eligibility for participation in the training was therefore limited to city teams made up of these different types of actors – a principle successfully tested in the Latin American course.

As with the Latin American version, competitive selection of teams for the Anglophone Africa course was based on draft proposals that teams submitted on urban agriculture interventions (new or to be improved) in their city. These proposals had to demonstrate the backing of local government so there was some likelihood of implementation. The Technical Committee assessed submissions according to performance criteria, including feasibility, and guided a short list of successful teams in preparing for the course. The following city teams participated in the March 2004 course:

- Kenya – Nairobi and Kisumu;
- Ghana – Accra and Kumasi;
- Cameroon – Bamenda;
- South Africa – Msunduzi (Pietmaritzburg);
- Uganda – Kampala.

Essentially, the course prepared these teams to implement specific interventions aimed at strengthening and institutionalizing urban agriculture in local urban development, both through the improvement of their proposals and through the learning modules.

Differently from its Latin America predecessor, the Anglophone Africa course provided post-course seed grants. These were awarded to three of the city proposals considered to show particular innovation and potential to yield results:

- Bamenda, Cameroon for work on UA as a strategy for poverty alleviation;
- Kumasi, Ghana to develop guidelines for community-based waste management for UPA;
- Kisumu, Kenya for work on integrating agroforestry into urban food-production systems.

The funds were provided jointly by IDRC and Urban Harvest. The Msunduzi team had already received project support prior to its selection, and merely needed to improve its capacity for implementation.

The course produced a number of outputs. An interactive CD-ROM for distance learning applications was made available through the Internet and informed the subsequent development of online courses on urban agriculture by RUAF and Ryerson University in Canada. Videos produced for different learning modules were later adapted and used for other purposes, such as the one made with support from DFID-UK Livestock Production Programme (LPP) on the participatory revision of UA legislation in Kampala, Uganda. Course materials were also used elsewhere, such as in a Kenyan national course for urban farmers on marketing and health aspects. For one year after the completion of the course, Urban Harvest ran the Anglophone Africa Urban Agriculture Network using an e-mail list-serve, in part to follow up on the implementation of the eight projects developed by the city teams. But it also helped build links between various institutions on substantive topics, for example among those who wanted to learn more about pathogens that can occur on crops. Health and UA was an important new topic developed through the Anglophone Africa course and this was followed up by an Urban Harvest book on the subject (Cole et al. 2008).

However, Urban Harvest's plan to consolidate all the regional capacity generated through follow-up networking ultimately was not achieved, due to lack of resources or an appropriate institutional framework linking the partners. Nevertheless, there were ripple effects – or outcomes – from the Anglophone Africa Course, along the lines analyzed by IDRC's Evaluation Unit in its work on "Outcome Mapping". A number of these can be traced through the follow-up activities, not only by the city teams, but through the denser network of partnerships created through the collective that worked on the course.

For example, several institutions expanded their UA mandate, supported by staff members who had attended the training course. A research team from the Kenya Agricultural Research Institute (KARI) in Kenya developed a project on research information services for urban and peri-urban agriculture and environment, which was funded by IDRC. This consolidated and disseminated information on urban agriculture in Kenya's major cities: Nairobi, Mombasa and Kisumu. Through a joint workshop by KARI and CIP-Urban Harvest on UA policy challenges, government representatives were sensitised on UA issues and KARI committed itself to incorporating UA into its research agenda. Finally, in early 2009 the Kenyan Government convened a task force to write its urban agriculture policy (see Chapter 15 below). A youth representative invited to attend the course as an observer shared the knowledge acquired on bio-intensive gardening (Getachew 2002, 2003) with a women's group in Nakuru, Kenya, to produce indigenous vegetables. The latter led to an Urban Harvest follow-up project there, first supported by IDRC. Growing numbers of women's groups produced and sold indigenous vegetables and continued to train other women's groups. The supply of the vegetable seeds has been successfully sustained through alliances with urban farmer groups in Kibera, Nairobi and Muranga, as well as the University of Nairobi.

In Kisumu, Kenya, the Kisumu Municipal Council, supported by the Lake Victoria Region Local Authorities Cooperation (LVRLAC), established an urban gardening and composting demonstration site in Kisumu and integrated urban agriculture into the City Development Strategy. The grant from the Anglophone Africa course was used to collect baseline data and make a rapid assessment of market chains and trade in agroforestry products in Kisumu city (see Chapter 13). Proposal writing skills developed in the course helped the Municipality secure funds from SIDA for an environmental assessment of the impact of solid waste disposal in Kisumu, while post-course networking led Kisumu and Nakuru to link their follow-up on waste management at the December 2004 stakeholder meeting in Nakuru (see Chapter 11). This in turn was facilitated by the Nairobi and Environs Food Security, Agriculture and Livestock Forum (NEFSALF) in collaboration with Urban Harvest (see Chapter 15).

In addition to the poverty survey in Bamenda, Cameroon, the Institute for Agricultural Research and Development (IRAD) held a follow-up seminar and included urban and peri-urban agriculture in its "participatory diagnostic survey", an exercise which helps set the national research agenda.

In Msunduzi, South Africa, the project "African Roots: Food garden initiatives and propagation of indigenous plants, (linked to HIV-AIDS strategies)", already funded and well integrated in local government programming, set up a stakeholder forum. This forum engaged different government departments, NGOs, community-based organizations (CBOs), research institutions and traditional authorities in developing a participatory plan of action for five zones where sites for the cultivation of medicinal plants were identified. People living with HIV/AIDS and local leaders and politicians took part in this planning. A total of 82 existing community gardens were identified and involved in the project, as were entry points in schools, clinics, drop-in centres, farmers associations, women's groups, youth groups and childcare centres. Knowledge acquired on nutritional and medicinal plant products and organic farming techniques – such as trenching, composting, mulching and rainwater harvesting – was further disseminated through demonstration sites and consolidated into a database, while a project team member joined a networking forum to link rural and urban initiatives on indigenous plants for HIV/AIDS-affected communities in KwaZulu-Natal Province.

In Accra, Ghana, the Metropolitan Director for Urban Agriculture in the Ministry of Food and Agriculture (MOFA) was briefed on the course, then became involved in both the follow-up project and the UA gender workshop held in Accra in September 2004. There was also follow-up on health aspects, with wide outreach through the largest urban farmers' association in Ghana, addressing risks and the potential of wastewater farming. In Kumasi the course project team used its grant to link up with the Ghana Environmental Protection Agency and IWMI to develop national compost guidelines for District Assemblies in Ghana. A participating NGO, Centre for the Development of People (CEDEP) incorporated course materials into its programme on sustainable livelihoods in peri-urban areas, while the Kumasi team's municipal representative influenced the city's Planning Committee and politicians on the development of UA policy guidelines.

Perhaps the most extensive follow-up has been in Kampala, Uganda, where two initiatives built on the learning were gained by that city's team through participation in the course. The proposal to address solid waste management through urban agriculture and other recycling activities was further developed as part of the IDRC Focus Cities Research Initiative that began in mid-2006 (see below). The Kampala team also developed and implemented the IDRC and McGill University-funded project "Making the Edible Landscape", integrating urban agriculture into urban design. The project was led by Kampala City Council and involved Makerere University, Kampala District Farmers' Association, RUAF and the local research coordinating organization (KUFSALCC) established through earlier Urban Harvest-supported research (Lee-Smith et al. 2008). "Making the Edible Landscape" explored ways to integrate urban agriculture as a permanent land use within new urban residential developments and the City Council allocated an area of land for low-income housing combined with agricultural space around the houses.

Confronting Urban Poverty and Environmental Burdens – the UPE Focus Cities Research Initiative

The Urban Poverty and Environment (UPE) Program Initiative launched by IDRC in 2005 aimed at easing environmental burdens that exacerbate poverty by strengthening the capacity of the poor to equitably access environmental services, reduce environmental degradation and vulnerability to natural disasters, and enhance use of natural resources for food, water, and income security. Lack of adequate solid waste management undermines human health and eats up financial resources through increased medical costs. Lack of piped drinking water can also lead to disease and drain financial assets through purchase of water from trucks. The Focus Cities Research Initiative (FCRI) is the major program of UPE and involves research on key themes associated with alleviating poverty through addressing environmental burdens, IDRC, 2008. These themes include water supply and sanitation, urban agriculture, solid waste management, vulnerability to natural disasters and housing. Land tenure and gender are cross-cutting issues. With any of these themes as the entry point, focus city projects establish linkages between them. They are conducted by research teams that include local government officials and staff, NGOs working within the community and research institutions.

The Kampala Focus City Project centres on building a sustainable and cohesive community through waste recycling, agro-enterprises and flood mitigation. Known by its popular title: "Sustainable Neighbourhoods in Focus (SNF)", it builds on earlier partnerships and projects involving Kampala City Council, Makerere University, Environmental Alert, the Ministry of Agriculture, Animal Industries and Fisheries, Urban Harvest and RUAF, working with the Kawaala–Kasubi neighbourhood to address vulnerability to flooding and high poverty levels through improved solid waste management and recycling and agricultural-based enterprise development.

"Making the Edible Landscape" and "Sustainable Neighbourhoods in Focus" represent a broadening of focus from the urban agriculture agenda to engage with the way urban natural resources need to be managed within the built environment and how better management can not only reduce poverty levels but also improve the quality of life in cities.

In Senegal, the Dakar Focus City Project – *The Mbeubeuss Landfill: Exploring Options to Protect Health, the Environment, and Livelihoods* – is conducting research with local stakeholders to inform policies for safe and sustainable uses for the landfill site. This site receives about 500 000 tons of waste each year and is the only facility of its kind in Dakar. Many local people live in and around the dumpsite and derive their living from scavenging and recycling different types of wastes (plastics, metals, glass, compost, etc.). They also earn income from creatively reusing other materials. Urban agriculture in the area is an important and stable source of local income; however, working and farming on the dumpsite expose residents to health risks. High levels of chemical and biological contaminants threaten land and ground water near the landfill. The focus city team, coordinated by IAGU, is exploring appropriate local options to reduce health impacts, improve the environment and create employment.

Conclusion

Some 15 years ago, the institutional system for development research in urban agriculture was embryonic. No targeted funding program existed, nor was there any decent public program specifically supporting this research. No qualified training mechanism was available, nor was there any dedicated dissemination vehicle.

Original research was largely carried out by social scientists as individual efforts, for the sake of knowledge and mostly as academic pursuits; studies described the extent, location, practice, benefits and constraints of urban agriculture.

But at the time, one of the authors of this chapter wrote

> A rapid shift can be expected throughout the 1990s toward more multidisciplinary and policy-oriented efforts, through regional and global networking, as several research centres, public agencies, nongovernmental organizations (NGOs), and external support agencies collaborate to create a more enabling environment for the proper management of agriculture and food production in and around cities (Mougeot 1994).

This chapter has submitted evidence to the effect that this prognosis became reality to a significant extent in Sub-Saharan Africa during the 1990s and the first decade of the new millennium.

The progress achieved can be attributed to a confluence of several key ingredients. What is remarkable about this confluence is not so much that it makes sense in theory but that it became possible in practice. Firstly, of fundamental importance was stable and dedicated programming on urban agriculture over at least a decade on the part of all three core partners with an international mandate. Guaranteed internal staffing over that period by these partners was essential: much institution-building for this new area had to take place within the core partners themselves, between them and with their respective networks. This required not only dedicated staff but, more importantly, commitment by institutional management to funding them. Of strategic importance was the steadfast commitment of one public research-funding partner, IDRC; this enabled IDRC and the two other partners to leverage other resources from bilateral agencies and some UN programs, such as the Sustainable Cities Program and the Urban Management Program of UN-Habitat-UNDP. Dilution of funding across a wide agenda or premature shifts in programming away from urban agriculture would have prevented partners from achieving many of the results and synergies needed for the group to advance both research and policy in this new area.

Secondly, as regards subject matter, the pre-existing in-house stock of expertise and experience held by all core partners served as a foundation on which they could develop new programs in urban agriculture, making a more credible and less risky case for programming. Thirdly, the ability of the core partners to mobilize their respective Sub-Saharan Africa networks of partners was critical to regional ownership; forums, workshops and seminars among funding agencies (through the Support Group on Urban Agriculture), among members of particular networks (such as the CGIAR) and regional networks composed of research, city and civil society actors, helped the core partners and their associates to identify links between urban agriculture and pre-existing agendas in the region, as well as inducing peer pressure to experiment with new strategies to address shared challenges.

Fourthly, regarding partnership, one cannot underestimate the ability of key individuals within each core partner to remain in leadership positions as an ingredient for nurturing relationships of trust for long-term inter-institutional collaboration.

Fifthly, each of the core partners was comfortable in acknowledging the others' niche capacities and sought opportunities for collaboration that would exploit complementarities (IDRC's multi-stakeholder policy-oriented research; CGIAR's agricultural research expertise; RUAF's approach to community planning and information dissemination). Finally, a gradual approach to collaboration among the core partners, rather than systematic coordination at all cost from the outset, ensured a level of comfort from which a growing convergence developed over time; this was sped up by joint initiatives, which in turn ended up influencing all partners' subsequent programming.

Yet, the prognosis remains a work in progress in the region. Of course, we can expect the three core partners to champion an urban agriculture agenda for some years to come. And there is evidence that, thanks to the peer pressure they induce, recent regional action-research activities have been more influential on local policies than the more locally confined projects were in the 1990s. Dialogue between researchers and policy-makers has been greatly enhanced in the region. For instance, multi-actor panels on city initiatives in urban agriculture were staged by the three core partners at the World Urban Forum in Vancouver in 2006. Those panels spoke on the various ways in which research has served local policy innovation, how urban policy has gained from supporting proper urban agriculture, and how a collective sharing of local innovations can influence their adaptation and adoption by other cities. There is also evidence that regional and international activities have raised the profile and credibility of participants back in their home country or city. The information afforded by the post-Nairobi course tracer of impacts among participants and their cities shows the worth of tailoring course contents to local demand and of investing in follow-ups. Not least, thanks to the Urban Agriculture Magazine published by RUAF and its regional centres in Sub-Saharan Africa, the growing community of practice has a powerful tool to disseminate its activities and their outcomes and connect with other relevant communities. Given the growing visibility of research and policy on African urban agriculture, more and more students are attracted to the field and more international funding supports graduate research by the new generation that was available to African nationals some 15 years ago.

For all the progress achieved, daunting institutional capacity challenges remain in Sub-Saharan Africa. While an international distance-learning certification course is being piloted at Ryerson University in Canada with the collaboration of the three core partners, and while cells of research do exist in several African universities, the region still does not have a regional program or facility of its own to form professionals in urban agriculture. Also, over a decade of research has informed policy changes in Sub-Saharan Africa but these are now begging for monitoring and evaluation attention. What difference has research really made to recent or ongoing policy-making on urban agriculture? How well are these new policies performing? The next, and final, chapter explores this question in relation to Urban Harvest-supported research. The city focus approach chosen by both the IDRC's UPE and the RUAF programs is also being implemented in a much more networked environment than was the case for single-city projects back in the mid-90s. One should expect that learning and innovation will be accelerated as a result – an expectation which should be reviewed in the future.

Finally, following an emphasis on global and regional initiatives which largely involved researchers and public officials, greater emphasis needs to be placed on raising the still more limited engagement of small producers themselves, their organizations and their links with other urban actors, as well as ways for making more effective their participation in policy making affecting their livelihoods. Identifying, analyzing, comparing and assisting their role in policy making, such as in the Nairobi case described in this book's concluding chapter, is an area of work that is now receiving and should continue to receive greater attention throughout the region. The quest for urban agriculture to significantly reduce food insecurity and environmental degradation in urban areas of the region cannot dispense with small urban producers playing a stronger role in the governance of this sector.

References

Cole, DC, Lee-Smith, D & Nasinyama, GN 2008, *Healthy City Harvests: generating evidence to guide policy on urban agriculture,* CIP/Urban Harvest and Makerere University Press, Lima, Peru.

FAO 2007, *The urban producer's resource book: a practical guide for working with low income urban and peri-urban producers' organizations,* Rome, 137 pp. plus appendices.

Frayne, B 2005, 'Survival of the poorest: migration and food security in Namibia', in Mougeot, LJA (ed) 2005, *Agropolis: the social, political and environmental dimensions of urban agriculture,* IDRC, Ottawa & Earthscan, London.

Gabel, S 2005, 'Exploring the gender dimensions of urban open-space cultivation in Harare, Zimbabwe', in Mougeot, LJA (ed) 2005, *Agropolis: the social, political and environmental dimensions of urban agriculture,* IDRC, Ottawa & Earthscan, London.

Getachew, Y 2002, 'The living garden: a bio-intensive approach to urban agriculture in Ethiopia', *Urban Agriculture Magazine* no. 6, RUAF, Leusden, Netherlands.

Getachew, Y 2003,'Micro-technologies for congested urban centers in Ethiopia', *Urban Agriculture Magazine,* no. 10, RUAF, Leusden, Netherlands.

Hovorka, A 2005, 'Gender, commercial urban agriculture and urban food supply in Greater Gabarone, Botswana', in Mougeot, LJA (ed) 2005, *Agropolis: the social, political and environmental dimensions of urban agriculture,* IDRC, Ottawa & Earthscan, London.

Hovorka, A, de Zeeuw, H & Njenga, M (eds) 2009, *Women feeding cities: gender mainstreaming in urban agriculture and food security,* CTA, RUAF, Urban Harvest, Practical Action Publishing, Warwickshire, UK.

IDRC (International Development Research Centre), 2005, *Urban poverty and environment prospectus 2005–2010,* Ottawa, ON.

IDRC (International Development Research Centre), 2009, *Urban poverty and environment program,* Ecopolis Brochure, Ottawa, ON, www.idrc.ca/upe-ecopolis

IDRC (International Development Research Centre) 2008, *Addressing the millennium development goals, one neighbourhood at a time,* Urban poverty and environment program, Focus City Research Initiative Ottawa, ON.

King'ori, P 2004, *Assessment of urban and peri-urban agriculture research in the Centres of the Consultative Group on International Agriculture Research (CGIAR) in Sub-Saharan Africa,* Urban Harvest Working Paper Series, no. 1, International Potato Center (CIP), Lima, Peru. 35 pp.

Lee-Smith, D & Cole, D 2008, 'Can the city produce safe food?', in Cole, D, Lee-Smith, D & Nasinyama, G (eds) *Healthy city harvests: generating evidence to guide policy on urban agriculture,* CIP / Urban Harvest and Makerere University Press, Lima, Peru.

Lee-Smith, D, Azuba, MS, Musisi, JM, Kaweesa, M & Nasinyama, GW 2008, 'The story of the health coordinating committee, KUFSALCC and the urban agriculture ordinances', in Cole, D, Lee-Smith, D & Nasinyama, G (eds) *Healthy city harvests: Generating evidence to guide policy on urban agriculture,* CIP / Urban Harvest and Makerere University Press, Lima, Peru.

Mougeot, LJA 1994, 'Leading urban agriculture into the 21st century: renewed institutional interest', in Egziabher, AG, Lee-Smith, D, Maxwell, DG, Pyar Ali Memon, PA, Mougeot, LJA & Sawio, CJ (eds) *Cities feeding people: an examination of urban agriculture in East Africa,* p. 105, IDRC, Ottawa, ON, 146 pp.

Mougeot, LJA (ed) 2005, *Agropolis: the social, political and environmental dimensions of urban agriculture,* IDRC, Ottawa & Earthscan, London.

Nabulo, G, Oryem-Origa, H & Diamond, M 2006, 'Assessment of lead, cadmium, and zinc contamination of roadside soils, surface films and vegetables in Kampala City, Uganda', *Environmental Research,* vol. 101, pp. 42–52.

Nabulo, G, Oryem-Origa, H, Nasinyama, G & Cole, D 2008a, 'Assessment of Zn, Cu, Pb and Ni contamination in wetland soils and plants', *International Journal of Environmental Science and Technology,* vol. 5, no. 1, pp 65–74.

Nabulo, G, Oryem-Origa, H, Nasinyama, G, Cole, D & Diamond, M 2008b, 'Assessment of heavy metal contamination of food crops in wetlands and from vehicle emissions', in Cole, D, Lee-Smith, D & Nasinyama, G (eds) *Healthy city harvests: Generating evidence to guide policy on urban agriculture,* CIP/Urban Harvest and Makerere University Press, Lima, Peru.

Redwood, M (ed) 2009, *Agriculture in urban planning: generating livelihoods and food security,* IDRC, Earthscan, London, 248 pp.

Sebastian, R, Lubowa, A, Yeudall, F, Cole, DC & Ibrahim, S 2008, 'The association between household food security and urban farming in Kampala', in Cole, D, Lee-Smith, D & Nasinyama, G (eds) *Healthy city harvests: Generating evidence to guide policy on urban agriculture,* CIP/Urban Harvest and Makerere University Press, Lima, Peru.

Serani S, Nasinyama GW, Nabulo, G, Lubowa A & Makoha, M 2008, 'Biological hazards associated with vegetables grown on untreated sewage-watered soils in Kampala', in Cole, D, Lee-Smith, D & Nasinyama, G (eds) *Healthy city harvests: Generating evidence to guide policy on urban agriculture,* CIP/Urban Harvest and Makerere University Press, Lima, Peru.

Smit, J, Ratta, A & Nasr, J 1996, *Urban agriculture: food, jobs and sustainable cities,* UNDP Habitat II series, New York, NY, 302 pp.

Yamamoto, S 2008 'Estimating children's exposure to organic chemical contaminants', in Cole, D, Lee-Smith, D & Nasinyama, G (eds) *Healthy city harvests: Generating evidence to guide policy on urban agriculture,* CIP/Urban Harvest and Makerere University Press, Lima, Peru.

Yeudall, F, Sebastian, R, Cole, DC, Ibrahim, S, Lubowa, A & Kikafunda, J 2007, 'Food and nutritional security of children of urban farmers in Kampala, Uganda', *Food & Nutrition Bulletin,* vol. 28, no. 2, pp. S237–S246.

Yeudall, F, Sebastian, R, Lubowa, A, Kikafunda, J, Cole, DC & Ibrahim, S 2008, 'Nutritional security of children of urban farmers', in Cole, D, Lee-Smith, D & Nasinyama, G (eds) *Healthy city harvests: generating evidence to guide policy on urban agriculture,* CIP/Urban Harvest and Makerere University Press, Lima, Peru.

Chapter 15
The Contribution of Research–Development Partnerships to Building Urban Agriculture Policy

Diana Lee-Smith and Gordon Prain

Introduction

This final chapter looks at the experiences of the research–development partnerships established in the different towns and cities where Urban Harvest worked from 2002 to 2006. Picking up the discussion of "policy and institutional dialogue and change" in Chapter 2, it draws conclusions about the extent and type of changes that emerged and explores what research–development collaboration might contribute to future policy change. The case study locations have different agro-climatic conditions, production systems, histories and social circumstances that make comparison difficult. However, there is much to be gained by reflecting on the similarities that do exist, including what took place in the different institutional contexts. In order to draw some conclusions we apply here principles developed in policy science and participatory research and development in agriculture, as discussed in Chapter 2 above.

As a prelude to the research undertaken in the three countries described in this book, a regional stakeholder meeting was held in Nairobi in 2000, involving representatives of different organizations from Sub-Saharan African cities. This was seen as a first step in building platforms for research–development collaboration in those cities. Such platforms are groups of collaborating local stakeholders engaged in the research–development process. Three broad areas of concern were identified with respect to urban agriculture:

- Specialized and mixed systems involving crops (including tree crops) and livestock and their relation to food security and income generation, with an emphasis on livelihoods;
- Crops, livestock and health, including recycling of organic wastes; and
- Policy and institutional issues of urban agriculture.

D. Lee-Smith (✉)
Mazingira Institute, Box 14550, Nairobi, Kenya
e-mail: diana.leesmith@gmail.com

These eventually evolved into the research areas described in Chapter 2. On the basis of themes and guided by local conditions and priorities, city stakeholders convened by one of the 15 CGIAR research centres[1] were invited to submit proposals to Urban Harvest. The research implemented by each resulting multi-stakeholder consortium forms the substance of this book. Whereas Chapter 2 examined what we learned from the research results, here we seek to document how the implementation of the projects by multi-stakeholder teams in the case cities interacted with existing local and international institutions, as they worked together or alongside.

In no case did the Urban Harvest supported research take place on a *tabula rasa* with no previous activity concerning urban agriculture. In one case, Yaoundé, Cameroon, a careful documentation of the context and institutional history formed part of the research at the instigation of local partners, one of whom had been investigating the subject for many years. The research outputs were then presented to a platform of stakeholders – identified by the research – who proceeded to formulate a plan to act on them. In Kampala, Uganda, intensive research on urban agriculture was conducted during the 1990s, and local government and local NGOs both had long-standing programs. The urban centres where research was conducted in Kenya have all had active research programs including, in the case of Nakuru, international research collaboration, which was an important basis for the research reported here.

The following sections of this chapter explore how, in each country and urban centre, researchers worked with decision-makers and producers to try and build relevant knowledge and institutions. In each case the aim was to set the direction for policy and other changes focused on urban agriculture as a key strategy for poverty alleviation and food security. The first section describes the formation of the Kampala Urban Food Security, Agriculture and Livestock Coordinating Committee (KUFSALCC) and the next the emergence of the Nairobi and Environs Food Security, Agriculture and Livestock Forum (NEFSALF), including its interaction with other towns. After that there is a brief section on the different setting in Yaoundé, Cameroon, already described in Chapter 5 above. While the direction for institution-building was also set there, it was not as dynamic as in the other two cases.

Finally, the chapter assesses the potential for new policy initiatives at national level in these countries and others. This assessment is made in the context of ongoing activities in the region supported by various agencies including IDRC, and draws some general conclusions about the potential of these for improving urban food security, urban natural resource management, and improved national agricultural productivity and market orientation.

---

[1] The Consultative Group on International Agricultural Research (CGIAR) not only supports the 15 Centres, it also initiates cross-cutting "system-wide and eco-regional programs (SWEPs)" to address themes of relevance to different Centres and to non-CGIAR stakeholders. Urban Harvest is one of these SWEPs.

Urban Agriculture in Uganda: Participatory Review of Bylaws and the Emergence of a Novel Research–Development Platform

Kampala as an Innovative City for Urban Farming

Kampala's relatively high-rainfall environment creating wetlands, as well as its history of agriculture in the ancient Kingdom of Buganda, have both favoured its urban agriculture. The practice expanded during Uganda's civil war and again due to structural adjustment policies during the 1980s and 1990s. However, there was no official policy recognition of the practice until Kampala's Structure Plan of 1994. This plan was anyway not widely supported or implemented, though it generated much public debate. During the late 1980s and early 1990s that debate was stirred when Maxwell's research involving Makerere University and Environmental Alert, a local NGO dealing with urban agriculture, established local agriculture's links to urban food security and nutrition. The research showed that urban agriculture was important in the livelihoods of 36 percent of Kampala households, consistent with data from neighbouring Kenya and Tanzania, where parallel studies were supported by the International Development Research Centre (IDRC) (Maxwell 1994; Atukunda 1998; Hooton et al. 2007; Lee-Smith et al. 2008).

Uganda's national constitution and decentralization policy favour a participatory approach to governance. Kampala City Council (KCC) is also a District Council which has a Department of Agriculture. Elected urban politicians, including the Mayor and his cabinet of city ministers, are close to issues affecting their voters, including urban agriculture. They are also close to urban agriculture in the literal sense that many raise their own crops and animals (Lee-Smith et al. 2008, p. 224). Also, they have the support of technical officers relocated from central government to KCC, including from the Ministry of Agriculture, Animal Industries and Fisheries (MAAIF). This opportunity was used creatively to develop urban agriculture and food supply; partnerships were formed with different departments of Makerere University and the National Agricultural Research Organization (NARO) to hold high-level awareness-raising workshops and encourage research on heavy metal contamination in urban farming. Extension services were provided to farmers in some urban areas through collaboration with Environmental Alert and other NGOs, while KCC officers reported upward on the need for a legal framework on urban agriculture (Hooton et al. 2007; Lee-Smith et al. 2008). Kampala's Department of Agriculture also established a typology of urban farming areas within its boundaries and this formed a framework for the detailed examination of urban and peri-urban agriculture presented in Chapter 6 of this book.

Institution-Building for Linking Research to Policy

The Urban Harvest-supported research on "Strengthening Urban Agriculture in Kampala", which began in early 2002, involved three research centres of the

Consultative Group on International Agricultural Research (CGIAR) as well as national and local researchers from several institutions, civil society groups and the Kampala City Council. The Council's Agriculture Department took an active role, especially in the field work and links to the communities, while the Ministry of Agriculture, Animal Industries and Fisheries (MAAIF) took part through its representative at City Hall. The National Agricultural Research Organization (NARO), which moved from a position of scepticism about agriculture in Kampala to recognizing its key livelihood role through participation in the regional stakeholder meeting held in 2000, also made an important contribution to this project.

What brought these partners together was the aim of better documenting the livelihoods, production and marketing aspects of urban agriculture in Kampala and thereby identifying opportunities for innovation. KCC participated through its technical staff in the Agriculture Department rather than as a policy-making body. Part of the reason for the limited attention to the policy dimension of urban agriculture at first was the leadership exercised by a CGIAR Center with a technology and enterprise innovation perspective, rather than an orientation toward policy change. There was a marked shift toward policy engagement when a parallel project on the health impacts of urban agriculture started up shortly afterward however. Many of these same actors also became involved directly or indirectly, but because of the importance attached to the health aspects of farming by policy-makers, the range of stakeholders also broadened from that of the livelihoods study.

Politicians as well as technical officers from the City Council became actively involved in guiding the direction of the health research and even in its implementation. After helping to identify the health benefits and risks of urban agriculture in the city, several stakeholders also agreed to take part in a committee to steer the direction of the research. This was an important step in institutionalizing the linkages between the separate people and organizations concerned about urban agriculture (Cole et al. 2008, Chapters 2 and 12).

The committee was formalized in June 2002 as the UA Health Coordinating Committee (HCC) and members took on different tasks. Apart from committee meetings, the researchers in the team met frequently at Makerere University's Faculty of Veterinary Medicine (whose head also chaired the committee). An urban agriculture resource centre set up there with Internet connection provided a physical as well as institutional space. Some of the earliest deliberations of the Committee involved farmers' concerns about the draft City Ordinances on urban agriculture that were being prepared by KCC. Pressures built up for the HCC to ensure the research contributed to that policy process and that the project team would collaborate on the proposed legal reforms (Lee-Smith et al. 2008).

One result of this collaboration between researchers and policy-makers was that the outputs of the Urban Agriculture Health Project included public health messages and policy guidelines as well as scientific reports (Cole et al. 2008). Further, the HCC added its voice to the pressure that was building for participatory review of the draft ordinances, and had the clout to influence this because of its membership. City Minister Winnie Makumbi and Council Committee Chair Rebecca Mutebi, who were both HCC members, also conferred with the Mayor and other Council members, who consequently changed from being opponents of urban farming to being supporters, at least of its proper regulation. This dynamic also played out in international gatherings dealing with urban food and agriculture (Lee-Smith et al. 2008).

The Urban Agriculture Ordinances

In 2001 KCC had drafted new Bills for Ordinances dealing with urban agriculture, livestock, meat, milk and fish handling based on model bylaws. Their rather prohibitive implications for urban agriculture disturbed local residents who were involved in farming. KCC eventually agreed in principle to a participatory review of the draft ordinances, thus acknowledging a clause in Uganda's constitution that encourages participatory law review including at grassroots level. Nevertheless, there was still quite a lot of internal opposition to urban food production and the idea of regularizing it. In 2003 Makerere University's Department of Veterinary Public Health and Preventive Medicine, on behalf of HCC, secured a grant from the UK's Department for International development (DfID) to support such a participatory, public review of the Ordinances. Funds and resources were also provided by Environmental Alert, KCC and Urban Harvest.

The review consisted of consultative workshops held across Kampala in August and September 2003 to debate the draft Ordinances. With farmers and local leaders contributing from the neighbourhoods and a large gathering of many stakeholders at District (City) level, numerous new ideas and recommendations emerged. Due to the presentation of old and new research, including the UA Health Project, there were proposals, for example, on measures to reduce toxic contamination of sites used for urban agriculture. Recommendations were made on various technical changes and on how to structure the draft Ordinances. Important suggestions from senior government level included the idea to prepare simple guidelines to accompany the proposed new Ordinances to facilitate their understanding and uptake, and for the launch of a national policy process on urban agriculture (KUFSALCC & Urban Harvest 2004; Lee-Smith et al. 2008).

KCC requested HCC to provide technical and editorial assistance to incorporate the recommendations into the draft Ordinances and these then passed into the review and approval process of the Council. After making its input, Council approved the draft bills in December 2003. Urban Harvest provided some of this technical assistance and again with DfID help supported the production of guidelines and pilot testing of the proposed Ordinances. Since pilot testing involved research in the service of City Council operations, HCC (which included high level KCC membership) began to think about broadening its mandate. The new laws on urban agriculture, livestock, fish, milk and meat were formalized in the Uganda Gazette in December 2006, the first example of participatory law-making under Uganda's 1995 Constitution (Lee-Smith et al. 2008).

A Novel Research–Development Platform for Urban Agriculture: KUFSALCC

Kampala's major achievement of drafting urban agriculture law with public participation was an important impetus for urban agriculture institution-building based on the HCC. But there were other, internal, bureaucratic factors. The research on "Strengthening Urban Agriculture in Kampala", led by the International Center on Tropical Agriculture (CIAT), was winding up while the health project was continuing. The research teams, which collaborated closely, both wanted to turn their results into useful outputs and make forward linkages to research use and implementation. This was of particular interest to the group working with schools, as described in Chapter 7 above, as well as the health researchers (Lee-Smith et al. 2008).

Collectively, they decided to broaden the mandate of HCC and formally constitute it as an independent non-governmental research organization, but with the same broad stakeholder membership, including City Council. So by the end of 2004 a new organization, the Kampala Urban Food Security, Agriculture and Livestock Coordinating Committee (KUFSALCC) was formally registered as an NGO under the laws of Uganda. It was now able to raise and manage funds according to the priorities identified by the team. Its priorities included monitoring the Ordinances and developing and publishing the guidelines for their implementation. These set out accurately the key elements and requirements in the form of glossy pamphlets in everyday language. Most importantly, these guidelines contained a policy statement about KCC's positive and supportive stance toward urban food production, in the form of a preamble. This policy statement, emerging from the participatory consultations, was however deleted from the laws as passed and in all subsequent information issued by KCC, leaving only the regulations and accompanying penalties.

A recent study of how the City Ordinances on urban agriculture came about concluded that a number of factors converged to make this innovation possible, namely the farmers themselves, civil society action, influential research, and key decentralization policies and actions. The main factor however was seen as the motivation, coordination and collaboration of a set of actors from different institutions, namely those that constituted KUFSALCC (Hooton et al. 2006, pp. 10–12).

Assessment and Future Directions

Hooton and colleagues (2007), in assessing the likely long-term impact of the important innovation that Kampala's new laws on urban food production represent, point out that only time will tell whether the initiative will have a lasting, or even a beneficial effect. The Ordinances contain provisions for all urban agriculture activities to have permits or licenses from KCC. While the consensus of the consultative process was that existing farm activities simply needed to be listed by KCC in order to qualify for a permit, while food handlers and livestock keepers would eventually upgrade to a license by following all the detailed provisions of the law, it is not certain how this will be interpreted or implemented in practice (Stren 2008).

For one thing, the assessment of feasibility carried out by the KUFSALCC demonstrated that the resources needed for a comprehensive record of all urban farming activities were enormous. In the absence of this comprehensive list of urban farmers implied by the Ordinance on Urban Agriculture, it is possible that harassment of farmers and rent-seeking behaviours by city officials may persist or even increase as a result of the regulatory requirement.

The documented preference of some KCC and government officials to contain and restrict urban agriculture because of perceived health threats (Lee-Smith et al. 2008), along with the City's dropping of the positive policy preamble to the Ordinances, also suggest that they may be implemented in a restrictive rather than supportive manner. Gore (2008) discusses the inherently political nature of public policy-making, with power relations among various actors playing out in different decision-making arenas. We can see an example of this when Kampala City Council allocated a hundred million USh (US$ 45 000) of its annual Local Government Development Grant (LGDP) to facilitate the popularization of the new Ordinances in collaboration with Environmental Alert in 2008. Rather than presenting a positive view of urban agriculture however, the posters mainly indicated prohibitions and penalties (Lee-Smith et al. 2008).

On the positive side, KUFSALCC's monitoring showed that pilot testing the Ordinances created awareness among farmers who learned about measures to protect their health and showed willingness to comply with the various provisions as long as they could afford them. The motivation and dedicated collaborative effort by numerous individuals in different institutions working with a common purpose was what ensured that the urban agriculture Ordinances were passed. The organization that they formed, KUFSALCC, helped them accomplish that purpose, although it did not maintain a dynamic presence subsequently. The later decline – or even demise – of KUFSALCC can be attributed to competition with KCC for funds and influence combined with lack of sustained interest and support among other partners. It was certainly an innovation in itself, being a hybrid NGO with government participation and a research body with policy advisory status in relation to local government. Such institutions are needed in order to develop effective structures of governance in African countries such as Uganda (Gore 2008).

Despite KUFSALCC's decline, it is as yet too soon to assess the impact of these developments. In early 2009, the Ministry of Agriculture, Animal Industries and Fisheries (MAAIF) finally followed up on the suggestion made several years earlier for the initiation of a process to develop a national policy on urban agriculture. A MAAIF official and member of KUFSALCC had taken the initiative to develop a draft policy even though there had been no official endorsement or follow up. In 2009 a budget was allocated to this process within the 5-year Development Strategy and Investment Plan, thanks to the determination of the staff member and lobbying by mayors of many of the smaller cities in Uganda.

The existence of the aforesaid committed group of individuals across many sectors is a necessary but not sufficient condition for such policy development. Political will is also needed within the decision-making arena. And in that sense Kampala has made a historic move in having legislation that acknowledges the existence of urban agriculture as a phenomenon to be regulated. This achievement may be added to its list of others, including the development of urban agriculture land-use typologies and having an urban Agriculture Department that actively collaborates with civil society organizations to bring services to city farmers.

Urban Agriculture in Kenya: Policy-Making from the Bottom and the Emergence of NEFSALF

Kenya's Absence of Urban Policy

Kenya's neglect of its rapidly growing urban centres – particularly the capital Nairobi – for the four decades after its independence in 1963 has been extensively documented (Stren & White 1989; UNICEF 1992; Lamba 1994; Lee-Smith & Lamba 1998, 2000). Apart from the underlying context of rapid demographic growth unaccompanied by a parallel economic growth, Nairobi's problems have been attributed to a crisis of governance, meaning the relationship between civil society and the state, between rulers and the ruled, the government and the governed (Gore 2008; McCarney et al. 1995). Failure to come to terms with reality in the form of informal sector economic activity was accompanied by problems of overlapping and conflicting structures of local government. These were compounded by the heightening of ethnicity as a factor in the competition for political power. In combination, these factors led to central government starving the capital city of power and resources with lower income groups particularly affected (Lee-Smith & Lamba 1998, 2000).

The city's housing demand was for instance estimated as 20 000 units per annum with only 5000 being built and the accumulating shortfall leading to an estimated 60 percent of residents living in informal settlements without services (Lee-Smith & Lamba 1998; Lamba 1994). Since those peoples' access to land was highly restricted, over half the population was living on less than 6 percent of the habitable land with no services, with resulting health and environmental problems (Lee-Smith & Lamba 2000; Kessides 2006). Ten years later the situation is much worse, given ever-increasing numbers on the same amount of land. The population of Kibera, the largest slum in Nairobi, is thought to be living at a density of 250 000 persons per km^2 (Muraguri 2008).

Those involved in the various competing arms of government and others among the middle class and elites have tended to exploit the system of informality by illicit trading of scarce resources and services for private gain, many using government as a private resource rather than an instrument for governing. The complete failure to incorporate these areas housing the majority of the city's population into the formal structures of governance had by 1998 led to a wide range of problems from health to security issues and ultimately social breakdown. Some approached this crisis by trying to build structures of governance from the bottom up. While some such community-based organizations addressing land, housing and transport issues could be found, albeit harassed and suppressed, there were no such structures addressing food security or urban farming by the year 2000 (Lee-Smith & Lamba 2000).

Despite the change in government in 2002, few policy changes had emerged by 2008 to benefit lower income groups in towns and cities in Kenya, although the policy environment opened up and became more tolerant of public participation and a number of government bills were in preparation. However, divergent approaches within government were in evidence, with the Ministry of Housing and Nairobi's Water Board demonstrating approaches to slum improvement with public participation, but the new Ministry of Nairobi Metropolitan Development condemning slum upgrading and calling for a policy of slum elimination (Muiruri & Kaseve 2008; Muraguri 2008; Kilonzo 2008). And despite incisive analyses of the chaotic systems of governance and widespread efforts at more decentralized decision-making systems – including the Local Authorities Service Delivery Action Plans (LASDAP) and Ward Committees – the legal frameworks to strengthen these and make them effective were still absent (Mitullah 2008; Okello & Mboga 2008; Kibinda 2008). These institutional innovations that practiced stakeholder participation in planning and allocation of public funds were also challenged in implementation through being spread too thinly and having to combat a climate of rampant corruption (Okello & Mboga 2008).

Urban Agriculture in Kenya

Kenya has a history of independent scholarly research on urban agriculture but until recently there was no official public sector support. A national sample of urban households in Kenya in the 1980s revealed that 64 percent of urban households grew crops, and 29 percent did so in the urban area where they lived (Lee-Smith & Memon 1994). The figures for Nairobi were 65 percent of households growing crops but only 20 percent in the city. About half of all urban households in Kenya (including Nairobi residents) kept livestock, but only 17 percent kept them in the towns where they lived, and even less (7 percent) in Nairobi. Nevertheless, as noted above in Chapter 10, there were 23 000 cattle in Nairobi at the time of the 1985 survey, and 20 years later the Kenyan Ministries of Agriculture and Livestock recorded 24 000 dairy cattle alone in the city, suggesting an overall increase in cattle numbers (Lee-Smith et al. 1987; Mukisira 2005).

In Nakuru at the turn of the century 35 percent of households were farming in town, as described in Chapter 11 above, 27 percent of all households growing crops and 20 percent keeping livestock in town (Foeken & Owour 2000; Foeken 2006). It was found that poorer sections of the urban population (who have less access to land) were proportionally under-represented among urban farmers than those who are better off, a trend particularly true for livestock keepers as confirmed by the data in Chapters 11 and 12 above. A separate study of a very low-income slum in Nairobi from the 1990s also confirmed that only 4 percent of households grew urban crops and 6 percent kept livestock for the same reasons (Lee-Smith 1997). In Kisumu in the 1980s 30 percent of urban households kept livestock and the same proportion grew crops (Lee-Smith et al. 1987). The figures were found to be higher for peri-urban households (60 percent engaging in crop or livestock keeping or both) in a 2002 survey cited in Chapter 13 above.

Shifts Toward Policy on Urban Agriculture in a Negative Environment

The policy environment in Kenya, like that of Uganda, was unfavourable to urban agriculture, but the governance structures favouring decentralization and participation were also missing in Kenya, so almost no moves supportive of urban agriculture took place during the 1990s. Outside of the leadership vacuum of government however, there were several urban environmental management initiatives that included urban agriculture, including Nakuru Municipal Council's "Local Agenda 21" and the "Greentowns" movement in many secondary towns, both with international and civil society support. The local authorities in Nairobi and Kisumu remained hostile, although some NGOs supported small urban farming projects in informal settlements. Meanwhile general dissatisfaction with the system of governance – manifest as political opposition – was growing, culminating in the change of government in December 2002.

When the first meeting of urban agriculture stakeholders was convened by the CGIAR in Nairobi in August 2002, it was in this tense political atmosphere. There was significant attendance by key individuals from the public sector, including Nairobi City Council. An Urban Harvest-supported research project emerged involving two CGIAR Centres headquartered in Nairobi, the International Livestock Research Institute (ILRI), which led the project, and the World Agroforestry Centre (ICRAF). Kenya Agricultural Research Institute (KARI) played an active role in the research, along with the Greentowns NGO. Links to public sector institutions remained fragile for the next 2 years however. The research, on recycling nutrients from organic wastes in Nairobi, is described in Chapter 10. Although the head of KARI called for a stakeholder workshop to develop a national policy on urban agriculture when the project was just beginning, it took time for a broader transformation in institutional attitudes to occur and for KARI to lead such a workshop toward the end of the project in 2004 (Ayagah et al. 2005).

A Policy Forum Built from the Grassroots: NEFSALF

Meanwhile, the Nairobi and Environs Food Security, Agriculture and Livestock Forum (NEFSALF) – bringing public, private and community sector stakeholders together in a policy platform – was convened by Mazingira Institute, a Kenyan NGO which had also carried out urban agriculture research in the 1980s. Emerging out of a focus on improved livelihoods through urban livestock keeping, NEFSALF's first action in January 2004 was to elaborate a "sectoral mix and cooperation model" to address the governance crisis through a bottom-up approach, building institutions through consultation rather than conflict. Aimed at facilitating group organization and network building at community level, the forum quickly grew, attracting many farmers and their associations. The farmers formed a network of associations with a gender-balanced executive and procedures and set their priorities, which included the need for skills training. Courses began in response, with input from the Ministries of Agriculture and Livestock which had also been attending the forum (Mazingira Institute 2005).

The forum became a venue for researchers as well, because of the early focus on improved livestock keeping, and farmers flocked to join because of the numerous knowledge benefits. Apart from seminars and training, the NEFSALF newsletter came out several times a year, starting in January 2005. At that time there were 80 members, belonging to 20 farmers' groups with up to 20 participants. By 2008 there were 693 members belonging to about 50 such groups. In 2006 the forum was having such a significant impact that the Ministry of Agriculture selected Nairobi Province to launch the second phase of its National Agriculture and Livestock Extension Programme (NALEP). Both the Minister for Agriculture and the Deputy Director of Extension called for new, supportive policy and regulatory frameworks for urban and peri-urban agriculture as an important means of improving livelihoods. NEFSALF officers were appointed to a Provincial Steering Committee to oversee implementation of NALEP (Mazingira Institute 2006).

Links Between NEFSALF and Urban Harvest-Supported Research

NEFSALF retained a close contact and integration with the Urban Harvest-supported research on recycling organic wastes through composting and use of manures in urban agriculture around Nairobi. The stakeholder group that formulated the study's conclusions contained in Chapter 10 above included farmers and public sector representatives from NEFSALF. And as a follow-up in December 2005, NEFSALF held farmer training sessions on composting and nutrient cycling, with inputs from the research team. Fifty-nine farmers and 11 agriculture and livestock extension staff were trained on urban agriculture and organic waste re-use including marketing (See Chapter 11).

Women and men from NEFSALF-affiliated farmers' groups also actively participated in formulating the recommendations of the workshop held by KARI with Urban Harvest support in 2004 (Ayaga et al. 2005). Despite these gains and the positive changes in KARI, the prospects for follow-up research activities involving Nairobi City Council remained poor, given the political circumstances in the capital described above. A better setting for building a research–development platform involving local government representation was Nakuru, in Kenya's Rift Valley

Province. There the Municipal Council (MCN) was keen to build on its earlier involvement in Local Agenda 21 planning and action as well as the considerable research that had just been completed (Foeken 2006). Urban Harvest collaborated with NEFSALF to support the stakeholder meeting there in December 2004. The aim was to do further research and development work in Nakuru, including examining the links between crop–livestock production and waste management, as described in Chapters 11 and 12. The Nakuru farmers expressed interest in developing a similar approach to NEFSALF, and the meeting also resolved to initiate work on a review of bylaws affecting urban agriculture, based on what was learned about the participatory law review in Kampala.

Following a resolution by Nakuru councillors in May 2005 to develop urban agriculture bylaws, MCN's Department of Environment began a consultative law-making process to enable and regulate farming within the municipality, including farmer consultations at ward level throughout the town. Research project workshops in Nakuru helped create awareness of agriculture as a productive sector of the urban economy and encouraged dialogue on bylaw development.

As these developments moved forward the agenda for making research-policy linkages in Kenya, it was apparent that the leadership focus was mainly in civil society, with responsive stakeholders in the national research body and certain parts of central government. The role of the CGIAR centres was quite limited, despite the cross-cutting facilitation of Urban Harvest.

Prospects for an Urban Agriculture Policy in Kenya

In Kenya's turbulent institutional environment with conflicting interests and political power struggles, it is difficult for an urban agriculture policy to be formulated and then implemented. Different ministries of central government take different positions as do different urban local governments, and Nairobi is a particular arena of conflict. Nonetheless there are positive signs. Picking up on recommendations from the KARI-led 2004 policy workshop supported by Urban Harvest, a task force to develop an urban agriculture policy in Kenya was in fact promulgated in 2007, led by the Ministry of Agriculture's Provincial Agriculture Board and KARI. In early 2009, this task force met to begin its work of drafting. A major factor here may be the influence of the former KARI Director who then became the Permanent Secretary in the Ministry of Agriculture. Further, the Ministry of Lands and Settlement proved to be receptive to some of the research lessons for legalizing urban agriculture and organic waste management, which emerged from the experiences reported in Chapters 10 and 11, and these have been incorporated in Kenya's new Land Policy adopted in 2010 (Ministry of Lands and Settlement 2006).

It is interesting to observe that the moves of some in the public sector to support policy reform suggests their willingness to follow the lead of, and even encourage, grass roots and civil society actors. The strong push for recognition coming from the large numbers of urban farmers' associations facilitated by the NEFSALF structure is a noteworthy development for both Nairobi and Kenya. It is a case of building structures of governance from the bottom up as anticipated by the 1990s analysis of the city's governance crisis, which continues at least a decade later.[2]

Only time will tell whether the growing political clout of such bottom-up institutions as NEFSALF can have any real impact in terms of policy changes that will benefit the urban poor through increased access to resources and improved means of livelihood. The incorporation of civil society actors from NEFSALF in the Provincial Steering Committee overseeing the National Agriculture and Livestock Extension Programme (NALEP) and the creation of a permanent NALEP stakeholder forum are also signs of positive institution-building for urban agriculture (Mazingira Institute 2006). Further, the Ministries introduced a system of loans for small urban and peri-urban enterprises and disbursed the first ones in late 2008. Proper regulation is an important aspect of better governance and the development of bylaws for urban agriculture in Nakuru is encouraging, although they are yet to be gazetted. As discussed in the case of Kampala, however, it remains to be seen how these are implemented in practice. The new regulations could be interpreted with political will to enhance peoples' livelihoods or simply as a mechanism for harassment and continuing exploitation of farmers due to systemic corruption.

Cameroon: Weak Institutions in an Urbanizing Agricultural Society

The findings of the institutional study that formed an essential part of the Urban Harvest-supported research in Cameroon are only briefly discussed here since the study itself is presented as Chapter 5 above by Bopda and Awono. Like Kampala and Nairobi, Yaoundé as an African capital developed from a colonial urban settlement super-imposed upon a landscape used by local people for food production. Legal frameworks and institutional structures were inherited from the colonial power. With a landscape and climate very similar to Kampala's (hilly terrain in a rainy tropical climate – though without being on a lakeshore), Yaoundé is a suitable place for agriculture.

The resulting state of affairs in the capital is currently "a huge number of farmers about whom little is known and who lack organization, consumers who are only poorly aware of the issues involved, a negative official perception of urban agriculture, an insufficient involvement of the state and local government, a very limited involvement of civil society, and an environment that both enables and constrains the activity" (Bopda & Awono, Chapter 5 above). Their chapter pinpoints the dynamics whereby the indigenous institution of the chiefdoms is responsive to the urban farmers but that this is also linked to corruption. Nevertheless, they perceive how,

---

[2]There was a massive breakdown in national government and public order in early 2008 following a disputed election in December 2007.

despite also being part of the official structures inherited from colonialism, the chiefdoms mediate farmers' interests, taking up the slack of the discontinuity between what officialdom tries to maintain as an ideal urban structure and the reality on the ground.

The research in Yaoundé considerably advanced our understanding of its urban farming systems and included a sophisticated institutional history and analysis. The research team comprised individuals from a broad range of international and local institutions, governmental, non-governmental and university-based, engaged in research, as well as a representative of the Yaoundé local government. During the research period however, this grouping was not able to facilitate the development of an action-oriented platform such as KUFSALCC or NEFSALF. The top–down system of government in Cameroon and the confusing dual system of authority described in Chapter 5 are probably both factors constraining organizational development: for example, civil society organizations have only been legally operating since the end of the 1990s.

Control of the citizenry in Cameroon through the centralized operation of an adapted traditional structure based on chiefs is typical of many African countries. It is one of the overlapping systems of government still existing in Kenya where it likewise operates as a system of both land allocation and corruption (Lee-Smith & Lamba 1998). In all cases it was a direct creation of the colonial rulers, who used it as a way of managing the indigenous populations. Only in places like Uganda where there were well-established kingdoms with their own systems of government was the alternative approach of indirect rule applied. When representative systems of government with free elections were introduced after independence in the 1960s, the colonial structures remained in place, leading to the complicated and conflicting systems of administration described in Cameroon.

The public meeting organized at the end of the urban agriculture research in Yaoundé in February 2004 set as one of its main priorities the establishment of a stakeholder forum. Its purpose was stated as the development of a strategy to address food security and other concerns of farmers, and the need for health and environmental protection (see Chapter 5 above). Thus despite the less free conditions for organizing, the conclusion drawn by stakeholders when they did meet was to move in exactly the same direction as those in Kampala, Nairobi and Nakuru. Due to lack of resources, this agenda has yet to move forward however.

As in the other urban centres, the stakeholders who met in Yaoundé wanted better recognition and regulation of urban food production, and for existing institutions of government to incorporate it within their responsibilities. They also wanted waste management and urban agriculture to be addressed together by the authorities, and it was clear that this was because the farmers themselves were already doing this to the extent they could, but needed help. And, also like those in the other urban centres, Yaoundé stakeholders wanted training and capacity building for farmers, including building on whatever existing extension services were being provided. In Yaoundé, as in Kenya, central government extension services were already providing extension services in town.

Yaoundé households rely a great deal on women's labour to produce food and this has been shown to be done in an exploitative way in the past (Bopda & Awono above Chapter 5), with elite politicians proclaiming in the 1970s the cultural expectation that women's duty was to feed the urban population. Meanwhile male chiefs were reported to "marry" hundreds of "wives" in order to profit from food produced by them (Hovorka & Lee-Smith 2006, pp. 127–128; Guyer 1987, p. 118). More than any other city in the region reporting these statistics, women heavily predominated among both commercial and subsistence producers of urban crops in the Yaoundé of 2003 as well as among food traders (Bopda et al. above, Chapter 5). It is possible that the low status and dismissal of urban agriculture as unworthy of public policy attention is linked to the correspondingly low status of women's activities.

Conclusions: The Potential for Urban Agriculture Policy and Institutional Change in the Region

Comparing the Three Countries Studied

Each of the country cases examined in this book casts a different light on a common effort to develop or adapt institutions that will come to terms with the on-the-ground reality of urban agriculture. The elements that are not in doubt, and indeed are common to all the cases, are the scale of the phenomenon of urban agriculture, its essential nature as a livelihood strategy of urban households, the complex array of actors that have a stake in it, and the absence of policies and institutions that currently address it.

Here we attempt to draw conclusions about the prospects for urban agriculture policy and institutional change based on our analysis of these cases, both in a broad conceptual framework and more specifically in relation to the investment in research and the building of research–development platforms. Taking a leaf from the companion book to this one (Cole et al. 2008) that addresses the public health aspects of urban agriculture, we have situated our examination of the development of urban agriculture policy and institutions in a governance context.[3] We suggest, along with Gore (2008), that a political science perspective that examines the power relations between the various actors involved needs to be applied to urban agriculture. Efforts to build stakeholder platforms as a prelude to policy and institution building should be looked at in relation to a broad set of variables, including power relations. As a first step in this direction, we examine some of our case study material in relation to a few variables at macro- and micro-level.

---

[3] That is, we have tried to gain an understanding of the relationships between civil society and the state, between rulers and the ruled, the government and the governed, in relation to urban agriculture.

Aspects of Comparison at the Macro-Level

Interdisciplinary, inter-agency platforms were developed in Kampala and Nairobi but not in Yaoundé. Despite similarities in terrain, the indigenous agricultural practices of crop and livestock farming, and colonial structures of governance inherited by elected governments following independence in the 1960s, a few key factors seem to have created some interesting differences among the three places. Not least of these are the variations in political developments since independence around four decades previously.

Whereas all three countries have been ruled by autocratic leaders who inherited the mantle of authoritarian colonial government and generally intensified centralized control, Uganda experienced at least a decade of civil war and Kenya at least a decade of civil turmoil before governance structures began to be reformed following changes of regime – in the 1980s in Uganda and after 2002 in Kenya. This pattern has not occurred in Cameroon. Another difference is the British system of legislation and local government inherited by Kenya and Uganda compared to the French system inherited by Cameroon.

All three countries have land tenure, ownership and control systems that are a mix of indigenous occupation patterns overlaid with colonial settlement and legal regimes derived from Europe. The latter have become increasingly established over time through a process of modernization. Where these systems overlap and intersect the informal sector operates, with its adapted forms of traditional tenure accompanied by cash transactions (see Sjaastad & Cousins 2009 for a discussion of current problems with the formalization of land tenure). Here urban agriculture also operates, usually in a disadvantaged position even in relation to other informal uses of land.

However, the general pattern of land use varies widely in the three countries, including in their capital cities. Being the location of the Buganda kingdom, Kampala evolved with a dual system of local government, involving both municipal and royal jurisdictions, and dual land tenure, with traditional, royal tenure having much higher status than any such systems in the capitals of Cameroon or Kenya. Thus "mailo" land and allocation practices survived in a recognized form alongside other forms of tenure in Kampala (Maxwell 1994). Similarly, urban agriculture was associated with royal power and status in Kampala (David et al. above, Chapter 6).

Such differences may account in part for the more dispersed and integrated pattern of both farming and of housing settlements of the rich and poor found in Kampala as compared to Nairobi's pattern of informal settlement and farming being squeezed into tightly controlled un-serviced areas. Yet Yaoundé's housing and urban farming pattern is much more like Kampala's than Nairobi's. While this may be in part attributed to geography, both Kampala and Yaoundé being built on hills separated by (partially farmed) wetlands while much of Nairobi is on a plain, there are also economic and political factors to consider.

Nairobi has seen much more foreign direct investment following on from its history as a favoured location for European settlement, and this may have contributed to increased alienation of land from poor indigenous populations as formalization of land tenure proceeded more rapidly than in the other two cities following independence. It is worth noting that, whatever their origins, conflicts over land and corruption in land transactions have been much more pronounced in Nairobi, and Kenya in general, than in the other two countries, leading to a Commission investigating illegally acquired land in 2004 (Government of Kenya 2004). After failure to act on this, there were further land conflicts following the disputed Kenyan election of 2007.

Some points of comparison between the three countries examined in this book, with respect to factors that may influence the development of policies and institutions on urban agriculture, are summarized in Table 15.1 below.

We can see from this comparison, which mostly draws on material from various chapters in this book and its companion volume (Cole et al. 2008), that Kampala exhibits a number of variations from the other two cities that may have influenced its earlier development of UA policies and institutions. In particular, the early emergence of civil society and engagement of local government stand out. Then, there is the respect for traditional land systems, not to mention the traditional role of urban farming in the Buganda kingdom. While research played a role in policy and institutional development in Kampala (Hooton et al. 2007; Lee-Smith et al. 2008), it seems to have not yet been as influential in Nairobi. It is too early to judge the impact of the very small investment in UA research in Yaoundé.

Table 15.1 Potential factors influencing the development of urban agriculture policy and institution-building at macro-level

Variable / city	Kampala	Nairobi	Yaoundé
Informal settlement	Mostly dense but dispersed	Very dense and highly concentrated	Mostly dense but dispersed
Evictions / harassment of informal dwellings	Patchy, recent	Extensive since independence	Sporadic since independence
Percent households farming in town	About 49%[a]	20%[b]	35%[c]
Control of land allocation	Formal but traditional systems respected	Formal corrupted, informal controlled by chiefs	Formal corrupted, informal controlled by chiefs
Civil society	Emerged late 1980s after civil war	Repressed till 1992, harassed till 2002	Repressed till late 1990s
Urban farming	Ignored till 1990s	Repressed or ignored till 2000s	Repressed or ignored till 2004
UA research	Since 1980s, substantial investment	Since 1980s, intermediate investment	1998–2000s, small investment
Central government engages with UA	2000	2006	2004
City government engages with UA	1993	Not yet	2004

[a] ref. Chapter 6 Table 6.7 and Cole et al. (2008); this figure is not based on a city-wide sample but selected case study zones
[b] 1985 data, Lee-Smith et al. (1987)
[c] unknown source quoted in Smit et al. (1996)

Aspects of Comparison at the Project, or Micro-Level

Further lessons emerge when we look at the relative performance of the research investments made in the three places. We touch first on research investment in urban agriculture in general, and then on the way it supported research–development platforms as a strategy for institutional and policy change.

While both Kenya and Uganda were two of the three countries where IDRC invested in research on urban agriculture in the 1980s, there was no such investment in Cameroon, although the French agency for agriculture research supported a few smaller studies on aspects of urban farming at that time (Bopda & Awono, Chapter 5). None of this early research engaged multiple stakeholders or research-policy dialogue however. Stakeholder-based research emerged later on, and has been developed by both Urban Harvest and RUAF, as mentioned in the previous chapter.[4]

As described above, the research–development platforms evolved differently in the cases examined in this book, responding to local circumstances. While it was an approach facilitated by Urban Harvest, it was the local stakeholders who determined the direction, and they in turn were responding to local conditions. And while Urban Harvest initiated research by consortia led by CGIAR research institutions, these rightly allowed other institutions to take leadership when carrying forward research implementation or connection to policy-making. In the case of Kampala and Nairobi, independent bodies, namely KUFSALCC and NEFSALF, were established, the former directly as a result of the Urban Harvest project and the latter in parallel to it. It is in the nature of such platform-building that whether an institution emerges, or what type it is, are not predetermined. What is important is whether a stable institutional and policy framework for urban agriculture emerges (Table 15.2).

It seems likely that research investment played a significant role in Kampala, Nairobi and to a lesser extent in Nakuru. Furthermore, in all three locations support for dialogue platforms was available both via research funds and through direct support. This was not the case in Yaoundé. If the funds raised in addition to the Urban Harvest support in Kampala and Kenya had also been found to invest in Yaoundé, the results might have been different. Such funds might have been put to good use by motivated local researchers and government officials. Building research–development platforms needs a lot of work and requires personnel that need to be paid for. KUFSALCC began to weaken with the reduction in Urban Harvest support although IDRC funding continued for Kampala City Council.

---

[4] Interestingly, Mazingira Institute, the grantee for the IDRC supported research on urban agriculture in Kenya in 1985 and facilitator of NEFSALF, had in fact previously pioneered stakeholder-based research in a study of Kenya's informal transport system, also supported by IDRC. That study had considerable policy impact (Stren & White 1989) despite being done in an era when civil society was severely restricted. It would also be interesting to compare institutional and policy change in the countries where RUAF supported action research and policy platforms with the countries dealt with here, where the research was more extensive.

Table 15.2 Urban agriculture policy and institution-building at micro-level; the impact of research-policy platforms

Variable	Kampala	Nairobi	Nakuru	Yaoundé
Research-policy platform				
Platform	KUFSALCC	NEFSALF	Links to NEFSALF	None
Initiated by	Civil society	Civil society	Civil society	n/a
CG project-facilitated?	Yes	No	Links facilitated	n/a
Community engagement?	No	Yes	Yes	n/a
Current activity	Dormant	Very active	Dormant	n/a
UA research Investment 2000–2006	> US$500 000	US$50–100 000	US$100 000 approx.	US$20 000
Investment in platform	US $20 000 p.a. up to 2005	US $20 000 p.a. continuing	None	None
Policy development				
Process	Policy drafted	Task force set up	Task force set up	None
Local government	Active	None	Active	Active
Central government	Leads process	Leads process	Leads process	n/a
National research body	Active	Active	Active	Supportive
Civil society	Active	Main advocate	n/a	Limited / none
Community engagement	None	Very active	Indirect, through research	None
CG role	Initial advice	Policy workshop	Policy workshop	Policy workshop
Investment by	None	Urban Harvest	Urban Harvest	None
Legal reform				
Process	Completed 2006	None	Drafting complete	None
Local government	Led process	n/a	Leads process	n/a
Central government	Active	n/a	Supportive	n/a
National research body	Supportive	n/a	Supportive	n/a
Civil society	Very active	Advocacy	Supportive	n/a
Community engagement	Very active	Advocacy	Advocacy	n/a
CG role	Very active	n/a	Very active	n/a
Investment by	DFID, Urban Harvest	none	Urban Harvest	None

Apart from the investment in research and platforms, another notable difference in these cities is the participation and roles of civil society and community-based organizations. It has to be asked whether the active presence of the community and facilitating role of civil society in Nairobi has not contributed to the greater viability and sustainability of the platform, compared to the situation in Kampala, where the community was not represented in the platform and where an external organization (Urban Harvest) and local university faculty with multiple, competing commitments were the lead agencies.

Prospects for UA Policies and Institutions in the Second Decade of the 21st Century

Apart from the obvious effects of widespread economic downturn and food price rises mentioned earlier, our analysis suggests that a number of factors may influence the development of urban agriculture policies and institutions in the coming years. Primarily, the decentralized model of government that entailed having agriculture as part of the structure of urban local government, as in Kampala, appears to have played a major part in that city, as did the encouragement of participatory governance. Therefore, it might be expected that countries having such decentralized government would be more likely to develop policies and institutions facilitating urban food production.

Likewise, it appears that political shifts toward more inclusive and democratic governance, including the development of a range of civil society organizations, are conditions that favour the development of such policies and institutions.

Finally, the role of research does appear to facilitate the emergence of urban agriculture policies and institutions. Despite the recalcitrance of Nairobi City's government, the presence of a substantial body of research on the subject and the active building of research-to-policy platforms may have been key factors influencing such developments in all three capital cities examined. Thus the role of both IDRC and Urban Harvest in supporting this type of research seems to have been influential and augurs well for future interventions of the same type in other towns and cities. In turn, the preparedness of these towns and cities to cope with hunger, food shortages and economic downturn may well benefit from such policies and institutions.

References

Atukunda, G 1998 Urban agriculture in Kampala, Uganda: reviewing research impacts. *IDRC Cities Feeding People Report 29B*, IDRC Canada. http://www.idrc.ca/en/ev-8249-201-1-DO_TOPIC.html [Accessed 10 January 2008].

Ayaga, G, Kibata, G, Lee-Smith, D, Njenga, M & Rege, R 2005. *Prospects for urban and peri-urban agriculture in Kenya*, Urban Harvest-International Potato Center, Lima.

Cole, DC, Lee-Smith, D & Nasinyama, GN 2008. *Healthy City Harvests: generating evidence to guide policy on urban agriculture*, CIP/Urban Harvest and Makerere University Press, Lima Peru.

Foeken, D 2006, *"To Subsidise my income" urban farming in an East-African Town*, Brill, Leiden, Boston, MA.

Foeken, D & Owuor, S 2000, *Urban farmers in Nakuru, Kenya*, ASC Working Paper No. 45, Africa Studies Centre, Leiden, and Centre for Urban Research, University of Nairobi.

Gore, C 2008, 'Healthy urban food production and local government', in Cole, D, Lee-Smith, Di & Nasinyama, G (eds) *Healthy city harvests: Generating evidence to guide policy on urban agriculture,* CIP/Urban Harvest and Makerere University Press, Lima Peru.

Government of Kenya 2004, *Report of the Commission of Inquiry into the Illegal/Irregular Allocation of Public Land,* the Government Printer, Nairobi.

Guyer, J (ed) 1987, *Feeding African cities: studies in regional history,* Manchester University Press, Manchester.

Hooton, N, Nasinyama, G, Lee-Smith, D, Njenga, M, Azuba, M, Kaweesa, M, Lubowa , A, Muwanga, J & Romney, D 2006, *Innovative policy change to support urban farmers in Kampala: what influenced development of the new city ordinances on urban agriculture?* Paper presented to the Innovation Africa Symposium, Kampala.

Hooton, N, Nasinyama, G, Lee-Smith, D & Romney, D in collaboration with Atukunda, G, Azuba, M, Kaweeza, M, Lubowa, A , Muwanga, J, Njenga, M & Young, J 2007, *Championing urban farmers in Kampala; influences on local policy change in Uganda,* ILRI/ODI/KUFSALCC /Urban Harvest Working Paper, ILRI Research Report No. 2. ILRI, Nairobi.

Hovorka, A & Lee-Smith, D 2006, 'Gendering the UA agenda', in van Veenhuizen, R (ed) *Cities farming for the future: urban agriculture for green and productive cities,* RUAF Foundation, IDRC & IIRR, Ottawa, ON.

Kessides, C 2006, *The urban transition in Sub-Saharan Africa. Implications for economic growth and poverty reduction,* Cities Alliance, SIDA, World Bank, Washington, DC.

Kibinda, PM 2008, *Nairobi City Council and the Metropolitan Growth Strategy* paper presented at a seminar on urban development in Kenya: towards inclusive cities?, French Institute for Research in Africa (IFRA), Nairobi.

Kilonzo, HM 2008, *The Nairobi Metropolitan Growth Strategy and the role of the new Ministry* paper presented at a seminar on urban development in Kenya: towards inclusive cities?, French Institute for Research in Africa (IFRA), Nairobi.

KUFSALCC & Urban Harvest 2004, *Report of a Participatory Consultation Process on the Kampala City Draft Bills for Ordinances on Food Production and Distribution,* KUFSALCC, Kampala & Urban Harvest, Nairobi.

Lamba, D 1994, *Nairobi's Environment: a review of conditions and issues,* Mazingira Institute, Nairobi.

Lee-Smith, D 1997, *"My house is my husband": a Kenyan study of women's access to land and housing,* Thesis 8: Department of Architecture and Development Studies, Lund University, Sweden. p. 116.

Lee-Smith, D & Lamba, D 1998, *Good Governance and Urban Development in Nairobi,* Mazingira Institute, Nairobi.

Lee-Smith, D & Lamba, D 2000, 'Social transformation in a post-colonial city: the case of Nairobi', in Polese, M and Stren, R (eds) *The social sustainability of cities: diversity and the management of change,* University of Toronto Press, Toronto, ON.

Lee-Smith, D, Manundu, M, Lamba, D & Gathuru, PK 1987, *Urban food production and the cooking fuel situation in urban Kenya – national report: results of a 1985 national survey,* Mazingira Institute, Nairobi.

Lee-Smith, D & Memon, PA 1994, 'Urban agriculture in Kenya', in Egziabher, AG, Lee-Smith, D, Maxwell, DG, Memon, PA, Mougeot, LJA & Sawio, CJ (eds) *Cities Feeding People: an Examination of Urban Agriculture in East Africa,* IDRC, Ottawa, ON.

Lee-Smith, D, Semwanga, MA, Muwanga Musisi, J, Kaweesa, M & Nasinyama, GW 2008, 'The story of the health coordinating committee, KUFSALCC and the urban agriculture ordinances', in Cole, DC, Lee-Smith, D and Nasinyama, GN (eds) *Healthy city harvests: Generating evidence to guide policy on urban agriculture,* CIP/Urban Harvest and Makerere University Press, Lima Peru.

Maxwell, DG 1994 'The household logic of urban farming in Kampala', in Egziabher, AG, Lee-Smith, D, Maxwell, DG, Memon, PA, Mougeot, LJA & Sawio, CJ (eds) *Cities feeding people: an examination of urban agriculture in East Africa,* IDRC, Ottawa, ON pp. 45–62.

Mazingira Institute 2005, *NEFSALF Bulletin* 1, Nairobi.
Mazingira Institute 2006, *NEFSALF Bulletin* 8, Nairobi.
McCarney, P, Halfani, M & Rodriguez, A 1995, 'Towards an Understanding of Governance', in Stren, R & Kjellberg-Bell, J (eds) *Perspectives on the City*, Centre for Urban and Community Studies, Toronto, ON, pp. 91–142.
Ministry of Lands & Settlement 2006, *Draft National Land Policy*, Ministry of Lands, Nairobi.
Mitullah, W 2008, 'Urban development in Kenya: institutional framework and stakeholders', paper presented at a seminar on urban development in Kenya: towards inclusive cities?, French Institute for Research in Africa (IFRA), Nairobi.
Muiruri, JM & Kaseve, C 2008, 'Water supply to the urban poor in Nairobi', paper presented at a seminar on urban development in Kenya: towards inclusive cities?, French Institute for Research in Africa (IFRA), Nairobi.
Mukisira, E 2005, 'Opening Speech', in Ayaga, G, Kibata, G, Lee-Smith, D, Njenga, M & Rege, R (eds) *Prospects for urban and peri-urban agriculture in Kenya*, Urban Harvest-International Potato Center, Lima.
Muraguri L 2008, *An overview of the Kenya slum upgrading program* paper presented at a seminar on urban development in Kenya: towards inclusive cities?, French Institute for Research in Africa (IFRA), Nairobi.
Okello, SO & Mboga, H 2008, *Assessment of LATF and LASDAP activities: the point of view of local authorities* paper presented at a seminar on urban development in Kenya: towards inclusive cities?, French Institute for Research in Africa (IFRA), Nairobi.
Sjaastad, E & Cousins, B 2009, 'Formalisation of land rights in the South: an overview' in *Land Use Policy*, 26:1, www.sciencedirect.com [Accessed 15 January 2009].
Smit, J, Ratta, A & Nasr, J 1996, *Urban agriculture: food, jobs and sustainable cities*, UNDP Habitat II series, New York, NY, 302 pp.
Stren, R 2008, 'Foreword', in Cole, DC, Lee-Smith, D & Nasinyama, GN (eds) *Healthy city harvests: generating evidence to guide policy on urban agriculture*, CIP/Urban Harvest and Makerere University Press, Lima Peru.
Stren, RE & White, R 1989, *African cities in crisis: managing rapid urban growth*, Westview, Boulder, CO.
United Nations Children's Fund (UNICEF) 1992, *Situation analysis of women and children in Kenya*, UNICEF, Nairobi.

Acronyms

Acronym	Meaning
AFID	Agency for Inter-regional Development
ANT	Anthracene
ASF	Animal Source Foods
AVD	Association of Volunteers for Development
AVRDC	The World Vegetable Center
BOD5	Biochemical Oxygen Demand, 5-day
BTB	Bovine Tuberculosis
C	Carbon
C.V.	Coefficient of Variation
c/l	Crop/Livestock
C:N	Carbon:Nitrogen
Ca	Calcium
CAID	Centre for Accompaniment of Initiatives for Development
CARBAP	African Research Centre on Bananas and Plantains
CBO	Community-Based Organization
CBS	Central Bureau of Statistics
Cd	Cadmium
CEDEP	Centre for the Development of People
CFP	Cities Feeding People
CFU	Colony-Forming Unit
CGIAR	Consultative Group on International Agricultural Research
CIAT	International Center for Tropical Agriculture
CIDA	Canadian International Development Agency
CIKSAP	Centre for Indigenous Knowledge Systems and by-Products
CIP	International Potato Center
CIPRE	International Centre for Promotion of Economic Recovery
CIRAD	International Centre for Research on Agriculture for Development
COD	Chemical Oxygen Demand
CR	Chromium
CRP	C-Reactive Protein
CUY	Urban Community of Yaounde
CUY	Urban Community of Yaounde
DfID	Department for International Development
DGIS	Directorate General for International Cooperation
DRSP	Document de Strategie de Reduction de la Pauvrete
DSCN	Direction de la Statistique et de la Compatibilite Nationale
EDI	Estimated Daily Intake

Acronym	Meaning
ELISA	Enzyme-Linked ImmunoSorbent Assay
ENPS	Ecole Nationale Polytechnique Superieure
ENSP	Higher National Polytechnic School
ESARO	East and Southern Africa Office
ETC	Educational Training Consultants
FAO	Food and Agriculture Organization of the United Nations
FC	Fecal Coliform
FCFA	Francs Communauté Financière Africaine
FCRI	Focus Cities Research Initiative
FGD	Focus Group Discussion
FLA	Fluoranthene
FS	Fecal Streptococcus
GAN	Green Africa Network
GDP	Gross Domestic Product
GRES	The Research Group on Social and Environmental Issues
GTZ	Gesellschaft für Technische Zusammenarbeit
ha	hectare
HCC	Health Coordinating Committee
HFS	Household Food Security
Hg	Mercury
HH	Household
HIA	Health Impact Assessment
IAGU	African Institute for Urban Management
ICDP	Indeno[$1,2,3$-cd] pyrene
ICLARM	World Fish Center
ICRAF	World Agroforestry Centre
IDRC	International Development Research Centre
IIRR	International Institute for Rural Reconstruction
IITA	International Institute of Tropical Agriculture
ILRI	International Livestock Research Institute
INC	National Institute of Cartography
INIBAP	International Network for the Improvement of Banana and Plantain
IPGRI	International Plant Genetic Resources Institute
IRAD	Institute for Agricultural Research and Development
IRR	Internal Rate of Return
ISRA	Senagalese Institute of Agricultural Research
ITDG-EA	Intermediate Technology Development Group - East Africa
IWMI	International Water Management Institute
JICA	Japan International Cooperation Agency
K	Potassium
KARI	Kenya Agricultural Research Institute
KCC	Kampala City Council
KGTPA	Kenya Greentowns Partnership Association
KSh	Kenyan Shilling
KUFSALCC	Kampala Urban Food Security, Agriculture and Livestock Coordination Committee
LASDAP	Local Authorities Service Delivery Action Plans
LC	Local Council
LGDP	Local Government Development Grant
LPP	Livestock Production Programme
LVRLAC	Lake Victoria Region Local Authorities Cooperation

Acronyms

Acronym	Meaning
MAAIF	Ministry of Agriculture, Animal Industries and Fisheries
M:F	Male:Female
MCN	Municipal Council of Nakuru
MDP	Municipal Development Partnership
MDP-ESA	Municipal Development Partnership for Eastern and Southern Africa
MES	Metal Sulphide
MFPED	Ministry of Finance, Planning and Economic Development
Mg	Magnesium
MINAGRI	Ministry of Agriculture
MINEFI	? Ministry of Economics, Finance and Industry CHECK (In Table 5.7) or better in French Ministere de l'Economie, des Finances et de l'industrie
MINEPIA	Ministere De L'Elevage, Des Peches Et Des Industries Animales
MINPAT	Ministère du Plan et de l'Aménagement du Térritoire
MINREST	Ministere de la Recherche Scientifique et Technique
MINUH	Ministry of Urban Affairs and Habitat
MOFA	Ministry of Food and Agriculture
MPND	Ministry of Planning and National Development
MUM-FAMrisk	Multimedia Urban Model, Family Risk
MUM-risk	Multimedia Urban Risk Model
N	Nitrogen
n	Number
NALEP	National Agriculture and Livestock Extension Programme
NMC	Nakuru Municipal Council
n/e	Not established
n/f	Not found
NAARI	Namulonge Agricultural and Animal Research Institute
NARO	National Agricultural Research Organization
NCC	Nairobi City Council
NEFSALF	Nairobi and Environs Food Security, Agriculture and Livestock Forum
NEMA	National Environmental Management Act
NGO	Non-Governmental Organization
NH4	Ammonium
NO3	Nitrate
NPV	Net Present Value
O2	Oxygen
ODA	Overseas Development Aid
OECD	Organization for Economic Co-operation and Development
P	Phosphorus
p	Significance
PAH	Polycyclic Aromatic Hydrocarbon
Pb	Lead
PHE	Phenanthrene
PMA	Plan for the Modernization of Agriculture
PO4	Phosphate
Ppkm2	Persons per square kilometer
ppm	Parts per million
ppt	Parts per thousand
PRGA	Participatory Research and Gender Analysis
PRSP	Poverty Reduction Strategy Paper
p-u	Peri-urban
PUA	Participatory Urban Appraisal

Acronym	Meaning
PYR	Pyrene
RUAF	Resource Centres on Urban Agriculture and Food Security
RVIST	Rift Valley Institute of Science and Technology
s.e.	Standard Error
sd	Standard Deviation
SIDA	Swedish International Development Agency
SNF	Sustainable Neighbourhoods in Focus
SP	Selling Price
SPSS	Statistical Package for Social Scientists
SSA	Sub-Saharan Africa
TB	Tuberculosis
u/p-u	Urban/Peri-urban
UA	Urban Agriculture
UBOS	Uganda Bureau of Statistics
UH	Urban Harvest
UK	United Kingdom
UMP-LAC	Urban Management Program's Latin America and Caribbean Office
UN	United Nations
UNDP	United Nations Development Program
UPA	Urban and Peri-urban Agriculture
UPE	Urban Poverty and Environment
US EPA	United States Environmental Protection Agency
USA	United States of America
USAIP	Urban Schools Agricultural Initiatives Project
USh	Ugandan shilling
VLA	Veterinary Laboratories Agency
WAZ	Weight for Age Z score
WHO	World Health Organization
ZBMI	Body Mass Index Z score
Zn	Zinc

Index

Note: Locators followed by 'f' and 't' refer to figures and tables respectively.

A
Aflatoxin M1, concentration response of, 234t
Africa, urban agriculture in, 13–14
   agriculture as urban livelihood strategy, 18–20
      contribution to household income, 20–23
      urban agriculture and markets, 23–25
   livelihoods, and markets, 15–16
      livelihoods in space, 16–18
   urban ecosystem health, 25–26
      benefits and risks of horticulture, 27–28
      benefits and risks of livestock-raising, 26–27
      benefits of waste recycling, 29–30
      policy and institutional dialogue and change, 30–32
      solid wastes, 28–29
Agricultural enterprises for income generation/household food security, 146t
Agricultural recycling of organic wastes in Nakuru, 29f
Agricultural research and rural bias, 1–4
Agriculture as urban livelihood strategy, 18–20
   contribution to household income, 20–23
   urban agriculture and markets, 23–25
Agriculture in African Urban institutions and policy, 4–6
Agroforestry
   and food products, 252–257
      fruits, 252–254
      medicinal plants, 256–257
      products transport, 257
      vegetables, 254–256
   and income security, 249
   and non-food products, 250, 257–259
      tree nurseries, 259
AGROPOLIS, 270–272

Animal husbandry, 44, 239
   men/women's perceptions of persons involved in, 240f
Animals, *see* Livestock

B
Biological hazards, 172–174

C
Cameroon, 37
   crop–livestock integration, 61–69
   institutional development of urban agriculture, 71–94
   with Kenya/Uganda, comparing urban agriculture of, 302
      macro-level comparsion aspect, 302–303
      micro-level comparsion aspect, 304–306
      UA policies and institutions, 306
   urban farming systems in Yaoundé, 39–57
   to Yaoundé, emigrants (%) from different parts of, 79
Cattle rearing systems, 242f
CGIAR centres' participation in urban harvest-supported Africa research, 273t
Chemical contaminants, 185
Chemical hazards, 27, 184
   and heavy metals, 51, 175, 185
Children
   exposed to combustion products, risks, 175–176
   nutrition in, 170–171
      household UA and, 170f
Cities, 3–5
   and European settlement, 5, 302
   food security in, 7, 53
   links to hinterland of, 28

Cities (*cont.*)
   and livestock, 61, 68, 101, 176–177, 185–186
      chicken rearing, 180–183
      dairying, 177–180
Civil society, 86–88, 303
Climate, 299
Community, 126, 135, 195–196
Compost, 29–30, 195–196, 202
   compared to manure, 196–198, 206
   marketing, 205–207
   pricing, 206
   production of, 206–207
   quality, 196, 204–205
Consultative group on International Agricultural Research, 1, 98, 142, 193, 269, 288–290
Contaminants
   chemical, 185
   dealing with
      biological hazards, 172–174
      children exposed to combustion products, risks, 175–176
      heavy metal contamination, 174–175
      nature of enquiry, 171–172
Crop–livestock integration in urban farming systems of Yaoundé, 61–62
   crop–livestock production in urban areas, 67–68
   findings on livestock enterprises, 63–65
   methods used for study, 62–63
   use of manure, 65–67
Crop–livestock–waste interactions in Nakuru's urban agriculture, 213–214
   methods used, 214
      data analysis, 215–216
      organic waste production and utilization, 215
      urban appraisals and household interviews, 215
   results
      characterizing sample, 216–217
      farmers' attitudes/use of crop inputs, 221–223
      household food consumption, 219
      manure production and use, 223–225
      policy influencing/technology transfer in Nakuru municipality, 226
      sources of manure for, 225–226
      types of livestock feeds/sources, 220–221
      urban farming plots, 218–219
      urban livestock production, 219–220
      use of raw organic household waste as animal feed, 221
Cropping system, 44, 106–108
Crops, 2, 43, 126
   destruction of, 258
   horticultural, 27, 48
   intercropping of, 107
   nutritional quality of, 48
   production, constraints, 108t
   staple, 4, 40, 54
   See also Food security

D
Dairy cattle
   and economic benefits, 8, 26
   and health risks, 25–26, 213, 231–232
Demographics
   population density, 17, 77, 143, 194, 198
   population growth, 42, 62

E
Economic characterization matrix, 159t
Employment, 20, 67, 78, 102, 177, 180, 237–238
Environment, 14, 21, 24, 139–140
   benefits of recycling to, 26
   management and planning for, 30, 68, 199–200, 214
   policies for, 72
   pollution of, 67–68, 231
   protection of, 92–93, 120, 280
Environmental Alert, 104, 111, 125, 281, 293

F
Farmers/farming
   evaluation/selection of enterprises with, 161t
   and food trading households (HHs), 116t
   objectives in selected urban and peri-urban zones, 115t
   wealth categories across urban to peri-urban continuum, 119t
Farming households along peri-urban to urban continuum, Kampala
   primary activities of, 21f
   sources of income, 22f
Farm inputs/marketing pathways of chicken producers, 181f
Food security, 7–8, 73, 104, 117, 119–121, 169, 183–184
   crops, 23, 146
   food supply, 43–44
   household, 26, 129, 145, 146t, 169–171, 185–186

Index 315

and livestock, 68, 148
local food production, 7–8, 22
and nutrition, 185–186
policy, 120–121, 251, 290
women, 117, 186
Formal produce markets, 142

G
Gender, 73, 107, 203, 223–225, 273
  benefits of livestock keeping in
    Nakuru, 239
  brucellosis, symptoms, 243t
  distribution of household head by, 238f
  health risks associated with livestock, 241f
  manure production/use, 227t
  ownership of urban farming plots, 220f
  perceptions of persons in animal
    husbandry, 242f
  perceptions of risks/benefits of urban
    farming, 239–241
  ranking of listed health risks in
    Nakuru-municipality, 240f
Governance, 7–8, 296–291, 304
  definition, 31
"Greening" of city, 26

H
Health
  human nutrition and, 14, 61
  impacts, identifying potential, 169–171
  remedies for different health
    problems, 257t
  risks
    from composting, 77, 203, 237
    dairy cattle and, 25–26, 215, 233–234
    disease associated with livestock
      keeping, 241f
    from livestock, 20, 178–179, 233–234
    from manure, 77, 203
    in Nakuru, gender ranking of
      listed, 240f
Health impact assessment of urban agriculture
  in Kampala, 169
  dealing with contaminants
    biological hazards, 174–176
    children exposed to combustion
      products, risks, 177–178
    heavy metal contamination, 176–177
    nature of enquiry, 173–174
  household food security and child
    nutritional benefits, 171–172
  findings on child nutrition, 173
  findings on household food security
    (HFS), 172–173

identifying potential health impacts,
  169–171
integrating findings into policy and
  practice, 185
managing healthy urban livestock,
  187–188
promoting food security and nutrition,
  185–186
reducing contamination, 186–187
managing urban livestock for health
  chicken rearing, 182–185
  dairying, 179–182
  nature of inquiry, 178–179
Health risks of urban dairy production in
  Nakuru, Kenya, 231–233
  approaches and methods used
    health risks examined, 233–234
    study population and data collected,
      234–237
  results
    farming household characteristics,
      237–239
    gendered perception of risks and
      benefits, 239–241
    practices in livestock husbandry,
      241–242
    selected health hazards investigated,
      242–246
Heavy metals, *see* Chemical contaminants
HIV/AIDS, 170, 172, 234, 253, 282
Horticulture, *see* Vegetables
Horticulture, benefits and risks of, 27–28
Household food security and child nutritional
  benefits, 171–172
  findings on child nutrition, 173
  findings on household food security (HFS),
    172–173
Household UA and child nutrition, 172f

I
IDRC and its partners in Sub-Saharan Africa
  2000–2008, 269–272
  AGROPOLIS, 272–274
  Anglophone Africa Course, 278–282
  RUAF foundation, 276–278
  UPE Focus Cities Research Initiative,
    282–283
  urban harvest, 274–276
Income, 20–23, 47–48, 104–106, 114–115,
  199, 235
  different groups involved in urban
    agriculture, 21
  diversification of, 239

Income (*cont.*)
  generating products, supply/
      constraints, 149t
  sources of, farming households,
      Kampala, 22f
  from urban agriculture, 20–23, 47–48
Informal food markets, 143
Institut de Recherche Agricole pour le
      Developpement, 49, 69
Institutional and regional context, 1
  agricultural research and rural bias, 1–4
  treatment of agriculture, 4–6
  urban agriculture research agenda for
      Africa, 6–9
Institutional development of urban agriculture
  background to institutional study in
      Yaoundé, 71–74
  history and development of urban Yaoundé,
      75–84
  institutional study and its results
    civil society, 86–88
    formal structures of government, 84–86
    market traders, views, 89–91
    Yaoundé's 1981 structure plan, 88–89
  methods, 74–75
International Centre for Tropical
      Agriculture, 9, 95
International Development Research Center, 3
International Institute for Tropical
      Agriculture, 9, 37
International Livestock Research Institute, 9,
      193, 195, 199, 274, 298
International Monetary Fund, 6

K
Kampala
  amount of cultivable land available to
      farmers in communities, 129
  farming households, primary activities
      of, 21f
  farm inputs/marketing pathways of chicken
      producers, 183f
  history of urban agriculture, 98–99
  sources of income of farming
      households, 22f
  urban agriculture profile, 97–121
Kampala, changing trends in urban agriculture
      in, 97–98
  agriculture as means of livelihood, 111–117
  background, 98–99
  description of urban agriculture
    Banda: new urban slum area, 101–103
    Bukesa: old urban area, 101
    Buziga: peri-urban area in transition to
        urban, 104
    Komamboga: peripheral peri-urban
        area, 104–106
  methods, 99–101
  policy implications, 117–121
  urban agriculture production systems, 106
    cropping systems, 106–108
    livestock systems, 109–110
    social capital and institutional support,
        110–111
Kampala City Council, 17, 19, 32, 98–99, 125,
      127, 136, 143, 147, 169, 172–173,
      183, 185, 273
Kampala, health impact assessment of urban
      agriculture in, 169
  dealing with contaminants
    biological hazards, 174–176
    children exposed to combustion
        products, risks, 177–178
    heavy metal contamination, 176–177
    nature of enquiry, 173–174
  household food security and child
      nutritional benefits, 171–172
    findings on child nutrition, 173
    findings on household food security
        (HFS), 172–173
  identifying potential health impacts,
      169–171
  integrating findings into policy and
      practice, 185
    managing healthy urban livestock,
        187–188
    promoting food security and nutrition,
        185–186
    reducing contamination, 186–187
  managing urban livestock for health
    chicken rearing, 182–185
    dairying, 179–182
    nature of inquiry, 178–179
Kampala, market opportunities for
      urban/peri-urban farmers, 141–144
  market study, 145–146
  methods, 144–145
    participatory urban appraisal, 145
    participatory evaluation of options,
        146–147
  results
    market survey, 151–153
    participatory evaluation of options,
        153–164
    participatory urban appraisal, 147–151

Kampala: schools as agents of urban
    agriculture, 123–124
  background and rationale for study,
    124–126
  households in school's communities buying
    seed, 130f
  methods used for research, 126
  schools as commercial seed
    producers, 133
    constraints, 135–136
    results, 134–135
    school-based commercial seed
      production, 133–134
  technology dissemination and extension
    assessment of crop enterprises and land
      availability, 129–130
    feasibility study and selection of
      schools, 127–128
    K132 Bean variety at Lubiri
      Nabagereka, 132
    Longe 5 Maize variety at Lubiri
      Nabagereka, 131–132
    Longe 5 Maize variety at valley
      St. Mary's, 132–133
    seed-demand assessment in schools,
      130–131
    sweetpotato at valley St. Mary's, 133
    using schools for agricultural
      extension, 131
Kampala Urban Food Security, Agriculture and
    Livestock Coordinating Committee,
    185, 290, 294
KCC urban agriculture classification
    system, 99t
Kenya, urban agriculture in, 193, 297–298
  benefits/risks of urban dairy production in
    Nakuru, 231–248
  with Camaroon/Uganda, comparing urban
    agriculture of, 303
  crop–livestock–waste interactions,
    215–229
  Kenya's absence of urban policy, 296–297
  NEFSALF, 298–299
  NEFSALF and urban harvest-supported
    research, 299–300
  prospects for an urban agriculture policy,
    300–301
  recycling nutrients from organic wastes,
    195–211
  shifts toward policy in negative
    environment, 298
  urban agroforestry products in Kisumu,
    251–264

Kisumu
  fruit type transactions in urban and
    peri-urban, 255t
  local vegetable species traded in
    markets, 257t
  market chains for agroforestry products
    in, 260f
  showing key open air markets, 254f
  transport cost for various products, 259t
Kisumu, urban agroforestry products in,
    251–252
  agroforestry food products
    fruit, 254–256
    local vegetables, 256–258
    medicinal plants, 258–259
    products transport, 259
  market chain analysis, 262–263
  methods, 253–254
  non-food agroforestry products, 259–261
  tree nurseries, 261–262
  study area, 252–253
KUFSALCC, 294–296

L
Land
  access to, 20, 110, 118, 199, 216, 276,
    296–297
  conversion of, 42
  eviction from, 40, 75–76, 305
  rented out by chiefs, 46, 102
  urban used for agriculture, 231
Livelihoods, 13–14
  framework, 15, 25
Livestock
  benefits of, keeping by gender in
    Nakuru, 239f
  crop-livestock interactions, 20, 93, 196
  and gender, 241
  health impacts of, 43, 98, 169–171
  health risks, by gender, 241f
  kept by farming households in
    Nakuru, 238f
  local races, 17
  managing urban
    chicken rearing, 182–185
    dairying, 179–182
    healthy, 187–188
    nature of inquiry, 178–179
  manure, compostion, 66t
  number of, per farm, 63t
  and nutrition, 26–27
  raising, benefits and risks of, 26–27
  and socio-economic profiles, 20

Livestock (*cont.*)
  types of, 109t
  and waste recycling, 29, 196

M
Manure
  animal, compostion, 66t
  contamination from, 52, 174
  gender and use of, 227, 229
  health risks from, 77, 203
  prices, 208, 210–211
  sourcing, 110, 223, 263
  types of, 206
  See also Compost
Market characterization matrix, 159t
Marketing chains, 207
Markets
  formal produce markets, 142
  informal food markets, 143
  local vegetable species traded in Kisumu, 257t
  occupation stated by traders in Yaoundé, 90t
  products and their purchasing conditions, 156t
  products in high demand in different outlets, 152t
  products in scarce supply in different outlets, 154t
  reason for scarcity of products, 155t
  selection of short listed products, 158t
  supermarkets, 144
  urban agriculture and, 23–25
  in Yaoundé by, different types of, 84
  Yaoundé, types of, 84t
Migration, 3, 76
Millennium Development Goals, 26, 57, 120, 123, 186

N
Nairobi
  chemical levels in compost/cattle manure, 206t
  demographics, 200
  organic waste recovery and recycling groups, 202t
  rural–urban manure and compost flows, 205f
  sites and population densities, 200f
  solid waste management in, 195, 274, 282–283
Nairobi City Council, 32, 201, 205, 210, 275, 298–299
Nakuru
  agricultural recycling of organic wastes in, 29f
  benefits of livestock keeping by gender in, 239f
  gender ranking of listed health risks in, 240f
  livestock kept by farming households, 238f
  urban farming plots, 221f
  uses of crop residues by farmers, 225t
Nakuru, benefits/risks of urban dairy production, 231–233
  approaches and methods used
    health risks examined, 233–234
    study population and data collected, 234–237
  results
    farming household characteristics, 237–239
    gendered perception of risks and benefits, 239–241
    practices in livestock husbandry, 241–242
    selected health hazards investigated, 242–246
Nakuru's urban agriculture, crop–livestock–waste interactions in, 215–216
  methods used, 216
    data analysis, 217–218
    organic waste production and utilization, 217
    urban appraisals and household interviews, 217
  results
    characterizing sample, 218–219
    farmers' attitudes, crop inputs, 223–225
    household food consumption, 221
    manure production and use, 225–226
    policy influencing and technology transfer, 228
    sources of manure for crop-only farmers, 227–228
    types of livestock feeds and their sources, 222–223
    urban farming plots, 220–221
    urban livestock production, 221–222
    use of raw organic household waste as animal feed, 223
NEFSALF, 298–299
  and urban harvest-supported research, links between, 299–300
Nutrient recycling, 210–211, 229
Nutrition, 43, 173–176
  in children, 27

Index 319

household UA and, 172f
contribution of perishable products to household, 47–48
See also Crops; Livestock

O

Organic waste, 28–30, 202–203, 217–218
  health risks
    from composting, 77, 203, 237
    from livestock, 20, 178–179, 233–234
    from manure, 77, 203
  in Nakuru, agricultural recycling of, 29f
  policy, 217
  uses of, 217–218
  See also Compost

P

Peri-urban to urban continuum, Kampala
  farming households (primary activities of) along, 21f
  sources of income of farming households along, 22f
Pesticide use by urban and peri-urban farmers in relation to land source, 52f
Pollution
  air, 108, 149, 171
  biological, 43
  estimated amount of effluent discharged by known industrial polluters, 50t
  point and non-point, 50
  traffic and effect on crops, 120
  water, 28, 37, 43
Poverty
  gender and, 251
  reduction through urban agriculture, 23, 68
Production characterization matrix, 160t

R

Recycling nutrients from organic wastes in Kenya's capital city, 195–196
  discussion of results
    compost and livestock manure marketing, 207–209
    compost quality, 206–207
    group compost operations, 203–206
    organic waste recovery and UA, 202–203
  methods used, 199–202
  Nairobi situation, 196–199
Research–development partnerships (contributing) to building UA policy, 289–290
  Cameroon, 301–303
  comparing Cameroon, Kenya, Uganda, 303

macro-level comparsion aspect, 304–305
micro-level comparsion aspect, 306–308
UA policies and institutions, 308
UA in Kenya, 297–298
  Kenya's absence of urban policy, 296–297
  NEFSALF, 298–299
  NEFSALF and urban harvest-supported research, 299–300
  prospects for urban agriculture policy, 300–301
  shifts toward policy in negative environment, 298
UA in Uganda
  institution-building for linking research to policy, 291–293
  Kampala as innovative city for urban farming, 291
  KUFSALCC, 294–296
  urban agriculture ordinances, 291
Resource Centres on Urban Agriculture and Food Security, 269, 275
Rural–urban linkages, 3, 53–56
Rural–urban manure and compost flows in and around Nairobi, 203f

S

Schools as agents of UA and seed production, 123–124
  background and rationale for study, 124–126
  methods used for research, 126
  schools as commercial seed producers, 133
    constraints, 135–136
    results, 134–135
    school-based commercial seed production, 133–134
  technology dissemination and extension
    community entry and selection of enterprises, 128–129
    crop enterprises and land availability, 129
    feasibility study and selection of schools, 127–128
    K132 Bean variety at Lubiri Nabagereka, 132
    Longe 5 Maize variety at Lubiri Nabagereka, 131–132
    Longe 5 Maize variety at valley St. Mary's, 132
    seed-demand assessment in schools, 129–130

Schools as agents of UA (*cont.*)
　　sweetpotato at valley St. Mary's, 132
　　using schools for agricultural
　　　　extension, 131
School's communities buying seed, proportions
　　of households in, 130f
Seeds, 255, 259
Slums, 76
Soil fertility management options, 222f
Solid waste, *see* Organic waste
Sub-Saharan Africa 2000–2008, IDRC and its
　　partners in, 267–270
　　IDRC'S regional courses and individual
　　　　research awards (AGROPOLIS),
　　　　270–272
　　relevant knowledge for action in
　　　　African urban agriculture – 2004
　　　　Anglophone Africa Course,
　　　　276–280
　　RUAF foundation, 274–276
　　UPE Focus Cities Research Initiative,
　　　　280–281
　　urban into international agricultural
　　　　research, 272–274
Sub-Saharan Africa, proportion of urban
　　population farming, 15f
Supermarkets, 142
　　demand high quality and consistent
　　　　supply, 155t

U

UA policy, research-development partnerships
　　(contributing) to building, 287–288
　　Cameroon, 299–301
　　comparing Cameroon, Kenya, Uganda, 301
　　　　macro-level comparsion aspect,
　　　　　　302–303
　　　　micro-level comparsion aspect,
　　　　　　304–306
　　　　UA policies and institutions, 306
　　UA in Kenya, 295–296
　　　　Kenya's absence of urban policy,
　　　　　　294–295
　　　　NEFSALF, 296–297
　　　　NEFSALF and urban harvest-supported
　　　　　　research, 297–298
　　　　prospects for urban agriculture policy,
　　　　　　298–299
　　　　shifts toward policy in negative
　　　　　　environment, 296
　　UA in Uganda
　　　　institution-building for linking research
　　　　　　to policy, 289–291

　　　　Kampala as innovative city for urban
　　　　　　farming, 289
　　　　KUFSALCC, 292–294
　　　　urban agriculture ordinances, 291
Uganda, 95
　　with Camaroon/Kenya, comparing urban
　　　　agriculture of, 301
　　changing trends in urban agriculture in
　　　　Kampala, 97–121
　　health impact assessment of urban
　　　　agriculture in Kampala, 167–187
　　identifying market opportunities for
　　　　farmers in Kampala, 139–165
　　schools as agents of UA extension and seed
　　　　production, 123–138
Uganda, urban agriculture in
　　institution-building for linking research to
　　　　policy, 289–291
　　Kampala as innovative city for urban
　　　　farming, 289
　　KUFSALCC, 292–294
　　urban agriculture ordinances, 291
United Nations Development Program, 3,
　　139, 268
Urban agriculture
　　and access to land, 118, 197, 214
　　benefits of, 168, 184
　　biological *vs.* chemical health hazards, 172t
　　crop choices, 161
　　crop–livestock systems, 227
　　for households involved in, 21t
　　income/food security balance, 26
　　livelihoods, and markets, 15–16
　　livelihoods in space, 16–18
　　as a livelihood strategy, 13, 18–20
　　marginalization of, 32
　　and nutrition security, 7, 26
　　policy and institution-building at
　　　　micro-level, 305t
　　research agenda for Africa, 6–9
　　Yaoundé, institutions involved in, 88t
Urban agriculture in Africa, 13–14
　　agriculture as urban livelihood strategy,
　　　　18–20
　　contribution to household income,
　　　　20–23
　　urban agriculture and markets, 23–25
　　livelihoods, and markets, 15–16
　　livelihoods in space, 16–18
　　urban ecosystem health, 25–26
　　　　benefits and risks of horticulture, 27–28
　　　　benefits and risks of livestock-raising,
　　　　　　26–27

Index

benefits of waste recycling, 29–30
  institutional dialogue and change, 30–32
  solid wastes, 28–29
Urban agriculture in Kampala, changing trends in, 97–98
  agriculture as means of livelihood, 111–117
  background, 98–99
  description of urban agriculture
    Banda: new urban slum area, 101–103
    Bukesa: old urban area, 101
    Buziga: peri-urban area in transition to urban, 104
    Komamboga: peripheral peri-urban area, 104–106
  methods, 99–101
  policy implications, 117–121
  urban agriculture production systems, 106
    cropping systems, 106–108
    livestock systems, 109–110
    social capital and institutional support, 110–111
Urban agriculture in Kenya, 295–296
  Kenya's absence of urban policy, 294–295
  NEFSALF, 296–297
  NEFSALF and urban harvest-supported research, 297–298
  shifts toward policy in negative environment, 296
  urban agriculture policy, 298–299
Urban agriculture, institutional development of (Yaoundé)
  background to institutional study in Yaoundé, 71–74
  history and development of urban Yaoundé, 75–84
  institutional study and its results
    civil society, 86–88
    formal structures of government, 84–86
    institutional mobilization, 88–89
    market traders, views, 89–91
  methods, 74–75
Urban agriculture in Uganda
  institution-building for linking research to policy, 289–291
  Kampala as innovative city for urban farming, 289
  novel research–development platform for urban agriculture: KUFSALCC, 292–294
  urban agriculture ordinances, 291
Urban agroforestry products in Kisumu, Kenya, 249–250

agroforestry food products
  fruit, 252–254
  local vegetables, 254–256
  medicinal plants, 256–257
  products transport, 257
  market chain analysis, 260–261
  methods, 251–252
  non-food agroforestry products, 257–259
  tree nurseries, 259
  study area, 250–251
Urban and peri-urban continuum/zones/farmers
  farming and food trading households (HHs), 116t
  farming objectives in selected, 115t
  market opportunities in Kampala, 139–142
  market study, 143–144
  methods, 142–143
  participatory urban appraisal, 143
  participatory evaluation of options, 144–145
  pesticide use by, 52f
  results
    market survey, 149–151
    participatory evaluation of options, 151–162
    participatory urban appraisal, 145–149
  wealth categories of farmers across, 119t
Urban cultivators, types, characteristics, 45t
Urban ecosystem health, 25–26
  benefits and risks of horticulture, 27–28
  benefits and risks of livestock-raising, 26–27
  benefits of waste recycling, 29–30
  definition, 25
  policy and institutional dialogue and change, 30–32
  solid wastes, 28–29
Urban farmers/farming
  crop production constraints identified by, 108t
  ownership of plots, 220
  plots by location, 219f
Urban farming systems in Yaoundé, 39–40
  household nutrition and perishable products, 47–48
  livelihoods of urban crop growers, 44–47
  methods used for five studies, 42–44
  origins and nature of urban agriculture, 40–42
  rural–urban linkages, 53–56
  seedling production for urban horticulture, 48–49
  water pollution, 50–53

Urban governance, *see* Governance
Urban Harvest, viii, 98, 125, 197–198, 272–274
  sustainable livelihoods framework, 15–15f
Urbanization, 3, 53, 73, 86
  livelihood strategies, 13, 15–16, 18–20, 119–120
  migration, 3
Urban population farming in Sub-Saharan Africa, 15f

V

Vegetables, 22–24, 39, 45–48, 142, 146t, 185, 254–256

W

Waste recycling, 29–30
Water pollution, 28, 37, 43
  levels of 12 Yaoundé streams, 51t
World Agroforestry Centre, 9, 136, 193, 197, 249, 296
World Bank, 2, 22, 61
World Health Organization, 51, 173, 256

Y

Yaoundé
  administrative levels and powers within Mfoundi Division, 85
  demographics, 76t, 77f
  emigrants (%) from different parts of Cameroon, 79f
  farmers as economically active population, 83t
  growth of, 82f
  history, 71–94
  institutions involved in UA in, 88t
  markets
    institutions dealt with by traders in, 91t
    occupation stated by traders in, 90t
    types of, 84t
  UA production in Yaoundé, 83
  water pollution levels of 12 streams, 51t
Yaoundé, crop-livestock integration, 61–62
  constraints/opportunities for crop–livestock production, 67–68
  findings on livestock enterprises, 63–65
  methods used for study, 62–63
  use of manure, 65–67
Yaoundé, urban farming systems in, 39–40
  livelihoods of urban crop growers, 44–47
  methods used for five studies, 42–44
  origins and nature of urban agriculture, 40–42
  perishable products to household nutrition, 47–48
  rural–urban linkages, 53–56
  seedling production for urban horticulture, 48–49
  water pollution, 50–53